Statistics in Sport

Arnold Applications of Statistics Series

Series Editor: **BRIAN EVERITT**
Department of Biostatistics and Computing, Institute of Psychiatry, London, UK

This series offers titles which cover the statistical methodology most relevant to particular subject matters. Readers will be assumed to have a basic grasp of the topics covered in most general introductory statistics courses and texts, thus enabling the authors of the books in the series to concentrate on those techniques of most importance in the discipline under discussion. Although not introductory, most publications in the series are applied rather than highly technical, and all contain many detailed examples.

Other titles in the series:

Statistics in Education Ian Plewis

Statistics in Civil Engineering Andrew Metcalfe

Statistics in Human Genetics Pak Sham

Statistics in Finance Edited by David Hand & Saul Jacka

Statistics in Sport

Edited by

Jay Bennett
Senior Consultant, Bellcore, New Jersey, USA

A member of the Hodder Headline Group
LONDON • NEW YORK • SYDNEY • AUCKLAND

To
Joseph and Anita Bennett
and
Marc and Helen Gaylburd
for their guidance, support and love

40115897

First published in Great Britain in 1998 by Arnold
a member of the Hodder Headline Group,
338 Euston Road, London NW1 3BH
http://www.arnoldpublishers.com

Copublished in North, Central and South America by
Oxford University Press Inc., 198 Madison Avenue,
New York, NY 10016

British Library Cataloguing in Publication Data
A catalogue record for this book is available from the British Library

Library of Congress Cataloging-in-Publication Data
A catalog record for this book is available from the Library of Congress

ISBN 0 340 70072 6

Publisher: Nicki Dennis
Production Editor: James Rabson
Production Controller: Helen Whitehorn
Cover designer: M2

Typeset in 10/11 pt Times by Academic and Technical, Bristol, UK
Printed in Great Britain by MPG Books, Bodmin, Cornwall

Contents

Contributors

Jay Bennett is a Senior Consultant in telecommunications reliability and traffic engineering at Bellcore. He has been Chair and Secretary-Treasurer of the Section on Statistics in Sport of the American Statistical Association. Jay received an A.B. Degree in mathematics from Colgate University and M.S. and Ph.D. degrees in statistics from Temple University.

Stephen R. Clarke is an Associate Professor in the School of Mathematical Sciences at Swinburne University. He has published about 50 articles in professional journals on applications in sport. Stephen received his Ph.D. by publication on Performance Modelling in Sport, and is a winner of the 1988 Operational Research Society President's Medal for a paper on cricket.

John Croucher is an Associate Professor of Statistics at Macquarie University, Sydney, Australia. He is the author of eleven books on management, statistics, crime and humour, plus over 60 papers in scientific journals. For seven years, John presented tips and statistical information on Australian football telecasts. His expertise is strategies in sport, on which he has lectured widely.

John A. Flueck is Visiting Professor at the Department of Management, University of Nevada-Las Vegas. He is a fellow of the American Statistical Association and co-established its Section on Statistics in Sports. John serves on the editorial board of *Physician* and *Sports Medicine*. Research interests include management science, operations management, quality and productivity improvement, environmental monitoring and statistical modeling of sports.

Pat Larkey is Professor of Decisionmaking and Public Policy at Carnegie Mellon University, where he has also been Head of the Department of Social and Decision Sciences. He is the author of numerous books and articles which have appeared in *Operations Research*, *Management Science*, *Statistical Sciences* and *Golf World*. Pat is currently Editor-in-Chief of *Policy Sciences*.

Yuanlong Liu is a Ph.D. candidate in the School of Human Kinetics at the University of British Columbia. His dissertation topic focuses on the statistical validity of using ratio score variables, and his broader research interests include statistical and measurement issues in health and physical education, and applied statistics in sports.

Tim McGarry is currently a Ph.D. candidate at the University of British Columbia where he is studying the control processes that underlie movement. Other active interests include the analysis of sport performance for signatures of athletic behaviors, as well as the analysis of sports tournament designs.

Carl Morris is Chair of Statistics at Harvard, having previously been at Rand and the University of Texas. A past Editor of *JASA* and of *Statistical Science*, Carl's research has emphasized hierarchical modelling, Bayesian and likelihood theory, exponential families and their applications. The statistical theory of sports and competition has been a long-term interest.

John Norman is Professor of Management Studies at the University of Sheffield, UK, and has published papers on decision-making in a variety of sports, including tennis, squash and orienteering. John has taught at universities in the USA, India, Japan and the UK. His academic speciality is operational research. John is a regular jogger and swinger (of Indian clubs).

Robert Schutz is a Professor, and past Director, of the School of Human Kinetics at the University of British Columbia. He has served as President of the Canadian Sports Association of Sport Sciences and Program Chair for the Statistics in Sports Section of the American Statistical Association. His publications deal with statistical and measurement issues in human kinetics research.

Ray Stefani earned BSEE (Notre Dame, 1962) and MSEE and Ph.D. degrees (Arizona, 1964 and 1971). He has been at Cal State Long Beach since 1971 as Professor of Electrical Engineering. Sport publications include the rate of improvement of Olympic winning performance, sports rating systems and making predictions for sports such as US football, Australian Rules football and soccer.

Hal S. Stern is Professor of Statistics at Iowa State University. His research interests include methods for assessing the fit of statistical models and Bayesian methodology. In addition to working on problems related to sports, he is interested in applications of statistics to the biological and social sciences.

Robert L. Wardrop received a BA in Mathematics from Oakland University, and a Ph.D. in Statistics from the University of Michigan. He is the author of a textbook, *Statistics: Learning in the Presence of Variation*. In 1996 he won the Chancellor's Award for teaching excellence at the University of Wisconsin-Madison, where he is a Professor of Statistics.

Bill Williams spent 20 years at Bell Laboratories, where he wrote on problems in sampling and statistical methods. Currently, Bill is on the faculty at Hunter College as Professor of Mathematics and Statistics with interests in the effect of language factors on mathematics learning and in hockey statistics. He is a Fellow of the American Statistical Association.

Dave Williams has played almost 200 games in the National Hockey League as a defenseman for the San Jose Sharks and the Anaheim Mighty Ducks. Currently, Dave is with the St Louis Blues and is captain of their American Hockey League affiliate in Worcester, Massachusetts. Dave graduated from Dartmouth College.

Preface

Sports are statistics. Behind the flash of a Michael Jordan dunk, the majesty of a Mickey Mantle home run, the soaring glory of a Tiger Woods drive, sports have a foundation built upon scoring systems – put simply, numbers. Virtually, all sports contests are unplayable without the generation of numbers. Goals must be counted in soccer, points are accumulated in basketball, runs are tabulated in cricket. While a winner in a race can generally be decided without a statistic, downhill skiing still must resort to timing to decide a victor. Sports in which form is a major element, such as figure skating and platform diving, must utilize numbers as expressions of subjective judgments.

Serious research in sports statistics has been going on for a long time. Predecessors to this book, such as *Management Science in Sports* and *Optimal Strategies in Sports*, were published in the 1970s. Research has been published in many journals with various interests and affiliations (see Schutz, *Amstat News*, April 1996, p. 22–23, or www.stat.duke.edu/~box/sis/pubs/journals.html). Still, however, no journal has dedicated itself to sports statistics. The diffusion of publications over many journals has made it difficult (if not impossible) to keep up with current research. Recently though, several journals have recognized the increasing interest in sports statistics. *Chance*, published by the American Statistical Association (ASA), has inaugurated a regular column by Hal Stern on sports statistics. *The Statistician* has a periodic section devoted to research papers on sports statistics.

A sign of the maturity of sports statistics has been the development of research bodies dedicated to sports statistics. The American Statistical Association created a Section on Statistics in Sports in 1992. The section, now numbering about 400 members, sponsors invited and contributed paper sessions at the annual Joint Statistical Meetings, publishes a proceedings of papers from the meetings, and maintains a web site that can be reached through the ASA site (www.amstat.org) or directly at www.stat.duke.edu/~box/sis. In 1993, the International Statistical Institute created a Sports Statistics Committee, which also publishes proceedings of their bienniel meetings.

Statistical analysis of sports provides inspiration for extensions of the same ideas to more general uses. In *The American Statistician* (November 1997), Frederick Mosteller pondered the lessons he has learned from his research in sports statistics. Bradley Efron and Carl Morris found that their application of shrinkage estimation to baseball statistics provided insight into applications to other sets of data. Stephen Jay Gould used his analysis of the disappearance of the .400 hitter in baseball as a springboard to issues in zoology. Bill James feels that since athletes have the best-kept records of performance among professional workers, these records could be used as a sample of society; the relationships of their athletic performance with other aspects of their lives could provide sociological insights.

In *Innumeracy* (Hill and Wang, 1988), John Allen Paulos recounts 'The earliest memory I have of wanting to be a mathematician was at age ten, when I calculated that a certain relief pitcher for the then Milwaukee Braves had an earned run average (ERA) of 135 ... Impressed by this extraordinarily bad ERA, I diffidently informed my teacher, who told me to explain the fact to my class ... When I finished, he announced that I was all wrong and that I should sit down. ERAs, he asserted authoritatively, could never be higher than 27.' (As baseball fans know, ERAs can be ∞.) Fortunately for Paulos (and to his teacher's disgruntlement), he was vindicated by the publication of the correct ERA of 135 in the *Milwaukee Journal*. Not all statisticians may have had such a dramatic and satisfying experience as this in their early lives, but it is unquestionable that many got their first taste of calculating and analyzing statistics from an interest in sports. For many, sports provided the first real application of mathematics beyond the contrived or less immediate examples too often given in textbooks. Schoolyard arguments over the relative merits of different athletes gave us the first opportunity to use the analysis of numbers as evidence in debate. As the sports themselves have provided generational continuity, the discussion of sports statistics allowed us in our youth to discourse as equals with our parents.

Sports statistics are a powerful tool to capture the attention of students. When compared with the typical examples provided in statistics textbooks (e.g. manufacturing processes, clinical trials), examples from sports statistics have two immediate advantages. First, the situation or context of the example requires less explanation – because of the relative familiarity with sports, the student can focus on the statistics rather than comprehending the background of the example. Second, the student is more likely to be intrigued to find the result. While other examples might be appealing at an intellectual level, the result of a sports problem or example has a greater chance of producing an intuitive or visceral reaction to the answer. In this difficult time of attracting students to mathematics and the sciences, instructors would do well to take advantage of these benefits afforded by sports statistics.

This book is composed of nine chapters, each dedicated to a particular sport, followed by four chapters, each dedicated to a theme which spans sports. The authors were charged with the task of providing a broad survey of professional and popular literature on their subjects with the option of delving more deeply into the topics of greatest current interest. A secondary goal of the book is to identify structure in the research and provide organizing themes in their presentation. This goal was inspired by the proposals of Robert Schutz to provide a framework for the study of sports statistics (The systematic study of 'Statistics in Sports': Do we need a framework? *1994 Proceedings of the Section on Statistics in Sports*, Alexandria, VA: American Statistical Association, pp. 16–20).

The first section provides a separate chapter on each of the nine sports covered (American football, baseball, basketball, cricket, soccer, golf, ice hockey, tennis, and track and field). While each chapter covers a different sport, the reader will identify recurrent themes across the chapters. For example, research in player value is found in the chapters on baseball, basketball, golf and ice hockey. Developing strategies with respect to the probability of winning is addressed in baseball, cricket, soccer, tennis, and track and field. Each chapter covers these themes with variations appropriate to the context of the sport. To heighten the reader's awareness of these interrelationships, the chapters contain pointers to other chapters which cover the material with a different orientation or degree of detail.

Certain sports with great fan interest (e.g. Australian Rules football, rugby, boxing, horse racing) are not highlighted in separate chapters. The selection of the sports covered was entirely my own. Each sport's selection was based on my understanding not only of the popularity of the sport but my perception of the depth of statistical research performed. I extend my apologies to all who find their favorite sport not covered.

The second section provides four chapters which examine general areas of research across sports. A general area of interest is rating teams or players and using these rating systems to predict outcomes. Another is the structure of tournaments in order to determine a champion and rank teams or players in an efficient and effective manner. In recent years, graphical techniques have been at the forefront of statistical research and their use in sports reflects this importance. Popular analysis of sports statistics has been plagued by overstatement of conclusions based on small samples; Bayesian statistics have been employed to study and ameliorate this problem. Each of these subjects is covered in a separate chapter.

The book is intended for an international audience. It is likely that any reader will have great familiarity with the majority of the sports covered and significantly less familiarity with the remaining sports. (Of course, which sports fall in the majority and minority categories depends greatly on the geographic location of the reader.) The authors have attempted to write the chapters in such a way that they should be understandable by most readers with a passing knowledge of the sport. Certain sports (particularly baseball and cricket) have sets of rules and terminology which defy simple explanation. In the case of baseball, an appendix has been provided with a brief description of the game. The reader is also advised to refer to the glossary which contains definitions from all sports necessary for following the chapters.

The book could be used at many levels in a course of instruction. On occasion, authors use some terminology which requires understanding of advanced statistical concepts, but, for the most part, the chapters should be easily understood by first-year college students. The book should prove useful as a supplementary text to introductory statistics, a primary text on statistics in a sports management program, and a text for first-year seminars popular in many liberal arts colleges. Of course, we hope that the sports fan who wishes greater depth in sports statistics analysis than can be provided by the newspaper, television, or sports radio will find the book of interest as well.

Perhaps the book's primary function is to provide a starting point for any serious researcher in sports statistics. Too often, researchers in the area have rediscovered results already published. This is understandable because research in sports statistics has been so diffused throughout the professional literature. Hopefully, this book will provide a common root from which statistical research in sports will grow and flourish.

The words 'he' and 'his' are used in this book for brevity and clarity only and no discrimination with regard to women's sports is intended or should be inferred.

Acknowledgments

I would like to thank Nicki Dennis for proposing the idea for the book and presenting me with the opportunity of assembling it. I would like to give a special note of thanks to Professor Frederick Mosteller and Professor James Albert for their invaluable advice and comments at critical junctures of the project. This project would not have been possible without the support of my wife, Lynn, who endured a long and very trying year as a statistical form of a 'football widow'. Most of all, I would like to extend my appreciation to the authors who gave of themselves so completely in their work.

JAY BENNETT
Bellcore

Part I
MAJOR SPORTS

1

American Football

Hal S. Stern

Department of Statistics, Iowa State University, USA

1.1 Introduction

1.1.1 A brief description of American football

There are a number of sports played across the world under the name 'football', with most of the world reserving that name for association football or soccer. Soccer is discussed in Chapter 5 of this volume; this chapter, however, considers American football, the basic structure of which is described in the next paragraph. Most of the material in this chapter is discussed in terms of the National Football League – the professional football league in the United States. Other versions of the game include college football in the United States and professional football in Canada. There can be substantial differences between different versions of American football, e.g. the Canadian and United States games differ with respect to the length of the field and the number of players, among other things. Despite these differences, the methodology and ideas discussed in this chapter should apply equally well to all versions of American football.

American football, which we shall call football in this chapter, is played by two teams on a field 100 yards long, with each team defending one of the two ends of the field (called goal lines). Games are 60 minutes long, and are broken into four 15-minute quarters. The two teams alternate possession of the ball and score points by advancing the ball (by running or throwing/catching) to the other team's goal line (a touchdown worth 6 points, with the additional opportunity to attempt a 1-point or 2-point conversion play), or failing that by kicking the ball through goal posts situated at the opposing team's goal (a field goal worth 3 points). The team in possession of the ball (the offense) must gain 10 or more yards in four plays (known as downs) or turn the ball over to their opponent. The ball is advanced by running with it, or by throwing the ball to a team-mate who may then run with the ball. As soon as 10 or more yards are gained, the team starts again with a first down, and a new opportunity to gain 10 or more yards. If a team has failed to gain the needed 10 yards in three plays then it has the option of trying to gain the remaining yards on the fourth play or kicking (punting) the ball to its opponent to increase the distance that the opponent must move to score points.

This very brief description ignores some important aspects of the game (the defense can score points via a safety by tackling the offensive team behind its own goal line, teams turn the ball over to their opponents via dropped/fumbled balls or interceptions of thrown balls) but should be sufficient for reading most of this chapter.

1.1.2   A brief history of statistics in football

As with the other sports in this volume, large amounts of quantitative information are recorded for each football game. These are primarily summaries of team and individual performance. For United States professional football these data can be found as early as the 1930s (Carroll *et al.*, 1988) and in United States collegiate football they date back even earlier. We focus on the use of probability and statistics for analyzing and interpreting these data in order to understand the game better, and ultimately perhaps to provide advice for teams about how to make better decisions.

The earliest significant contribution of statistical reasoning to football is the development of computerized systems for studying opponents' performances (e.g. Purdy, 1971; Ryan *et al.*, 1973). Professional and college teams currently prepare reports detailing the types of plays and formations favored by opponents in a variety of situations. The level of detail in the reports can be quite remarkable, e.g. they might indicate that Team A runs to the left side of the field 70% of the time on second down with five or fewer yards required for a new first down. These reports clearly influence team preparation and game-time decision-making. These data could also be used to address strategy issues (e.g. should a team try to maintain possession when facing fourth down or kick the ball over to its opponent) but that would require more formal analysis than is currently done – we consider some approaches later in this chapter. It is interesting that much of the early work applying statistical methods to football involved people affiliated with professional or collegiate football (i.e. players and coaches) rather than statisticians. The author of one early computerized play-tracking system was 1960s' professional quarterback Frank Ryan. Later in this chapter we will see contributions from another former quarterback, Virgil Carter, and a college coach, Homer Smith.

1.1.3   Why so little academic research?

Football has a large following in the United States and Canada, yet the amount of statistical or scientific work by academic researchers lags far behind that done for other sports (most notably, baseball). It is interesting to speculate on some possible causes for this lack of results. We briefly describe three possible causes: data availability, the nature of the game of football, and professional gambling.

First, despite enormous amounts of publicity related to professional football, it is relatively difficult to obtain detailed (play-by-play) information in computer-usable form. This is not to say that the data do not exist – they clearly do exist and are used by teams during the season to prepare their summaries of opponents' tendencies. However, the data have not been easily accessible to those outside the sport. The quality of available data is improving, however, as play-by-play listings can now be found on the World Wide Web through the National Football League's own site (www.nfl.com). These data, however, are not yet in a convenient form for research use.

A second contributing factor to the shortage of research results concerns the nature of the game. Examples of the kinds of things that can complicate statistical analyses include: scores occur in steps of size 2, 3, 6, 7, 8 rather than just a single scoring increment, the game is time-limited with time management an important part of strategy, and actions (plays) move the ball over a continuous surface with an emphasis on 10-yard pieces. All of these conspire to make the number of possible situations that can occur on the football field extremely large, which considerably complicates analysis.

One final factor that appears to have worked against academic research about the game itself is the existence of a large betting market on professional football games. A great deal of research has been carried out on the football betting market, including methods for rating teams (described later in the chapter) and methods for making successful bets (described in Chapter 12). Unfortunately, because of its applicability to gambling, a large portion of this research is proprietary and unavailable for review by other researchers.

The remainder of this chapter is organized as follows. The next three sections describe research results and open problems in three major areas: player evaluation, models for assessing football strategy, and rating teams. Following that is a section that touches on other possible areas of research. The chapter concludes with a brief summary and a list of references. One reference merits a special mention at this point; *The Hidden Game of Football*, a 1988 book by Bob Carroll, Pete Palmer and John Thorn, is a sophisticated analysis of the game by three serious researchers with access to play-by-play data. However, written for a popular audience, the book does not provide some of the details that readers of this book (and the author of this chapter) would find interesting. We refer to this book quite often and use CPT (the initials of the three authors) to refer to it.

1.2 Player evaluation

Evaluation of football players has always been important for selecting teams and rewarding players. Formally evaluating players, however, is a difficult task because several players contribute to each play. A quarterback may throw the ball five yards down the field and the receiver, after catching the ball, may elude several defensive players and run 90 additional yards for a touchdown. Should the quarterback get credit for the 95-yard touchdown pass or just the five yards the ball traveled in the air? What credit should the receiver get? We first review the current situation and then discuss the potential for evaluating players at several specific positions.

1.2.1 The current situation

Evaluation of players in football tends to be done using fairly naive methods. Football receivers are ranked according to the number of balls they catch.

Running backs are generally ranked by the number of yards they gain. Punters are ranked according to the average distance they kick the ball without regard to whether they are effective in making the opponent start from a poor field position. Kickers are often ranked by the number of points scored. The most complex system, the system for ranking quarterbacks, is quite controversial – we will review it shortly. Defensive players receive little evaluation in game summaries beyond simple tallies of passes intercepted, fumbles recovered, or quarterbacks tackled for a loss of yardage. Offensive linemen, whose main job is to block defensive players, receive essentially no formal statistical evaluation. Several problems are evident with the current situation: the best players may be misevaluated (or not evaluated at all) by existing measures, and it can be very difficult to compare players of different eras (because there are different numbers of games per season, different football philosophies, and continual changes in the rules of the game).

1.2.2   Evaluating kickers

The difficulty in apportioning credit to the several players that contribute to each play has meant that a large amount of research has focused on aspects of the game that are easiest to isolate, such as kicking. Kickers contribute to their team's scoring by kicking field goals (worth 3 points) and points-after-touchdowns (worth 1 point). On fourth down, a coach often has the choice of: (1) attempting an offensive play to gain the yards needed for a new first down; (2) punting the ball to the opposition; or (3) attempting a field goal. Evaluation of the kicker's ability will have a great deal of influence on such decisions. Berry and Berry (1985) use a data-analytic approach to estimate the probability that a field goal attempted from a given distance will be successful for a given kicker. They then propose a number of measures for comparing two kickers, e.g. the estimated probability of converting a 40-yard field goal. Interestingly, Morrison and Kalwani (1993) examine all professional kickers and conclude that binomial variability is sufficiently large that a null model that says *all* kickers are equally good (or bad) would be accepted. This suggests that rating kickers may not be a good idea at all. Of course, as they point out, it is also possible that some kickers are indeed better than others but that the 30 or so field goal attempts per season are not enough to detect the difference. In addition to comparing kickers, it can be valuable to explore the factors affecting the probability of success of a field goal. Bilder and Loughin (1998) pool information across kickers to determine the key factors affecting the success of a field goal. They find that yardage is most important, but that the score at the time of the kick matters, with field goals causing a change in the lead more likely to be missed than others. This effect is akin to the clutch (or choke) factor so heavily researched in baseball (see Chapter 2).

For a single team, an interesting set of questions concerns the effective use of that team's kicker. Irving and Smith (1976) built a detailed model of the probability of a successful field goal for a single kicker. The result of their analysis, a plot showing the probability of a successful kick from any point on the field, was used by the coaching staff at the University of California, Los Angeles during the 1972–73 season, to assist in decision-making.

1.2.3 Evaluating quarterbacks

The quarterback is the player in charge of the offense, the part of the team responsible for trying to score points. He is certainly the most visible player and many argue he is the most critical player on the team. The main skill on which quarterbacks are evaluated is their ability to throw the ball to a receiver for a completed pass. A ball that is thrown and not caught is an incomplete pass and gains no yards. Even worse, a ball that is thrown and caught by an opponent is an interception and results in the opponent taking possession of the ball. The official National Football League system for rating quarterbacks awards points for each completed pass, intercepted pass (negative value), touchdown pass, and yards earned. Essentially, the system credits quarterbacks for their passing yardage and includes a 20-yard bonus for each completed pass, an 80-yard bonus per touchdown pass, and a 100-yard penalty per interception. The system has been heavily criticized for favoring conservative short-passing quarterbacks – after all, two five-yard completions are better rewarded than one 10-yard completion and one incomplete pass. CPT (recall that is Carroll, Palmer and Thorn, 1988) describe the existing system and propose a modest revision of similar form. CPT suggest that there should be no reward for completing a pass and that the touchdown and interception bonuses are too large. Their system appears to be a bit of an improvement, but still does not tie quarterback performance ratings to the success of the offense in scoring points or the success of the team in winning games.

1.2.4 Summary

Here we have briefly reviewed some of the difficulties with evaluating players, with a focus on kickers and quarterbacks. In this era of greater freedom in player movement from team to team, research regarding the value of a player or the relative value of two players will become even more crucial. Some of the problems associated with existing methods could be improved by careful application of fairly basic statistical ideas, e.g. examining the proportion of successful kicks rather than the total, examining the yardage contribution of receivers rather than just the number of catches, or considering the yards gained per attempt by running backs.

Two problems associated with evaluating players are more substantial and will be difficult to overcome, and so are candidates for future research work. The first is the problem of apportioning credit for a play among the various players contributing, e.g. the quarterback and receiver on a pass play. This might be resolved by more detailed record keeping, perhaps an assessment of how much yardage would have been obtained with an 'ordinary' receiver. Even then it is difficult to imagine how the contribution of the linemen might be incorporated. One possibility is a plus/minus system like that used in ice hockey (see Chapter 7), which rewards players on the field when positive events occur (points are scored) and penalizes players on the field when negative events occur (the ball is turned over to the opponent). The second problem with player evaluation is that the focus on yardage gained, although natural, means that more important concerns such as points scored and games won are not used explicitly in player evaluation. For example, all interceptions are treated the same, even those that occur on last-second desperation throws. As part of our discussion of football strategy in the next section, we will build up some tools that might be used to improve player evaluation.

1.3 Football strategy

1.3.1 Different types of strategy questions

As described in the introduction, professional and college football teams use data on opposing teams' tendencies to prepare for upcoming games. The data have generally not been used to address a number of other strategy questions that require more statistical thinking. Here we provide examples of some of these types of questions. The first issue concerns point-after-touchdown strategy: football teams have the option of attempting a near certain 1-point conversion after each touchdown (probability of success is approximately 0.96) or attempting a riskier 2-point conversion (probability of success appears to be roughly 0.40–0.50). The choice will clearly depend on the score, especially late in the game. Porter (1967) constructs decision rules for end-of-game extra-point strategy, but his method does not make any suggestions for decisions earlier in the game. Another example of a strategy question concerns fourth-down decision-making. Teams have four downs to gain 10 yards or must give the football over to their opponents. On fourth down, a team must choose whether to try for a first down with the risk of giving the ball to its opponent in a good scoring position, attempt a field goal (worth three points) or punt the ball to its opponent so that the opponent's position on the field is not quite so good. Choosing among the three options clearly depends on the current game situation and also requires reasonably accurate information about the value of having the ball at various points on the field. A key feature of both the point-after-touchdown and fourth-down strategy questions is that the optimal strategy will almost certainly depend on the game situation as measured by the current score and time remaining.

Other strategy questions can be isolated from the game context in the sense that the optimal strategy does not depend on the game situation. For example, Brimberg *et al.* (1998) have analyzed the placement of punt-returners in Canadian football (the wider and longer field and more frequent punts make this a more important issue in Canada than it is in the US). They find that a single returner can perform nearly as well as two returners, and that if two returners are used they should be configured vertically, rather than the more traditional horizontal placement (i.e. at the same yard line).

In the remainder of this section we focus on strategy questions that need to be addressed in the context of the game situation (score, time remaining, etc). We consider two ways of assessing the current game situation and use them to develop appropriate strategies. First, we measure the value of having the ball at a particular point on the field by estimating the expected number of points that a team will earn for a possession starting from the given point. Decisions can then be made to maximize the expected number of points obtained by the team. Following that, we consider a more ambitious proposal, measuring the probability of winning the game from any situation. Strategies may then be

developed that directly maximize the probability of winning (the global objective) rather than maximizing the number of points scored in the short-term
(a local objective). Making decisions to maximize the team's probability of
winning the game would seem to be a superior approach, but we will see that
it turns out to be quite difficult to put this idea into practice.

1.3.2 Expected points

1.3.2.1 Estimating the expected number of points for a given field position

Carter and Machol (1971) use a data-based approach to estimate the expected
number of points earned for a team gaining possession of the ball at a given
point. Let $E(\text{pts}|Y)$ denote the expected number of points for a team beginning
a series (first down) with the ball Y yards from the opposing team's goal. The
natural statistical approach for evaluating the expected number of points is to
examine all possible outcomes of a possession starting from the given point,
recording the value (in points) of each outcome to the team with the ball and
the probability that each outcome will occur. In football, there are 103 possible
outcomes, four of which involve points being scored: touchdown (7 points,
ignoring for the moment questions about point-after-touchdown strategy),
field goal (3 points), safety (-2 points, i.e. 2 points for the opponent), and opponent's touchdown (-7 points, i.e. 7 points for the opponent). The remaining 99
outcomes cover the cases when the ball is turned over to the opposing team with
the opponent needing Z yards for a touchdown, with Z ranging from 1 to 99.
The opponent can expect to score $E(\text{pts}|Z)$ points after receiving the ball Z
yards from its target and hence the value of this outcome for the team currently
in possession of the ball is $-E(\text{pts}|Z)$.

There are two complications that must be addressed before the expected point
values can be determined. First, the probability of each of the outcomes is
unknown and must be estimated. Carter and Machol find the probability of
each outcome based on data from 2852 series (2852 sequences of plays that
began with first down and 10 yards to go). The second complication is that,
as derived above, the expected number of points for a team Y yards from the
goal depends on the expected number of points if its opponent takes possession
Z yards from the goal. In total, it turns out that there are 99 unknown values,
$E(\text{pts}|Y)$, and these can be found using the 99 equations that define the expected
values.

In fact, Carter and Machol choose not to solve this large system of equations
with their limited data. Instead, they combine all the series that began in the
same 10-yard section of the field (e.g. 31–40 yards to go for a touchdown).
Their results are provided in Fig. 1.1. The results can be summarized by
noting that it is worth about 2 points on average to start a series at midfield
(50 yards from the goal line), and every 13 yards gained (lost) corresponds
roughly to a 1 point gain (loss) in expected value. Following this rule of
thumb, we find that having the ball near the opposing team's goal is worth a
bit less than 7 points since the touchdown is not guaranteed, and having the
ball near one's own goal is worth a bit more than -2 points (the value of
being tackled behind one's own goal). Interestingly, it appears that starting
with the ball just beyond one's own 20-yard-line (80 yards from the goal) is a

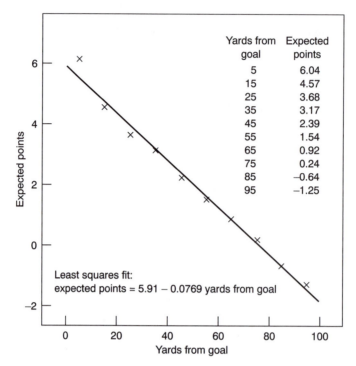

Figure 1.1   Expected points for a team with first down and 10 yards to go from various points on the field and the associated least squares line. Data are from Carter and Machol (1971).

neutral position (with zero expected points) and that is fairly close to the typical starting point of each game. The Carter and Machol analysis was carried out using data from the 1969 season. Football rules have been modified over time and it is natural to wonder about the effect of such rule changes. CPT redid the Carter and Machol analysis using 1986 data and obtained similar results.

1.3.2.2   Applying the table of expected point values

It is possible to use the expected point values of Fig. 1.1 to address some of the football strategy issues raised earlier. First, we describe Carter and Machol's use of their results to evaluate the football wisdom that says turnovers (losing the ball to your opponent by making a gross error) near one's own goal are more costly than turnovers elsewhere on the field. From the data in Fig. 1.1 one can see that a turnover at one's own 15-yard-line (85 yards from the target goal) changes a team from having an expected value of −0.64 to one of −4.57 (the opponent's value after taking possession is 4.57), a drop of 3.93 expected points. The same turnover at the opponent's 45-yard-line changes the expected points from 2.39 to −1.54, a drop of 3.93 expected points! Turn-overs are worth about 4 points and this value does not seem to depend on the location at which the turnover occurs.

Next we consider the question of appropriate fourth down strategy. Here is a specific example: consider a team at its opponent's 25-yard-line (25 yards from the goal) with fourth down and one yard required for a first down that will allow the team to maintain possession of the ball. Suppose the offensive team tries to gain the short distance required; then they will either have a first down still in the neighborhood of the 25-yard-line (expected value 3.68 points from Fig. 1.1) or the other team will have the ball 75 yards from their target goal (expected value to the team currently in possession is −0.24 points). Professional teams are successful on fourth-down plays requiring one yard about 70% of the time which means that they can expect $(0.7)(3.68) + (0.3)(-0.24) \approx 2.5$ points on average if they try for the first down. The result is recorded as an approximation – it ignores the possibility that the offensive team will gain more than a single yard but it also ignores the possibility that the team will lose ground if it turns the ball over. Field goals (kicks worth 3 points if successful) from this point on the field are successful about 65% of the time. Should the field goal miss, the other team would take over with 68 yards-to-go under current rules (expected value to the offensive team of approximately −0.65 points). Trying a field goal yields $(0.65)(3) + (0.35)(-0.65) = 1.7$ points on average. Clearly teams should go for the first down rather than try a field goal as long as these probabilities are reasonably accurate. In fact, for the specified field goal success rate, we find the probability of success required to make the fourth-down play the preferred option is about 0.50. CPT investigated a number of such scenarios and found that field goals are rarely the correct choice for fourth-down situations with six or fewer yards required for a first down (at least when evaluated in terms of expected points).

Expected points might also be used to evaluate teams' performances. Teams could, for example, be judged by how they perform relative to expectation by recording the expected number of points and the actual points earned for each possession. If the offensive team begins at their 25-yard-line (75 yards from the goal) and scores a field goal then they have earned 3 points, 2.76 more than might have been expected at the start of the possession. The contributions of the offense, defense, and special teams (punting and kicking) could be measured separately. It is more difficult to see how this can be applied to evaluating individual players since expected point values are only determined for the start of a possession. Our fourth-down example above indicates how we can assign values to plays other than at the start of a possession, but it becomes quite complicated when we try to assign point values to second- or third-down situations. If expected point values were available for every game situation, then it might be possible to give a player credit for the changes in the team's expected points that result from his contributions. Of course, partitioning credit among the several players involved in each play remains a problem.

1.3.2.3 Limitations of this approach

There are limitations associated with using expected point values to make strategy decisions. The first limitation is that the Carter and Machol (and CPT) expected point values are based on aggregate data from the entire league. Individual teams might have different expected values. For example, a team that prefers to advance the ball by running might have lower expected values from a given point on the field than a team that prefers to advance the ball by throwing. A second limitation is that, even if we accept a common set of expected point values, applying the expected point values to determine appropriate strategies requires assessing the probabilities of many different events. For example, if a team's kicker has an extremely high probability of success, or a team has an ineffective offense that is not likely to succeed on fourth down, then the fourth-down strategy evaluation we considered earlier might turn out differently. It is probably not appropriate to think of this as a problem; it merely points out that proper use of Fig. 1.1 requires the user to make determinations of the relevant probabilities.

A final limitation is that the expected points approach completely ignores two key elements of the game situation, the score and the time remaining. The correct strategy in a given situation should surely be allowed to depend on these important factors. As an extreme example, consider a team that trails by 2 points with 1 minute remaining in the game and faces fourth down on the opponent's 25-yard-line with one yard needed for a first down. An analysis based on expected points that suggests a team should try for the first down is clearly invalid because at that late point in the game, it is more important to maximize the probability of winning (achieved by trying to kick the field goal) than to maximize the expected number of points. We next consider an approach that treats maximizing the probability of winning as the objective and tries to take into account all the important elements of the game situation.

1.3.3   The probability of winning the game

1.3.3.1   *Estimating the probability of winning for a given situation*

Carter and Machol's expected number of points for different field positions can be used to make optimal short-term decisions when the time remaining is not a critical element (time remaining is important near the end of the first half and the end of the game). Decisions late in the game need to be motivated more by concerns about winning the game than about maximizing the expected number of points. This motivates an alternative approach to football strategy that requires estimating the probability of winning the game from any current situation. For the purposes of this chapter, we define the current game situation in terms of: the current difference in scores (ranging perhaps from −30 to 30), the time remaining (perhaps taking the 60-minute game to consist of 240 15-second intervals), position on the field (1 to 99 yards from the goal), down (1 to 4), and yards needed for a first down that will allow the team to maintain possession (ranging perhaps from 1 to 20). The win probabilities can be estimated for each game situation in a number of different ways. We describe two basic approaches: an empirical approach similar to that used by Carter and Machol, and an approach based on constructing a probability model for football games. Either approach must deal with the enormous number of possible situations. Using the values given above, there are more than 100 million possible situations.

An empirical approach   Conceptually at least, we can proceed exactly as Carter and Machol did and obtain probability estimates directly from play-by-play data. For any game situation, we need only record the frequency with which the situation occurs and the ultimate outcome (win/loss) in each game. However, the number of possible situations is far too large for this approach to be feasible. After all, there are more than 100 million situations and only 240 National Football League games per season with 130 plays per game. CPT perform an analysis of this type by restricting attention to the beginning of a team's possession (situations with first down and 10 yards to go), taking the current difference in scores to be between −14 and 14, and taking the time remaining to consist of 20 3-minute intervals. These modifications reduce the number of situations to a more manageable number, $29 \times 20 \times 99 = 57\,420$. They use two seasons' data to obtain estimates of the probability of winning the game for each of the 57 420 situations. For example, the probability that a team beginning the game with first down at its own 20-yard-line ultimately wins the game, is 0.493 according to CPT. By way of comparison, a team starting with first down at its own 20-yard-line but trailing by 7 points with 51 minutes remaining in the game has a probability of winning equal to 0.281. Unfortunately, there is no description of how the win probabilities were actually estimated from the data so it is difficult to endorse them completely. More important, it is not possible to obtain win probabilities for situations that are not explicitly mentioned in the book.

In order to pursue this approach further, we approximate the win probability function derived by CPT using some simple statistical modeling. Using 76 win probability values provided in the book, we derive the following fairly simple logistic approximation that gives the probability of winning, p, in terms of the current score difference, s, the time remaining (in minutes), t, and the yardage to the opposing team's goal, y, at the beginning of a team's possession

$$\ln\left(\frac{p}{1-p}\right) = 0.060s + 0.084\frac{s}{\sqrt{t/60}} - 0.0073(y - 74) \qquad (1.1)$$

where 'ln' is the natural logarithm. This approximation is motivated by the Stern (1994) model that relates the current score and time remaining to the probability of winning a basketball or baseball game. Note that the logistic equation empirically establishes a team's own 26-yard-line ($y = 74$ yards from the goal) as a neutral field position at the start of a possession. For the two situations described in the preceding paragraph, this approximation gives 0.489 and 0.249 (compared to the CPT values, 0.493 and 0.281). Table 1.1 compares the values obtained by CPT and those obtained by the logistic approximation for a number of situations. (The final column of Table 1.1 will be discussed later.)

The probability of winning is shown graphically in Figs 1.2(a)–(c) for selected values of the score difference, s, time remaining, t, and yards from the goal, y. Figure 1.2(a) shows the importance of the time remaining. Even relatively modest score differences become significant as the time remaining decreases towards zero. Figure 1.2(b) indicates that for the logistic approximation the effect of field position is (for the most part) independent of the score difference and time remaining. Figure 1.2(c) shows once again the effect of time remaining, with the curve corresponding to less time remaining being steeper near the zero score difference. Figure 1.2(c) also illustrates a weakness of the logistic approximation. Because it is not derived expressly for football, the logistic approximation does not account for the 3- and 7-point scoring increments, i.e. the curves in

Table 1.1 Win probabilities from Carroll *et al.* (1988), along with two alternatives described in the text

Score difference	Yards from goal	Time remaining (minutes)	Estimated win probability		
			CPT	Logistic	Dynamic programming
0	80	60.0	0.493	0.489	–
0	80	23.6	0.490	0.489	–
−7	80	47.4	0.274	0.246	–
−7	67	21.0	0.218	0.205	–
−7	74	13.5	0.153	0.161	–
−8	94	10.5	0.025	0.097	0.097
5	67	7.7	0.842	0.820	(0.811, 0.816)
−5	74	6.6	0.178	0.174	(0.267, 0.270)
5	78	2.4	0.945	0.915	(0.981, 0.989)
−5	58	1.8	0.069	0.071	(0.037, 0.152)
5	50	1.3	0.990	0.964	(0.998, 1.00)

Fig. 1.2(c) are continuous. The logistic approximation treats the difference between a 4-point deficit and a 6-point deficit as being no different than the difference between a 7-point deficit and a 9-point deficit. Clearly the latter difference is much more significant, since the 9-point deficit will require the team

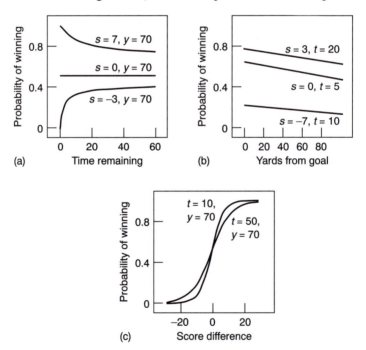

Figure 1.2 Probability of winning as a function of the score difference, s, the time remaining (in minutes), t, and yards from the goal, y, using the logistic approximation: (a) probability as a function of time remaining for three selected score/yards-from-goal combinations; (b) probability as a function of yards from goal for three selected score/time-remaining combinations; (c) probability as a function of score difference for two time-remaining/yards-from-goal combinations.

that is behind to get at least two scores to tie or win, whereas the 7-point deficit can be made up with a single score. In contrast, both a 4-point and a 6-point deficit can be overcome by a single score. A second weakness of the logistic approximation is that when the score difference is equal to zero the probability of winning does not depend on the time remaining (this is also visible in Fig. 1.2(c) as curves with different amounts of time remaining intersect when the score difference is zero). It seems likely that the probability of winning would be higher for a team with $s = 0$, $y = 1$, $t = 1$ (excellent field position near the end of a tied game) than for a team with $s = 0$, $y = 1$, $t = 59$ (excellent field position very early in a tied game). Before using the win probabilities to find answers to our strategy questions, we consider another approach to estimating the win probabilities.

Dynamic programming   Dynamic programming is a technique that can be used to find optimal strategies and simultaneously derive the probability of winning from a given situation under optimal play. We first describe a decision-theoretic formulation of football that allows us to apply dynamic programming. Let us take the two teams in the game to be Team A and Team B. As before, we consider the current situation, or state (as it is generally called in dynamic programming), of the football game as being given by: the difference in scores, the time remaining, the position on the field, the down and the yards needed for a first down. In addition, we will need to keep track of which of the two teams has possession of the ball so that we add this to the definition of the state. Each state is associated with a value that can be thought of as defining the objective of the game, e.g. we might take the value of a state to be the probability that Team A wins starting in the given state. From any state, the two teams have a limited number of actions from which they must choose. Although there is considerable flexibility in defining this set of actions, for now we will restrict our attention to the choices available to the team in possession of the ball. Their possible actions include run, short pass, long pass, punt and field goal. Not every action is reasonable from every state (e.g. we would not try a field goal on first down from our own five-yard-line), but any reasonable model will avoid choosing these suboptimal actions. Team A should choose the action at each point in the game that will give them the highest probability of winning (i.e. they try to maximize the expected value of the next state) and Team B should choose the action that will give Team A the lowest probability of winning (i.e. they try to minimize the expected value of the next state). We require the distribution of possible outcomes for each of the possible actions (a difficult task that we will return to shortly) to solve for the optimal action in a given state. Dynamic programming is an algorithm for finding the optimal action for every state and determining the value of being in that state (this is the probability that Team A wins from that state).

Dynamic programming starts at the end of the game (no time remaining) by defining any state in which Team A is ahead of its opponent as having value one, and any state in which Team A is behind as having value zero. Ties can be given the value one-half. These values, corresponding to the probabilities of Team A's winning the game, are obvious because there is no time remaining in the game. Now, given that we know the value of every state at the end of the game, we can back up one time unit (15 s is the specification used here) and determine the optimal strategy for any state with one time unit remaining. First, we evaluate the expected probability of winning under each action by averaging over the distribution of possible outcomes. Team A should choose the action that gives it the highest expected probability of winning the game. Team B, when it is in possession of the ball, should choose the action that gives Team A the lowest expected probability of winning the game. After determining the optimal strategy and value for every state with one time unit remaining, we can continue to move backwards from the end of the game. We find the optimal strategies for the states at time t by averaging over the results that we have already found for future states. Dynamic programming is a powerful computational algorithm for solving complex decision problems like this one.

It remains only to describe how we determine the distribution of outcomes under any action. In theory, this distribution could be obtained by a careful analysis of detailed play-by-play data. Here, a small sample of play-by-play data was used to suggest an approximate distribution. To illustrate, Table 1.2 gives the distribution for a run play, a short pass play, and a long pass play. Each row of the table gives one possible outcome (yardage gained and an indication of whether the ball has been turned over to the opponent) and the probability that it occurs. These probability distributions were constructed to match known features of the true distributions, e.g. the probability of a lost fumble is 0.015 and the probability of an intercepted pass is 0.04 (results cited in CPT and, more recently, in Brimberg and Hurley, 1997). Note that passes may result in an interception or fumble so the probability of a turnover is 0.055 when averaged over all pass plays. Similarly, the mean gain on a run is just under four yards. The remaining details of the distribution represent a crude estimate based on limited data. The distribution of possible outcomes

Table 1.2 Assumed distribution of outcomes for run plays, short pass plays, and long pass plays. Each play is assumed to consume one 15 s time unit

			Distribution of outcomes for various actions					
Run play			*Short pass play*			*Long pass play*		
Yards	Turnover	Probability	Yards	Turnover	Probability	Yards	Turnover	Probability
−4	0	0.020	−5	0	0.030	−10	0	0.045
−2	0	0.060	−5	1	0.010	−10	1	0.005
−1	0	0.065	0	0	0.400	0	0	0.595
−1	1	0.005	3	0	0.065	0	1	0.055
0	0	0.145	5	1	0.025	18	0	0.195
0	1	0.005	6	0	0.140	27	0	0.080
1	0	0.125	8	0	0.130	50	0	0.020
1	1	0.005	8	1	0.010	99	0	0.005
2	0	0.110	12	0	0.075			
3	0	0.090	16	0	0.055			
4	0	0.070	20	0	0.040			
6	0	0.090	35	0	0.017			
8	0	0.060	99	0	0.003			
10	0	0.050						
15	0	0.085						
30	0	0.010						
50	0	0.004						
99	0	0.001						

for punts and field goals was created using a similar procedure. The details of the distributions for these two actions are not provided here.

There is plenty to criticize here, e.g. the use of only a single distribution for all run plays, the use of only two passing distributions (short and long), the discrete approximations to phenomena that are nearly continuous in nature, and the complete exclusion of defensive actions. However, the biggest difficulty with this approach to determining optimal strategy is computational. The state space is enormous and, to this point, it has only been possible to solve for optimal strategy during the last 10 minutes of the football game. In addition, the strategy findings appear to be quite sensitive to the specified distributions which (in theory) reflect the relative abilities of the two teams. Distributions that are inaccurate may lead to unintended consequences. For example, an earlier version of the distributions in Table 1.2 led to the conclusion that all teams should always choose to throw long passes (unless ahead and trying to run out the clock). Even with these limitations, the optimal strategies obtained from this model are useful. For one thing they suggest that 2-point conversions after touchdown should be attempted more often than they are in practice. This is based on the current rate of success in United States professional football (approximately 0.50 for the 2-point conversion and 0.96 for the 1-point conversion).

The expected win probabilities produced by the dynamic programming approach are included in Table 1.1 for comparison with the other methods. Intervals are given when the time remaining is between two time units. The dynamic programming results are similar to those obtained by CPT; however, some substantial differences do occur. It appears that the dynamic programming approach allows for a greater probability of come-from-behind wins (likely due to some favorable features of the distribution of outcomes assumed for long passes).

The potential of dynamic programming was realized long ago. The annotated bibliography of the book on sports statistics edited by Ladany and Machol (1977) includes a reference to Casti's (1971) technical report which apparently outlines a similar approach. More recently, Sackrowitz and Sackrowitz (1996) developed a dynamic programming approach to evaluating ball control strategies in football. Their work is similar to that described here except that team possessions are analyzed rather than individual plays. They define a limited set of offensive strategies for a team (ball control, regular play, hurry-up) and assign a distribution for time used by each strategy and a probability of scoring a touchdown for each strategy. Their finding is that a team should not change its style of play for a particular opponent.

1.3.3.2   *Applying the estimated win probabilities*

We can now return to the types of strategy considerations that were evaluated earlier using expected points. For this discussion, we use the logistic approximation to the win probability (because the CPT results are not available for all the situations in which we are interested). We do not use the dynamic programming results because it is evident that more work is required to make this approach feasible. It should be noted, however, that the dynamic programming approach is a promising one for addressing detailed strategy questions.

Table 1.3 Change in win probability due to a turnover for several different scores, field positions, and time remaining

Score difference	Yards from goal	Time remaining (minutes)	Win probability		
			Before turnover	After turnover	Decrease
0	25	45	0.589	0.502	0.087
0	50	45	0.544	0.456	0.088
0	85	45	0.480	0.394	0.086
3	25	5	0.804	0.742	0.062
3	50	5	0.773	0.705	0.068
3	85	5	0.725	0.650	0.075

Recall that Carter and Machol (1971) found that the effect of a turnover did not depend on the location on the field where the turnover occurred. It seems likely that the time remaining in the game will make a difference with respect to this issue. Table 1.3 gives the probability of winning before and after a turnover at several different locations at two different points in the game. Early in the game we find that the Carter and Machol result holds, but later in the game the location of the turnover on the field does matter. Turnovers near your own goal late in a close game are more costly than turnovers near midfield, as intuition might suggest.

Interestingly, the optimal fourth-down strategy also depends on the time remaining. Early in the game, win probabilities support the recommendation derived using expected points – teams should go for the first down rather than kick a field goal. However, optimal late-game strategy appears to be sensitive to the model used for estimating win probabilities. The logistic approximation does not inspire great confidence so we do not provide the numerical details here.

Win probabilities might also be used to evaluate team performances. The offensive part of a football team could, for example, be judged by their net effect on the team's win probability. CPT propose win probabilities for precisely this purpose and work through three games in detail. The CPT approach only estimates win probabilities at the start of each possession so that it would be difficult to use them for evaluating individual plays or players. If win probabilities were available for every possible situation, as they would be if dynamic programming were used to estimate them, then it might be possible to give a player credit for the changes in the team's win probability that result from his contributions. This approach could also be used to assess the effectiveness of running plays and passing plays or the effect of penalties by summing the changes in win probability associated with all plays of a given type. Once again the difficult problem of partitioning credit among the several players involved in each play requires some thought.

1.3.3.3   Limitations

Conceptually, win probabilities come closest to providing the ideal information needed to make effective strategy decisions. One limitation of this approach is that, as with expected point values, the win probabilities are estimated from aggregate data (using either the CPT or dynamic programming approach) and thus may not be relevant for a particular team or game. The win probability for Team A in a particular situation may be different than if Team B were in the same situation. It still seems that a set of 'average' win probabilities would be a useful decision-making tool.

A more important issue at this point in time is the difficulty in obtaining credible estimates for the win probabilities. There are problems with both the empirical approach of CPT and the dynamic programming approach that we considered. Large amounts of data are required to apply the empirical approach of CPT and to expand the number of situations for which win probabilities are defined. We must also decide how many different situations to address. For example, in professional football in the United States the home team is usually thought to have a 3-point advantage, or put another way, the home team wins approximately 59% of all games. Should we compute separate win probabilities for the home and visiting team for each state? Dynamic programming, our second approach to estimating win probabilities, has great potential but also requires additional data. Data are needed to construct realistic distributions for the various plays/actions. In addition, it would be good to expand the model to include both offensive and defensive choices of actions at each state. This would make things more realistic than the offense-only model considered here. During games, teams try to outguess each other, so that the offense will try to use a run play when the defense expects a pass play. Incorporating offensive and defensive actions would require the distribution of outcomes for each offensive action under a variety of assumptions about the defensive team's strategy. Unfortunately, this would take our fairly large dynamic programming problem and make it even more complex.

Some researchers have worked in the opposite direction, constructing simpler models that can yield informative results on particular questions, e.g. Brimberg and Hurley (1997) describe a simple model of football and use it to assess the effect of turnovers on the probability of winning a game.

1.4   Rating of teams

Owing to the physical nature of football, teams usually play only a single game each week. This limits the number of games per season to between 10 and 20 games (depending on whether we are thinking of United States college football, United States professional football, or Canadian professional football). The seasons are not long enough for each team to play every other team. Typically teams are organized in leagues or divisions within which all teams play each other once or twice; however, these teams will play different schedules outside the division. Because teams play unbalanced schedules, an unequivocal determination of the best team is not possible. Play-off tournaments are used to determine champions in professional football but not in major United States college football. There are more than 100 college teams competing at the highest level and a unique champion is not determined on the field of play. The performances of the best teams are judged by a poll of coaches or sportswriters to identify a champion. It is natural to ask whether statistical methods can be used to rate teams and identify a champion. Even though professional football uses a play-off tournament to identify a champion, there is some interest in rating teams there as well, especially in the middle of the season. This is primarily because the question of how to find suitable ratings for teams is closely related to questions concerning prediction of game outcomes and preparation of a betting line. Prediction is covered in Chapter 12, so here we limit ourselves to a brief review of the work that has been carried out concerning the rating or ranking of football teams.

There has been interest in rating college football teams with unbalanced schedules for a long time. Dickinson (1941) describes an approach he used in the 1920s and 1930s which gave teams points for each game they won, with the number of points depending on the quality of the opponent. This is an example of a rating method that relies only on a record of which teams have defeated which other teams (with no use made of the game scores). Other examples of this type in the statistical literature include the methods of Bradley and Terry (1952) or Andrews and David (1990) for data consisting of contests/comparisons of two objects at a time. The National Collegiate Athletic Association (NCAA) is the governing body for college sports in the United States and is responsible for determining champions in a variety of sports. The NCAA relies on a measure of this type, the Ratings Percentage Index (a combination of a team's winning percentage, the average of its opponents' winning percentages (OWP) and the average of its opponents' OWPs), in a variety of sports but football is not one of them.

An extremely popular approach to rating teams makes use of the scores accumulated by each team during their games. Such ratings have become increasingly popular due to their relevance for prediction (see also Chapter 12). Most often these ratings approaches apply the method of least squares or related normal distribution theory to obtain ratings that minimize prediction errors (Leake, 1976; Harville, 1977, 1980; Stefani, 1977, 1980; Stern, 1995; Glickman and Stern, 1998). We briefly describe the basic idea of these approaches. Suppose that R_i is used to represent the rating for team i and R_j is the rating for team j. When team i plays team j the ratings would predict the outcome as $R_i - R_j \pm H$, where H is a home-field advantage measure (approximately 3 points in professional and college football in the United States) and the sign of H depends on the site of the game. If we use Y to represent the actual outcome when these teams play, then the prediction error is $Y - (R_i - R_j \pm H)$. Given the results from a collection of games, we can estimate the ratings to be those values that make the prediction errors as small as possible, e.g. least-squares ratings minimize $\sum (Y - (R_i - R_j \pm H))^2$. Ratings of this type appear in the *USA Today* newspaper during the college football season. Of course, it is not necessarily true that methods based on normal distributions are appropriate for analyzing football scores. Mosteller (1979) presents a 'resistant' analysis of professional scores to prevent unusual scores (outliers) from having a large effect. Bassett (1997) introduces the possibility of using least absolute values in place of least squares in order to minimize the effect of unusual observations. Rosner (1976) builds a model for rating teams or predicting outcomes that makes explicit use of the multiple ways of scoring points in football. Mosteller (1970) and Pollard (1973) provide exploratory analyses of football scores but do not focus on rating team performance.

1.5 Some other topics

Any presentation of the relationship of probability and statistics to football (or any other sport for that matter) will focus on those aspects of the sport that the author finds most interesting and promising. This section provides references to other work not discussed in detail. We also mention some problems that have not received much attention but might benefit from statistical analysis.

Professional football teams are constructed primarily by two means: teams draft players from college football teams and teams sign 'free agents' (players currently without a contract). Evaluating the contributions of players and placing an economic value on those contributions are obviously relevant to making personnel decisions. These issues have not yet received much attention. The player dispersal draft that allocates new players to teams has been around a long time but has also not received much attention. Price and Rao (1976) built a model for evaluating a variety of different player allocation rules. Other business and economic issues are addressed by Noll (1974). In one chapter of that edited volume, Noll carries out an analysis of attendance in many sports including football.

One strategy issue that is not appropriately addressed by any of the discussions here is the effective use of timeouts and other time management strategies. Carter and Machol (1971) discuss this issue briefly in their work on expected points. CPT also discuss the use of timeouts but both discussions are mainly qualitative. As regards time management strategies, Sackrowitz and Sackrowitz (1996) carry out an investigation of time management by asking whether altering one's strategy to use more/less time can increase the probability of winning.

1.6 Summary

Football teams have expressed a willingness to use statistical methods to learn from available data. Most teams keep detailed records of opponents' tendencies and use that information to plan strategy for upcoming games. In addition, Bud Goode has a long history of consulting for professional teams, identifying the key variables correlated with winning football games and then providing advice on how teams might improve their performance with respect to these variables (see, for example, Goode, 1978). The discussion here shows that more extensive use of statistical methods in football might provide an opportunity for enhanced player evaluation, and improved decision-making. In this era of greater freedom in player movement from team to team, research regarding the value of a player or the relative values of two players will become even more crucial. With respect to decision-making, the results here suggest that football coaches should attempt fewer field goals (worth 3 points) and instead take more fourth-down risks in pursuit of touchdowns (worth 6, 7 or 8 points). More complete results about player evaluation and optimal strategy will require more data and a more substantial research effort.

References

Andrews, D. M. and David, H. A. (1990) Nonparametric analysis of unbalanced paired-comparison or ranked data. *Journal of the American Statistical Association*, **85**, 1140–1146.

Bassett, G. W. (1997) Robust sports ratings based on least absolute errors. *The American Statistician*, **51**, 99–105.

Berry, D. A. and Berry, T. D. (1985) The probability of a field goal: rating kickers. *The American Statistician*, **39**, 152–155.

Bilder, C. R. and Loughin, T. M. (1998) It's good! An analysis of the probability of success for placekicks. *Chance*, **11**(2), 20–24, 30.

Bradley, R. A. and Terry, M. E. (1952) Rank analysis of incomplete block designs. I. The method of paired comparisons. *Biometrika*, **39**, 324–345.

Brimberg, J. and Hurley, W. J. (1997) The turnover puzzle in American football. Technical report, Royal Military College of Canada, Kingston, Ontario, Canada.

Brimberg, J., Hurley, W. J. and Johnson, R. E. (1998) A punt returner location problem. To appear in *Operations Research*.

Carroll, B., Palmer, P. and Thorn, J. (1988) *The Hidden Game of Football*. New York: Warner Books.

Carter, V. and Machol, R. E. (1971) Operations research on football. *Operations Research*, **19**, 541–545.

Casti, J. (1971) Optimal football play selections and dynamic programming: a framework for speculation. Technical report, Project PAR284-001, Systems Control, Inc., Palo Alto, CA.

Dickinson, F. G. (1941) *My Football Ratings – from Grange to Harmon*. Omaha, NE: What's What Publishing Co.

Glickman, M. E. and Stern, H. S. (1998) A state-space model for National Football League scores. *Journal of the American Statistical Association*, **93**, 25–35.

Goode, B. (1978) Relevant variables in professional football. *ASA Proceedings of the Social Statistics Section*, 83–86.

Harville, D. (1977) The use of linear model methodology to rate high school or college football teams. *Journal of the American Statistical Association*, **72**, 278–289.

Harville, D. (1980) Predictions for National Football League games via linear-model methodology. *Journal of the American Statistical Association*, **75**, 516–524.

Irving, G. W. and Smith, H. A. (1976) A model of a football field goal kicker. In *Management Science in Sports* (edited by R. E. Machol, S. P. Ladany and D. G. Morrison), pp. 47–58. New York: North-Holland.

Ladany, S. P. and Machol, R. E. (Eds) (1977) *Optimal Strategies in Sports*. New York: North-Holland.

Leake, R. J. (1976) A method for ranking teams: with an application to college football. In *Management Science in Sports* (edited by R. E. Machol, S. P. Ladany and D. G. Morrison), pp. 27–46. New York: North-Holland.

Morrison, D. G. and Kalwani, M. U. (1993) The best NFL field goal kickers: are they lucky or good? *Chance*, **6**(3), 30–37.

Mosteller, F. (1970) Collegiate football scores, U.S.A. *Journal of the American Statistical Association*, **65**, 35–48.

Mosteller, F. (1979) A resistant analysis of 1971 and 1972 professional football. In *Sports, Games, and Play: Social and Psychological Viewpoints* (edited by J. H. Goldstein), pp. 371–399. Hillsdale, NY: Lawrence Erlbaum Associates.

Noll, R. G. (Ed) (1974) *Government and the Sports Business*. Washington, DC: The Brookings Institute.

Pollard, R. (1973) Collegiate football scores and the negative binomial distribution. *Journal of the American Statistical Association*, **68**, 351–352.

Porter, R. C. (1967) Extra-point strategy in football. *The American Statistician*, **21**, 14–15.

Price, B. and Rao, A. G. (1976) Alternative rules for drafting in professional sports. In *Management Science in Sports* (edited by R. E. Machol, S. P. Ladany and D. G. Morrison), pp. 79–90. New York: North-Holland.

Purdy, J. G. (1971) Sport and EDP . . . It's a new ballgame. *Datamation*, **17**, 1 June, 24–33.

Rosner, B. (1976) An analysis of professional football scores. In *Management Science in Sports* (edited by R. E. Machol, S. P. Ladany and D. G. Morrison), pp. 67–78. New York: North-Holland.

Ryan, F., Francia, A. J. and Strawser, R. H. (1973) Professional football and information systems. *Management Accounting*, **54**(9), 43–47.

Sackrowitz, H. and Sackrowitz, D. (1996) Time management in sports: ball control and other myths. *Chance*, **9**(1), 41–49.

Stefani, R. T. (1977) Football and basketball predictions using least squares. *IEEE Transactions on Systems, Man, and Cybernetics*, **7**, 117–120.

Stefani, R. T. (1980) Improved least squares football, basketball, and soccer predictions. *IEEE Transactions on Systems, Man, and Cybernetics*, **10**, 116–123.

Stern, H. S. (1994) A Brownian motion model for the progress of sports scores. *Journal of the American Statistical Association*, **89**, 1128–1134.

Stern, H. S. (1995) Who's number 1 in college football? . . . and how might we decide? *Chance*, **8**(3), 7–14.

2

Baseball

Jay M. Bennett
Bellcore, USA

2.1 History

With recorded games dating back to 1839 and the development of a professional league in 1876, baseball is the oldest organized sport in North America. Over the intervening years, the game has spread beyond the boundaries of the United States to immense popularity in Canada, Mexico, Cuba, Central America, South America and Japan. However, statistical analyses have focused on play in Major League Baseball (MLB), primarily in the 'modern era', starting with the first World Series in 1903. At that time, Major League Baseball encompassed 16 teams in the National and American Leagues, each team representing a city in the United States. Today MLB preserves the two leagues with a total of 30 teams (including two in Canada). While the statistics and techniques described here can be (and are) applied to any baseball league, this chapter describes their application within Major League Baseball.

Statistics have always played a major role in spectator interest in baseball. According to Leonard Koppett (1991, p. 225), the Hall of Fame sportswriter, 'Statistics are the lifeblood of baseball'. To Bill James, 'Baseball statistics have the ability to conjure images' and 'tell stories' (James, 1985, p. 153). Perhaps the best expression of this union between statistics and storytelling is the box score invented by sportswriter Henry Chadwick. Through its enumeration of the baseball events credited to each player in the game, the devoted fan can get a reasonable picture of the game's drama. (See the appendix to this chapter for a sample box score from Cohen *et al.*, 1976.) Based on counts compiled in game box scores, summary statistics for batting, fielding, and pitching were developed throughout the 19th century. By the start of the modern era, in 1903, MLB had developed most of the standard statistics recognized by baseball fans today.

'In truth, baseball statistics are not statistics at all. They are accounting ledger entries, to which little or no statistical manipulation has ever been applied' (Kindel, 1983, p. 184). This statement had validity in the days when *The Sporting News* provided the most extensive coverage of baseball statistics, but, starting in the 1950s, statisticians like Frederick Mosteller (1952), John

Smith (1956) and Ernest Rubin (1958) became interested in applying statistical models to baseball. However, the most important work in this period was performed by George Lindsey (1959, 1961, 1963) who pioneered applying statistical models to address the questions of greatest enduring interest: optimal strategies and player performance.

With the advent of electronic calculators and computers in the 1960s, baseball saw an explosion of new interest in its existing statistics, ideas for new statistics, and the ways in which those statistics could be used. *Percentage Baseball* (Cook and Garner, 1966) brought widespread media attention to serious analysis of baseball statistics. Macmillan's 1969 publication of the first computer-generated baseball encyclopedia gave impetus to the movement. Historical baseball data were now among the most easily obtained massive data sets available to the public. Many papers, which took advantage of the availability of the data and computer technology, were published in the ground-breaking sports collections, *Management Science in Sports* (Machol *et al.*, 1976) and *Optimal Strategies in Sports* (Ladany and Machol, 1977). Members of the Society for American Baseball Research (SABR), established in 1971 and now numbering over 6000 members, devote much of their research efforts to analysis of baseball statistics.

The dominant figure who emerged from this period is Bill James. Like Chadwick, James' contributions to baseball statistics and their popularization have made him a significant figure in baseball history (Johnson and Ward, 1994). From 1977 through 1986, James published his *Baseball Abstract*, an annual collection of essays on the development and application of baseball statistics. In honor of SABR, James created the popular term 'sabermetrics' for the statistical analysis of baseball data.

The success of the *Baseball Abstract* stimulated a rush of similar annuals which went beyond the basic tabulations presented in *The Sporting News* publications. Most notably, from 1985 through 1993, the Elias Sports Bureau published an annual *Elias Baseball Analyst* in which it publicly made available the analyses it had been performing for its baseball team clients. Since 1990, STATS, Inc., another sports information services bureau, has published its *Baseball Scoreboard* series, which examines provocative statistical questions. John Thorn and Pete Palmer (1985) published the best single-volume history of baseball statistics, *The Hidden Game of Baseball*. In 1989, they incorporated non-traditional baseball statistics into the first edition of their baseball encyclopedia, *Total Baseball*, which recently was accepted as the official MLB encyclopedia.

Part of this explosion of interest has been made possible by the increasing collection of more detailed play-by-play (even pitch-by-pitch) records via computer which started with the EDGE 1.000 system developed by Richard Cramer in 1983 (Klein, 1983). Data have received greater dissemination via Internet web sites and CD-ROM encyclopedias. Fan fascination with baseball statistics continues to increase, even though baseball's position as the dominant American spectator sport has waned in recent years.

This chapter provides an overview of major topics addressed by statistical research in baseball. The subject of greatest interest by far is the value rating of individual players described in Section 2.2. The desire to compare players in different eras has led to further research in techniques to adjust for variations in the play of the game over baseball history (Section 2.3). Players and managers attempt to optimize performance by considering the many factors that affect the ability to score runs. Section 2.4 describes some of these factors while Section 2.5 considers how these factors and others could be used by managers and general managers in their respective strategic decision-making. Much of fan interest is derived not only as spectators of individual games, but also by tracking the accomplishments of players in achieving statistical milestones (Section 2.6). Section 2.7 is a potpourri of subjects that have been controversial side-issues or have just intrigued statisticians. Readers unfamiliar with baseball are advised first to read the appendix, which describes the basics of the game. The glossary also contains many definitions useful in grasping the game's terminology.

2.2 Player value

The initial driving impetus for the development of baseball statistics was the capability to measure individual player performance. 'As time passes, the evaluation of a player comes to rest more and more on his statistics. There is a simple reason for this, which is that everything else tends to be forgotten. His statistics remain exactly the same, and eventually the statistics become the central part of the player's image' (James, 1994, p. 322).

The discrete structure of baseball play provides a simple fundamental framework for statistical analysis. In contrast to such continuous-time games as soccer and hockey, the basic baseball event, the batter's opportunity to face the pitcher and the result of this encounter, is relatively easy to analyze. In addition, the line-up of batters enforces a more even distribution of opportunity among players than is encountered in most sports such as basketball. This relative equalization of opportunity invites comparison of batters and their contributions to the team.

Player statistics can be separated into offensive measures (for hitting and stealing) and defensive measures (for pitching and fielding). Several newer measures have attempted to incorporate both offensive and defensive measures into a single statistic to gauge the player's overall value. All of these measures are calculated from counts of baseball events in the player's record. Thorn and Palmer (1989) is an excellent source for the definition of baseball statistics. Albert (1998) provides an introduction to the basic traditional statistics and some of the newer 'sabermetric' measurements.

2.2.1 Offensive measures

2.2.1.1 Batting measures

The most commonly used statistic to measure a batter's effectiveness is the Batting Average (BA) – hits (H) divided by the number of at-bats (AB). BA (and all other statistics to be discussed) are compiled typically for each season and for a career. Newspapers and sports periodicals track the BA leaders in each league throughout each baseball season until its end, when the player with the highest BA is crowned as the league batting champion.

Table 2.1   Benchmarks for traditional measures since 1903

	Batting Average		On-Base Percentage		Slugging Percentage	
	Player	BA	Player	OBP	Player	SLG
Season record	Rogers Hornsby	.424	Ted Williams	.551	Babe Ruth	.847
Career record	Ty Cobb	.366	Ted Williams	.483	Babe Ruth	.690
High level		.300		.400		.500

A more sophisticated measure is the Slugging Percentage (SLG), the average number of bases reached per at-bat

$$SLG = (H + 2B + 2 \times 3B + 3 \times HR)/AB$$

This measure places greater value on power hitters who may not be among the league leaders in BA, but who get more extra base hits (doubles (2B), triples (3B) and, especially, home runs (HR)).

Another variation of BA is the On-Base Percentage (OBP)

$$OBP = \frac{H + BB + HBP}{AB + BB + HBP + SF}$$

Like BA, OBP is a measure of the probability of getting on-base; unlike BA, OBP includes offensive contributions from bases on balls (BB) and being hit by the pitcher (HBP), and counts sacrifice flies (SF) as missed opportunities.

These three measures of batting effectiveness are kept by Major League Baseball and reported in the media. Rogers Hornsby set the season record for BA in 1924, but could not top Ty Cobb's career BA record set from 1905 to 1928. In general, batters with a .300 season BA or higher are among the best in a season and those with .300 career BAs are among the all-time greats. Table 2.1 provides similar benchmarks for the other traditional measures as well.

The effectiveness of these traditional batting statistics has been questioned and many alternatives have been suggested. None has supplanted SLG, OBP and, especially, BA in the collective minds of the baseball public, but several have been recognized as valuable alternatives. The most successful alternatives are the Linear Weights and Runs Created models which were constructed from relationships of individual baseball events with runs scored. Unless a batter hits a home run, a run can only be scored by the combined efforts of multiple players. Because of the need for collaboration in scoring runs, these techniques have been developed by relating team totals in different offensive categories to team runs scored over a season.

Linear Weights model   The Linear Weights model has its origins with Lindsey (1963) who used his data on the frequencies of different batting situations and the expected number of runs scored in each situation to estimate the value of each hit in terms of runs scored. In order to compare his estimates to the weights used in the Slugging Percentage, Lindsey calculated the ratio of expected runs for each hit to the expected runs for a single. The results, as shown in Table 2.2, indicate that the Slugging Percentage gives too much weight to the triple (3B) and the home run (HR).

Palmer (Thorn and Palmer, 1989) extended Lindsey's concept. Using a computer simulation of all major league games played from 1901 to 1978, Palmer

Table 2.2   Expected runs per hit (Lindsey, 1963, p. 498)

	Type of hit			
	1B	2B	3B	HR
Expected runs	0.41	0.82	1.06	1.42
Ratio	1.00	1.97	2.56	3.42

generally confirmed Lindsey's results and expanded upon them to produce a formula for the expected number of Runs Above Average (RAA) performance which results from a set of batting events

$$\text{Runs Above Average} = 0.47 \times 1B + 0.78 \times 2B + 1.09 \times 3B + 1.40$$

$$\times \, HR + 0.33 \times BB - 0.25 \times (AB - H)$$

Runs Created model   The intuitive principle behind the Runs Created model had its origins with the Scoring Index proposed by Earnshaw Cook (Cook and Garner, 1966, p. 63): runs are the product of (1) the ability to get on base and (2) the ability to advance runners. Bill James' basic version of this principle (James, 1986, p. 279) states that

$$\text{Runs} = \frac{(H + BB)(H + 2B + 2 \times 3B + 3 \times HR)}{AB + BB}$$

where $H + BB$ is the on-base factor, total bases ($H + 2B + 2 \times 3B + 3 \times HR$) is the advancement factor, and $AB + BB$ is an opportunity factor.

Accuracy of models   Many other authors have proposed measures evaluating baseball players. James and Palmer are among the few who have tested their performance models with respect to a standard statistic, namely runs scored by teams in a season. Thorn and Palmer (1985) calculated the standard errors of estimates for team runs scored based on several offensive statistics (Table 2.3). Team BA provides the least accurate estimate of runs scored by a team in a season and team OBP provides little improvement. Team SLG is more accurate than the other measures, but the Linear Weights model halves the SLG standard deviation. For the Runs Created model, James (1986) estimated a level of accuracy approximately equal to that of the Linear Weights model. Bennett and Flueck (1983) performed a similar analysis of various models using data from the 1969–76 seasons.

Table 2.3   Accuracy of several offensive statistics in estimating team runs in a season, 1946–82 (Thorn and Palmer, 1985, pp. 58–59)

Team statistic	Standard deviation of team run residuals
Batting Average	54.8
On-Base Percentage	53.0
Slugging Percentage	39.9
Linear Weights	19.8

2.2.1.2 *Stealing*

Runners can advance without the aid of a batted ball by stealing bases. The runner can be caught stealing (CS) if he is tagged out before reaching the base. Traditionally, players are rated by the number of stolen bases (SB) and their Stolen Base Average (SBA)

$$SBA = SB/(SB + CS)$$

Palmer (Thorn and Palmer, 1989) incorporated base stealing into his Linear Weights model by adding Stolen Base Runs

$$SBR = 0.3 \times SB - 0.6 \times CS$$

to the runs created from batting. James also incorporated stealing into his Runs Created model

$$Runs = \frac{(H + BB - CS)(H + 2B + 2 \times 3B + 3 \times HR + 0.55 \times SB)}{AB + BB}$$

2.2.2 Defensive measures

2.2.2.1 *Pitching statistics*

Of course, many of the hitting statistics just discussed can be applied to pitchers, but the major traditional pitching statistics are the won–lost record and Earned Run Average (ERA).

The most quoted statistic for pitchers is the Earned Run Average, the average number of earned runs allowed by a pitcher per nine-innings

$$ERA = (earned\ runs/innings\ pitched) \times 9$$

A pitcher with an ERA below 3.00 is considered very good and ERAs below 2.00 are extraordinary. However, this statistic has been criticized as biased in favor of relief pitchers. Relief pitchers often enter the game in the middle of an inning; they are not charged for runners on base at the moment they enter the game and have less chance of allowing runs to score with one or two batters out in the inning. For this reason, ERAs for relief pitchers are considered to underestimate their 'real' ERAs comparable to those of starting pitchers. James (1977, p. 114) identified starter and relief pitcher counterparts on the basis of hits and walks allowed per inning; he found that starters had ERAs about 0.20 runs greater than their reliever counterparts.

The effectiveness of a pitcher may be affected by his battery mate, the catcher. The ability of a player to 'handle' a pitcher is often the paramount reason for using the player as a catcher even if he does not hit as well as other catchers on the team. To capture this ability quantitatively, Wright and House (1989, p. 22) have proposed calculating an ERA for catchers based on the runs scored while catching.

The won–lost record or a pitcher's Winning Percentage (WPct) has always been a suspect statistic for the pitcher's value. Fans with some justification feel that the statistic is biased by the quality of the team supporting the pitcher, especially with respect to run support (see Siwoff *et al.*, 1988, p. 161); some pitchers gain more wins not because they have allowed so few runs, but because their teams have scored more runs for them in support. Many techniques have been suggested to adjust the pitcher's Winning Percentage to account for run support. The most common technique has been to compare the pitcher's Winning Percentage with that of his team when he did not pitch. Examples of such techniques are described in Deane (1996) and Neft *et al.* (1974). Recently, Quality Starts (games in which the starting pitcher allows no more than two earned runs over at least six innings pitched) have been tracked to gauge pitcher performance apart from run support (Dewan *et al.*, 1990, pp. 184–185).

Because of the relief pitcher's increasing importance and his lack of opportunity in obtaining wins, his most important statistic is the Save, awarded when the relief pitcher completes a winning game without relinquishing a lead. Despite further constraints on the size of the inherited lead, the Save has drawn much criticism because of the relative ease with which it can be obtained (see Section 2.7.2 on progress of the score) and the inequities of getting Save opportunities. This criticism resulted in the Blown Save being given to a relief pitcher who fails in a Save opportunity. In addition, in order to reward middle relievers who have little chance for Wins or Saves, a Hold is credited to a pitcher who enters the game in a Save situation, retires at least one batter, and leaves the game with the inherited lead preserved (Dewan *et al.*, 1994, p. 171).

2.2.2.2 Fielding statistics

Fielding statistics have long been considered the weakest and least useful of baseball statistics. The key statistic for many years has been the Fielding Average (FA) which measures the probability of *not* making an error

$$FA = 1 - E/(A + PO + E)$$

based on the number of assists (A), putouts (PO) and errors (E).

Many analysts have recognized the weakness in this statistic. Because of official scoring criteria for errors, good fielders who are able to reach batted balls not reachable by lesser fielders are penalized with more errors and FAs lower than their lesser counterparts. A favored alternative is the Range Factor (RF)

$$RF = (A + PO)/G$$

the number of plays per game (G) by the fielder. Neft (1986) credits Al Wright with its invention as early as 1875. James (1976) popularized the Range Factor and provided contemporary RF percentiles for different fielding positions (Table 2.4). Several criticisms of the Range Factor (Siwoff *et al.*, 1985, p. 115) are as follows.

Table 2.4 Distributions of Range Factors by fielding position (James, 1976)

	Fielding position						
Percentile	*First base*	*Second base*	*Shortstop*	*Third base*	*Center field*	*Right field*	*Left field*
0.8	9.8	5.6	5.0	3.2	2.9	2.2	2.2
0.6	9.3	5.3	4.8	3.0	2.7	2.1	2.0
0.4	8.8	5.1	4.6	2.8	2.5	2.0	1.9
0.2	8.5	4.8	4.4	2.6	2.4	1.8	1.8

- It does not adjust for pitching staff characteristics (e.g. strikeout pitchers afford less fielding opportunities (and thus lower Range Factors) for their fielders).
- It is based on games rather than innings played.
- Putouts are often not made as a result of a fielder's range. The high RFs for first basemen in Table 2.4 are from putouts on throws from the other infielders; these putouts do not reflect the first baseman's skills in fielding batted balls.

Neft (1986) outlined RF adjustments in answer to some of these criticisms. Dewan *et al.* (1990, pp. 215–219) modified the RF concept by dividing the playing field into zones of responsibility for each fielder and calculating a Zone Rating for the proportion of balls hit into each zone which are converted into outs by the fielder.

2.2.3   Contribution to winning

Ultimately, the goal is to establish how much a player contributes to victory. Runs are an indirect measure from which Palmer and James have established models to estimate the player's contributions in terms of wins. Palmer (1982) estimates that a player contributes about one win above average for every 10 Runs Above Average. Thorn and Palmer (1989) present techniques for adjusting the 10 runs parameter for individual players.

James (1982) developed the Pythagorean Method to estimate a team's Winning Percentage based on the number of runs scored and allowed by the team

$$\text{WPct} = \frac{\text{Runs scored}^2}{\text{Runs scored}^2 + \text{Runs allowed}^2}$$

James found that the method estimates a team's wins in a season with a 4.30 standard deviation of observed minus predicted wins. James calculates a won–lost record (comparable to a pitcher's won–lost record) based on the premise that the player of interest is a team unto himself

$$\text{Player wins} = \text{Player games} \times \text{Player WPct} \quad \text{and}$$

$$\text{Player losses} = \text{Player games} - \text{Player wins}$$

where

$$\text{Player games} = (\text{AB} - \text{H})/25.5$$

and

$$\text{Player WPct} = \frac{(\text{Runs created}/\text{Player games})^2}{(\text{Runs created}/\text{Player games})^2 + \text{Average team runs per game}^2}$$

Mills and Mills (1970) took the novel approach of skipping the intermediary step of run estimation and used the probabilities of winning in each baseball situation (score, inning, outs, baserunners) to measure directly player contributions to winning on each play. For example, if a player at-bat produced a hit that raised his team's probability of winning from 0.5 to 0.6, the player received Win Points proportional to half the change in probability (0.05) and the pitcher who gave up the hit received Loss Points proportional to the other half (0.05). For each player, these Win and Loss Points were accumulated over games and combined into a Player Win Average (PWA). This system replaced the standard multivariate measures with a single value for evaluating and comparing all players (batters, starting pitchers and relief pitchers). Unfortunately, while the Mills brothers provided many win probability estimates in their analysis of the 1969 World Series, a complete set was not published.

Bennett and Flueck (1984) devised a technique for estimating these win probability values based on data collected from the 1959 and 1960 seasons by Lindsey (1961, 1963). They also introduced a PWA variant called Player Game Percentage (PGP), which is the contribution of each player to his team's probability of victory. Besides applying PGP to select the Most Valuable Players in recent World Series (Bennett, 1994), Bennett (1993) used PGP in his analysis of Shoeless Joe Jackson's batting record in the 1919 World Series. In 1921, Jackson was banned from MLB for his alleged participation in throwing the series. A resampling analysis repeatedly reshuffled Jackson's batting results with his batting situations and then re-evaluated Jackson's batting performance using PGP. Because Jackson's actual PGP rating was in the middle of the resampling PGP distribution, Bennett concluded that Jackson had batted to the full extent of his capabilities.

2.2.4 Examples

Table 2.5 presents examples of applying the Runs Created and Linear Weights models to the record seasons in Table 2.1. While both models give the same ordering in terms of runs produced, the Runs Created model places Ruth and Williams closer to each other and gives Williams a higher Winning Percentage (WPct) because of the lower League Runs/Game (LR/G) in 1941.[1]

2.3 Changes in the game over time

2.3.1 Normalization for eras

While the rules of baseball have not changed substantially throughout its history, styles of play, changes in equipment, and openness of competition have led to changes in scoring (Browning, 1980). In the National League, Batting Averages rose to record highs around 1930 and have generally declined since then (Fig. 2.1). The American League shows a pattern similar to that of the National League in Batting Average until the inception of the Designated Hitter (DH) in 1973, when BA rose above the National League level. The DH rule increased BAs by substituting a skilled batter for the pitcher (notoriously poor hitters) in the batting line-up, but not in the field. The adoption

---

[1] The calculations in Table 2.5 used the basic forms of the Runs Created and Linear Weights models as presented in Section 2.2.1. They differ slightly from published results derived from the complete forms which vary for different periods in baseball history and require additional data.

Table 2.5 Evaluation of season record performances using the Runs Created and Linear Weights models

| Player | Year | AB | H | 2B | 3B | HR | BB | SB | CS | Runs Created | | | Linear Wts. |
										RC/G	LR/G	WPct	RAA
Babe Ruth	1921	540	204	44	16	59	144	17	13	17.3	5.1	.920	135
Ted Williams	1941	456	185	33	3	37	145	2	4	17.2	4.8	.928	112
Rogers Hornsby	1924	536	227	43	14	25	89	5	12	15.1	4.5	.917	98

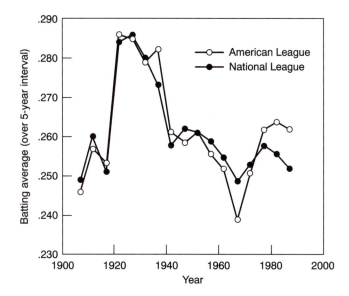

Figure 2.1 League batting averages over five-year intervals (data from Wright, 1994).

of this rule by the American League (and its disdain by the National League) remains the most controversial rule change in MLB.

In the National League, home run production per team per game rose steadily from less than 0.2 in the 1910s up to a peak over 0.9 in the 1950s. It has since dropped, although not so precipitously, to the 0.7 level (Wright, 1994). This change has often been attributed to a change in the resiliency of the baseball. Baseballs in MLB are not produced under specifications, such as those for the golf ball by the USGA and the PGA. The effect of the ball on play has drawn the attention of physicists and engineers (Adair, 1990; Watts and Bahill, 1990), more than statisticians. Cook and Garner (1966) concluded that the ball has changed substantially since 1901.

2.3.2 Are players today as good as yesterday?

Fans of the game have a great interest in comparing players from different eras. Because of changes in the game, this comparison involves more than a direct comparison of player statistics. The desire to compare players' performance across baseball history has led to much research in the normalization (or adjustment) of player performance for the era. Through the years, *Sports Illustrated* has published at least three articles (Weiskopf, 1977; Browning, 1980; Verducci, 1997) on adjusting Batting Averages based on the difference between the player's BA and the league BA in that year. Thorn and Palmer (1985) describe several approaches for normalization of annual statistics to adjust for different eras including:

- dividing the statistic by the league average;
- dividing the statistic by the value for the league leader; and
- calculating standard deviations from the league average.

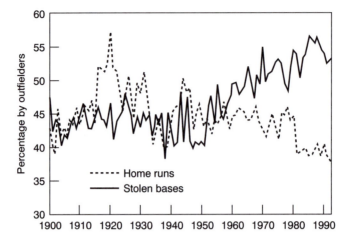

Figure 2.2   Outfielders' share of home runs and stolen bases throughout baseball history (data from Siwoff *et al.*, 1993).

Part of the difficulty in making such comparisons is that the needs of offensive players at different positions have changed. Siwoff *et al.* (1993) demonstrated that the outfielder has shifted from a power position to a speed position since the mid-1950s. Figure 2.2 indicates that outfielders' share of home runs has shrunk while their share of stolen bases has increased.

On the basis of the number of players per US male citizen between the ages of 20–39 years, Deane (1985) argued that the level of major league competition reached an all-time high in recent years. Cramer (1980) contended that there has been a steady increase in batting skill throughout history. His conclusions are based on a system of summing year-to-year changes in batting measures of players. Adams (1982) applied Cramer's yearly corrections to support a similar contention for pitching and fielding skills.

While most research has focused on the changing level of play over baseball history, Schutz (1995) found that the stability of batting performance in different eras has been declining steadily since 1928 (e.g. the year-to-year performance of batters was much more erratic in 1990 than in 1930). He also found greater year-to-year consistency in power measures than in Batting Average measures.

2.4   Situational effects on scoring runs

The simplest model for the process of scoring runs is a Markov process. In the basic models, each state in the process is defined by which bases are occupied by runners and the number of outs. The probabilities of transition from one state to another are based on measures of batting and pitching skills (e.g. probability of hitting a single, probability of hitting a home run). Stern (1997) presented a summary of the application of Markov models to baseball statistics.

While the Markov process seems on the whole to be a good model for a baseball game, there are indications of situational effects often not included in such models.

- Pitchers are liable to be less effective after their team has a big inning, scoring four runs or more (ERA 3.21 before and 3.83 after) and after an at-bat in which the pitcher got on base (ERA 2.79 before and 4.30 after) (Dewan *et al.*, 1994, pp. 154–155).
- A batter hits better in his plate appearance after being hit by the pitcher (Dewan *et al.*, 1994, p. 164).
- Pitchers hold an advantage when confronting a batter for the first time (Siwoff *et al.*, 1991, p. 71).
- Power pitchers (pitchers with high strikeout and walk rates) perform much better at night while finesse pitchers (pitchers with low strikeout and walk rates) demonstrate little difference in daytime and night-time performance (James, 1984, p. 260).

Unfortunately, in many cases, analysis of these effects can be difficult because of the small sample sizes for the situations from which the statistics are derived. Chapter 13 includes several applications of Bayesian statistics to this problem.

This section examines some analyses of major situational effects on transition probabilities.

2.4.1 The count of balls and strikes

The most elemental battle in baseball is the pitcher–batter confrontation. So it is natural that the count of balls and strikes should have a major effect on the game. The totals in each row of Table 2.6 enumerate the eventual result after each occurrence of a count in the 1989 season. The averages in each row summarize the expected value of that state (count) to the batter and the opposing pitcher. By comparing rows, we can see the importance of each pitch in influencing the outcome of each at-bat. For example, the difference between a ball and strike on the first pitch is dramatic (BA .267 versus .229, SLG .400 versus .331) especially for OBP (.372 versus .269). Table 2.6 demonstrates how batting improves (deteriorates) as balls (strikes) increase. The sharp drop in SLG at two strikes indicates that the batter may be cutting down on the power in his swing to improve his chances of making contact.

Table 2.7 shows how the batter changes strategy with the count. Note how the batter becomes more aggressive, swinging more often, with two strikes. In particular, foul balls are most probable at full counts (3 balls, 2 strikes). This strategy of fouling pitches off to get a good pitch to hit is apparently effective. Dewan *et al.* (1994, p. 103) provided an analysis of fouls hit on full counts. As fouls increase from 0 to 2 or greater, BA rises from .220 to .240, OBP rises from .455 to .474, and SLG rises from .341 to .394. Although these differences are not dramatic, they indicate a slight advantage to the batter as he fouls off more pitches.

Katz (1986) converted data similar to that collected by Dewan into a transition matrix and modeled plate appearances as a Markov process. His model was able to reproduce values for BA and OBP close to those shown in Table 2.6.

Table 2.6 1989 hitting performances after different counts (hitting by pitchers excluded) (Dewan et al., 1990, pp. 64–65, 257)

After the count		Averages			Totals								
Balls	Strikes	BA	OBP	SLG	AB	H	2B	3B	HR	HBP	BB	IBB	K
0	0	.257	.320	.381	138 500	35 621	6199	856	3060	854	12 515	1743	22 195
	1	.229	.269	.331	61 135	14 015	2401	321	1068	384	3103	219	14 594
	2	.177	.206	.249	22 580	3 998	648	102	255	157	702	16	9 120
1	0	.267	.372	.400	56 052	14 945	2608	390	1370	271	9412	411	7 601
	1	.236	.301	.347	51 984	12 277	2135	278	1030	271	4706	54	11 189
	2	.185	.230	.262	34 475	6 365	1055	143	438	204	1897	6	13 004
2	0	.285	.491	.436	16 367	4 668	836	141	452	53	6717	170	1 901
	1	.254	.382	.384	25 768	6 554	1146	172	617	79	5371	40	4 596
	2	.201	.294	.291	26 254	5 276	919	135	396	117	3423	9	8 934
3	0	.291	.729	.449	2 868	835	142	27	86	11	4721	59	363
	1	.278	.579	.437	7 705	2 144	393	63	235	23	5570	15	1 064
	2	.225	.459	.340	11 712	2 634	470	78	240	44	5095	3	3 402

Table 2.7 Batting strategy and contact at different counts, 1991–93 (Dewan *et al.*, 1994, p.102)

At count		Batter took pitch		Batter swung at pitch		
Balls	Strikes	Ball	Strike	Missed	Fouled	Hit into play
0	0	0.436	0.256	0.062	0.110	0.136
	1	0.442	0.107	0.098	0.159	0.195
	2	0.490	0.048	0.111	0.165	0.185
1	0	0.351	0.207	0.081	0.158	0.203
	1	0.367	0.110	0.104	0.186	0.233
	2	0.375	0.045	0.127	0.207	0.246
2	0	0.323	0.254	0.063	0.152	0.209
	1	0.292	0.100	0.098	0.223	0.287
	2	0.297	0.044	0.124	0.241	0.294
3	0	0.377	0.529	0.014	0.034	0.045
	1	0.273	0.144	0.068	0.198	0.316
	2	0.223	0.040	0.106	0.276	0.356

2.4.2 Handedness

Lindsey (1959) performed the first study to confirm what baseball people had suspected for years, that batters hit for a substantially higher BA against pitchers of opposite handedness (i.e. left versus right or vice versa) than against pitchers of the same handedness. After adjusting for handedness, Lindsey also demonstrated that the binomial distribution was an appropriate model for the distribution of hits (Fig. 2.3).

2.4.3 Stadium effect

Baseball stadiums have different sizes, shapes and environmental conditions (indoor, artificial turf, wind, humidity, altitude) which can affect the difficulty of scoring runs. Lowry (1992) described the physical characteristics of all MLB stadiums, past and present.

Any analysis of stadium effects must take into account the possibility of a home field advantage and remove it from the analysis. Hurley (1993) demonstrated that the home team has a statistically significant advantage; home teams win about 54% of all games and have won about 60% of play-off games from 1925 to 1989. Still, as discussed in Chapter 12 on predicting outcomes, baseball has the smallest home field effect among the major team sports.

Through the calculation of a Park Factor (PF) for each team's home stadium, Thorn and Palmer (1985, p. 86) contend that season scoring can be affected as much as ±20% by the stadium. In its basic form

$$PF \approx \frac{\dfrac{\text{Team runs allowed per home game}}{\text{Team runs allowed per away game}}}{\dfrac{\text{League runs allowed per home game}}{\text{League runs allowed per away game}}}$$

For example, the 1906 World Champion Chicago White Sox were the 'Hitless Wonders', primarily because their park had PF = 0.820, one of the lowest

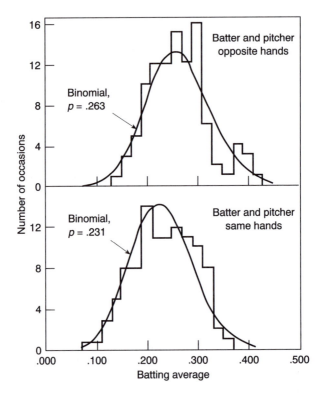

Figure 2.3 Actual distributions of batting average versus binomial prediction for 50 times at-bat (for same and opposite handedness) (redrawn from Lindsey, 1959).

since 1900. A contemporary stadium notorious for reducing batting performance is the Houston Astrodome. James (1985, p.154) describes how Jose Cruz's batting statistics would be at the Hall of Fame level of George Brett, Mike Schmidt and Eddie Murray if he had played for a team other than the Houston Astros.

Several studies have focused on the effects of artificial turf as opposed to natural grass. Siwoff *et al.* (1987, p.89) examined the percentages of hits by each team that went for doubles and calculated the ratio of home versus away percentages for each team. A similar analysis was performed on stolen base average. The summary of their results in Table 2.8 indicates that teams having home stadiums equipped with artificial turf have more extra-base hits

Table 2.8 Effects of stadium surface on extra-base hits and stolen base average, 1982–86 (Siwoff *et al.*, 1987, p.89)

Stadium surface	Number of stadiums	Average home/away ratio	
		Extra-base hit percentage	*Stolen base average*
Artificial turf	10	1.12	1.05
Natural grass	16	0.93	0.96

Table 2.9 Effects of game temperature on hitting, 1987–89 (Dewan *et al.*, 1990, p. 249)

Temperature (°F)	AB	BA	OBP	SLG	Games	R/G	H/G	HR/G
90+	23 640	.263	.329	.402	348	9.1	17.9	1.83
80–89	56 851	.263	.328	.403	827	9.1	18.1	1.85
70–79	93 829	.259	.323	.391	1383	8.6	17.6	1.69
60–69	51 721	.253	.320	.385	767	8.5	17.1	1.65
<60	36 380	.248	.318	.366	548	8.0	16.4	1.40

and greater stealing success at home than away; the effect is just the reverse for teams having home stadiums with natural grass. The analysis supports the belief that artificial turf provides a faster surface, which aids the offense.

2.4.4 Temperature

Table 2.9 indicates minor improvements in BA and OBP as the temperature rises. However, temperature appears to have a much more significant effect in improving power hitting; both SLG and home runs per game (HR/G) increase demonstrably with temperature. These effects could be attributed to stages of the baseball season, but Adair (1990) has stated that the baseball does travel farther at higher game temperatures.

2.5 Strategy

Baseball strategy can be divided into two components.

- On the field strategy conducted by a team's manager is designed to get the best out of the team's players in each game.
- Off the field strategy conducted by a team's General Manager (GM) is designed to acquire the best personnel for the team.

2.5.1 Manager strategy

Many of the results presented in the previous section could be used by the manager to identify situations of opportunity. In addition, baseball has several common tactics employed by managers as well as standard situations in which they are employed (referred to as 'the book'). As Bill McKechnie, one of the most successful managers from the 1920s through the 1940s, often said: 'If you take care of the percentages, the percentages will take care of you' (James, 1997, p. 73). Most research into the efficacy of these strategies has agreed on the basic conclusions: (1) the sacrifice bunt and intentional walk are of little value; (2) the stolen base can be of value only if the probability of success is sufficiently high (about 70%). The techniques used to reach these conclusions have varied from Markov model approaches (Trueman, 1976) to Monte Carlo simulations (Cook, 1977).

Lindsey (1963) used his analysis of run scoring in order to examine tactical strategy. These analyses were based on distributions of runs scored in the remainder of a half-inning conditional on the current out and base situation

Table 2.10   Distribution of runs scored in remainder of half-inning (Lindsey, 1963, Table I, p. 485)

Bases occupied	Outs	No. of situations	Probability of scoring runs				Expected runs
			0 runs	1 run	2 runs	>2 runs	
None	0	6561	0.747	0.136	0.068	0.049	0.461
	1	4664	0.855	0.085	0.039	0.021	0.243
	2	3710	0.933	0.042	0.018	0.007	0.102
1	0	1728	0.604	0.166	0.127	0.103	0.813
	1	2063	0.734	0.124	0.092	0.050	0.498
	2	2119	0.886	0.045	0.048	0.021	0.219
2	0	294	0.381	0.344	0.129	0.146	1.194
	1	657	0.610	0.224	0.104	0.062	0.671
	2	779	0.788	0.158	0.038	0.016	0.297
3	0	67	0.12	0.64	0.11	0.13	1.39
	1	202	0.307	0.529	0.104	0.060	0.980
	2	327	0.738	0.208	0.030	0.024	0.355
1,2	0	367	0.395	0.220	0.131	0.254	1.471
	1	700	0.571	0.163	0.119	0.147	0.939
	2	896	0.791	0.100	0.061	0.048	0.403
1,3	0	119	0.13	0.41	0.18	0.28	1.94
	1	305	0.367	0.400	0.105	0.128	1.115
	2	419	0.717	0.167	0.045	0.071	0.532
2,3	0	73	0.18	0.25	0.26	0.31	1.96
	1	176	0.27	0.24	0.28	0.21	1.56
	2	211	0.668	0.095	0.170	0.067	0.687
Full	0	92	0.18	0.26	0.21	0.35	2.22
	1	215	0.303	0.242	0.172	0.283	1.642
	2	283	0.671	0.092	0.102	0.135	0.823

(Table 2.10). Lindsey's approach was to consider each strategy in different out and base situations. The given strategy has value if the state resulting from the strategic move has a greater probability of scoring runs needed to overcome (or preserve) the lead in that inning. The next subsection provides an example of applying Lindsey's techniques to the sacrifice strategy.

2.5.1.1   Sacrifice

In a close game, the manager is often willing to give up an out in order to advance a runner to a position in which a key run can be more easily scored. The batter is instructed to bunt (tap the ball into the infield) and force the fielders to throw him out instead of the runner. Representative of most analyses is the conclusion by Trueman (1976) that the sacrifice always decreases the expected number of runs, but may improve the chance of scoring a key run if the batter can produce a successful sacrifice 70% of the time (50% for pitchers batting).

Lindsey (1963) arrived at similar conclusions. As an example of his analysis consider a runner on first base with no outs in a close game. The manager frequently will order the batter to sacrifice; if successful, this maneuver will advance the runner to second base, but the batter will generally be out. Using Table 2.10, we can see that with a runner on first base and no outs, the expected number of runs scored in the remainder of the half-inning is 0.813 and the probability of scoring no runs is 0.604. The standard result of the sacrifice hit produces a runner on second base and one out; only 0.671 runs are expected to be scored from this state and the probability of not scoring at all has increased slightly to 0.610. This seems to be a poor trade-off; however, further examination indicates that the probability of scoring *one* run has increased from 0.166 to 0.224. Thus, Lindsey's analysis of the sacrifice hit strategy in this situation would conclude that it is generally a poor strategy unless one run is desperately needed.

Lindsey's analysis is actually a bit more complex. In the example, he would also consider the possibility that the sacrifice hit was not successful (costing an out without advancing the runner) and the possibility that the bunt was a hit (placing runners on first and second with no outs). The skill of the batter as a bunter, relative to his skill as a batter swinging normally, becomes an important factor. Lindsey used similar techniques to analyze a variety of strategies.

2.5.1.2 Stolen base

If the runners on base are fast, the manager may not have to resort to the sacrifice in a close game. The runners can attempt to steal the next base. 'If this strategy is attempted in an effort to get one run across in a close game, the required success probability for the baserunner is in the neighborhood of 0.50, with a range of around 0.30 to 0.70, depending on the batter and line-up. If this strategy is utilized to increase the expected number of runs scored, the required success probability increases to around 0.65, with a range of approximately 0.40 to 0.85' (Trueman, 1976, p. 14). Lindsey (1963) generally supports Trueman's conclusions and provides a table for the critical successful steal probability to increase the chance of overcoming a lead in different out and base situations. The Linear Weights for stolen bases and caught stealing also support Trueman's result.

The manager may change his base stealing strategy depending on the count. Table 2.11 (Dewan *et al.*, 1994) indicates that the stolen base success rate is affected by the count. The manager of the fielding team can counter stolen base attempts by ordering a pitchout, a deliberate high pitch to the catcher giving him the best chance to throw out an anticipated base stealer. The pitcher can also throw to the fielder at an occupied base in order to pick-off a runner or prevent him from taking a big lead. Dewan *et al.* (1990, pp. 150–152) found the value of the pitchout to be inconclusive. However, a decrease in Batting Average and Slugging Percentage with an increase in pick-off throws indicate that pick-off throws may distract the batter (Table 2.12).

2.5.1.3 Intentional walk

The manager may order a batter to be given an intentional base on balls for two major reasons: to get to a weaker hitter in the line-up and to set up force-out and double-play situations. Various analyses tend to agree that the strategy has little

Table 2.11   Stolen base averages on different counts (1991–93)
(Dewan *et al.*, 1994, pp. 142–143)

| Count | | | | |
Balls	Strikes	SB	CS	SBA
0	0	2565	1166	.687
	1	1080	498	.684
	2	465	166	.737
1	0	1284	664	.659
	1	1048	590	.640
	2	802	345	.699
2	0	330	132	.714
	1	611	383	.615
	2	777	322	.707
3	0	29	6	.829
	1	213	136	.610
	2	443	420	.513

value. Trueman (1976) found that it increased the expected number of runs scored and only recommended its use to get to the ninth batter in the line-up. Lindsey (1963) found its best use was to set-up force plays when runners are at second and third bases with one out.

2.5.1.4   Positioning fielders

Managers often position fielders depending on the situation, such as moving the infielders in towards home plate for a likely bunt or moving the outfielders in towards the infield to prevent the winning run from scoring on a hit late in the game. Managers may also position outfielders depending on the tendencies of hitters and pitchers, the most famous being the shift (Fig. 2.4) developed by Lou Boudreau against Ted Williams in 1946 (Dickson, 1989). Williams, one of the greatest of all hitters, was a left-handed batter whose natural swing pulled the ball to right field. The figure shows how Boudreau shifted his fielders towards the right side to reduce Williams' chances of getting a hit with his natural swing, and possibly force him to change his swing and hit to left field with less power. Table 2.13 shows where different hits occurred in a small sample of 47 games involving the 1982 Kansas City Royals (James, 1983). Dewan *et al.* (1991) devised a system for recording the position of all batted balls and graphing them as an aid to managers. (See Chapter 11 on graphical techniques for an example.)

Table 2.12   Batter's performance when pitcher holds runner on first base, 1993 (Dewan *et al.*, 1994, p. 81)

Throws to first	BA	SLG
0	.287	.429
1	.279	.423
2 or more	.274	.395

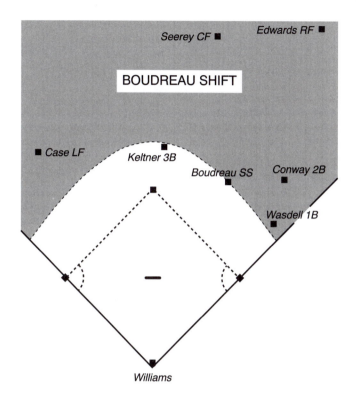

Figure 2.4   Williams shift developed by Lou Boudreau in 1946.

Table 2.13   The distributions of hit locations (James, 1983, p. 118)

Location	1B		2B		3B		HR	
	No.	*%*	*No.*	*%*	*No.*	*%*	*No.*	*%*
In the infield	91	15%						
Through the infield	219	36%	3	2%	1	5%		
Down the foul lines	36	6%	84	58%	3	14%	2	2%
In front of the outfielders	220	36%	5	3%	3	14%	1	1%
In the outfield gaps	47	8%	36	25%	13	59%		
Over the outfielders			18	12%	2	9%		
Over the outfield fence							104	97%
Total	613	100%	146	100%	22	100%	107	100%

2.5.1.5   *Batting order*

Each player upon entering the game is assigned a position in the team line-up which cannot be changed. Naturally, then, the composition and batting order of a team's starting line-up are major strategic decisions for the manager. Throughout its history, the batting order has evolved into a traditional pattern (Table 2.14) best summarized by Siwoff *et al.* (1989, p. 128): 'a base stealer to lead off, followed by a contact hitter and a high-average hitter; power in the

Table 2.14 1989 National League batting statistics by line-up position (per 600 plate appearances) (Dewan *et al.*, 1990, p. 279)

Line-up position	BA	OBP	SLG	HR	K	SB
1	.254	.323	.358	8	74	28
2	.261	.323	.368	9	77	17
3	.278	.349	.424	15	88	19
4	.266	.349	.453	22	95	9
5	.248	.307	.387	14	88	10
6	.250	.317	.374	12	87	10
7	.247	.303	.354	10	88	6
8	.233	.301	.321	6	85	6
9 (Pitcher)	.168	.218	.222	4	157	3

cleanup spot, then the extra base hitters; the automatic outs at the bottom of the order.'

While researchers have demonstrated a relative unanimity in the results of analyses of most managerial decisions, the batting order remains a contested issue. Freeze (1974) concluded that the best batting order could improve over the worst by less than three wins per season. Petersen (1977) found that a batting order of decreasing on-base average produced about a 1% improvement in scoring runs over a standard order. Cook (1977) supported the traditional batting order of two players with high on-base percentages, followed by the three players with the highest Slugging Percentages. Pankin (1992), using a Markov model to analyze batting orders, emphasized that the difference between high and low scoring line-ups is primarily increased scoring in the first inning. Seifert (1994) used a Monte Carlo simulation to demonstrate that the Minnesota Twins could have scored 12 more runs than they actually did in 1991 by using a line-up ordered by James' Runs Created formula, from high to low. The guiding principles behind this ordering are that the best offensive player gets the most at-bats and leads off about 40% of the innings.

Bukiet *et al.* (1997) utilized a unique transition matrix for each player in the line-up. For a given line-up, the transition matrix for each player is applied in turn in a Markov chain to produce a distribution of runs scored in nine innings. The optimal (worst) line-up is the one that produces the distribution with the highest (lowest) expected number of runs in nine innings. When applied to 1989 National League teams, optimal line-ups produced 30 to 50 more runs and 6 to 11 more victories per season than the worst line-ups. The study found that sluggers were not in the clean-up (fourth) position in most optimal line-ups and pitchers batted seventh or eighth (not last) in all but one optimal line-up.

Considering the continuing interest in batting orders, researchers have found surprisingly little improvement over the traditional order. All batting order models have assumed fixed performance levels for the players. Few analyses have addressed the possibility that batting position can affect player performance. For example, managers will try to place the best hitter in back of a good hitter to allow this 'protected' hitter to get better pitches to hit. Dewan *et al.* (1994, p. 129) provided some evidence that protecting a hitter has a positive effect on batting performance.

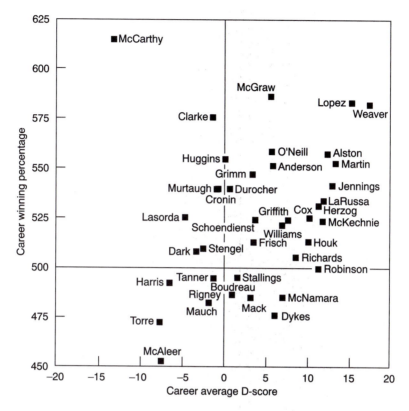

Figure 2.5 Career winning percentage (WPct) versus career D-scores (actual WPct minus expected WPct) for 40 managers with ten or more full seasons from 1901–92 (redrawn from Boynton, 1995).

2.5.1.6 *Manager evaluations*

Game strategy is only one factor in evaluating managers, but it is the focus of fan interest. Boynton (1995), Thorn and Palmer (1994) and Siwoff *et al.* (1993, p. 20) used similar techniques to evaluate managerial ability to get the most out of their teams' run performance, the sign of a good baseball tactician. Using the formulas each had developed for team Winning Percentage (WPct), managers were evaluated on the WPct their teams achieved above or below the WPct expected, based on the runs scored and allowed by their teams. Figure 2.5 summarizes Boynton's results for 40 managers with ten or more years of experience with respect to their career WPcts and their D-scores (actual WPct minus expected WPct). Managers at the top of the figure had high Winning Percentages, but managers to the right are those whose teams performed better than expected. Joe McCarthy, a great manager from the 1920s, 1930s and 1940s, has the highest Winning Percentage, but has a low D-score, indicating that he did not get the most from his team's run scoring and run prevention capabilities, according to Boynton's perspective. Boynton also found that about 2/3 of these managers experienced a decline in their D-scores over their careers. According to Boynton, the best manager, Earl Weaver, could add about three wins per year while Siwoff's best manager, Billy Martin, added about 4.5 wins per season. James (1997) disputed Boynton's system based on runs and proposed a technique based on the won/lost record over recent years; his system rated Joe McCarthy as the best manager.

2.5.2   General Manager

The major responsibilities of the General Manager (GM) are to draft amateur talent, to trade players, and to retain/acquire free agent talent. Since the advent of free agency in the 1970s, fan interest in off-season strategy at the GM level has increased. At times, fans appear to be more interested in off-the-field events than in the actual play of the game on the field. This phenomenon is reflected in the emergence of Rotisserie Baseball, as the single greatest use of baseball statistics today by fans. Rotisserie Baseball is a game based on daily baseball statistics in which each participant assumes the role of a general manager of a baseball team in a fictitious league composed of actual baseball players.

2.5.2.1   Draft

The baseball draft is much chancier than similar drafts in football and basketball. Most players require experience in the minor leagues before their skills are developed enough that a reasonable judgment on their professional capability can be made. Thomas (1994, 1996) developed a ranking system to gauge a team's success in the first round of the draft.

2.5.2.2   Trades

Several researchers have sought to measure the value of players gained and lost through trades. James (1982) developed a Value Approximation Method, but the technique does not have the statistical support earned by the Runs Created Model. In a related model, Lieff (1992) evaluated the skill of general managers in obtaining and retaining talent. The metric used was based on the percentage of Most Valuable Player (MVP) votes obtained by players through their careers. Teams gain points for MVP points of players on the team and lose MVP points for votes obtained by the player after leaving the team. Boyle (1996) analyzed making major overhauls in a team and found that it weakened good teams and had little effect on losing teams. Division and league champions' turnover averages 29% while other teams' turnover averages 52% (Siwoff *et al.*, 1993, p. 73).

2.5.2.3   Player age

Age is an important consideration in trading for a player. Siwoff *et al.* (1989, p. 90) contended that the peak age for a baseball player was 26 by examining the ages at which players had their best totals in home runs, stolen bases, hits and wins (for pitchers) (Fig. 2.6).

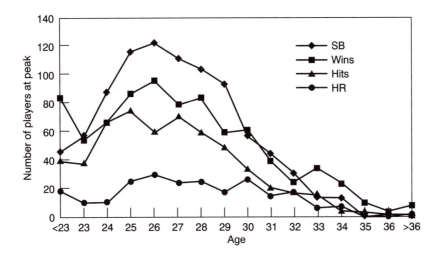

Figure 2.6   Number of players who peaked in HRs, Hits, Stolen Bases and Wins at different ages (players who retired before 1987 with at least 25 home runs, 25 stolen bases, 150 hits, or 15 wins in a season) (data from Siwoff *et al.*, 1989, p. 90).

Table 2.15   Change in team win percentage for different average ages of team starting players, 1920–41, 1946–77 (James, 1982, p. 191)

Average age	Number of teams			% Improved
	Improved	Declined	No change	
25	33	11	2	71.7%
26	61	40	3	58.7%
27	103	92	11	50.0%
28	115	112	8	48.9%
29	91	101	9	45.3%
30	41	64	3	38.0%

James (1982, pp. 192–193) came to a similar conclusion (age 27) by totaling the number of batting championships, 100-RBI seasons, and 20-win seasons at different ages in addition to an analysis using his Value Approximation Method. James also found that teams have less chance of improving as they age. Table 2.15 indicates that as the average age of a team increases, the chance of improving the won–lost record in the following year decreases. Note that the 50% point of improving is at James' peak age of 27.

2.6   Records

In few sports do numbers hold such significance as baseball. The meaning of a number such as 61 will bring instant recognition to the baseball fan.[2] Particular

[2] The highest number of home runs hit by a player in a season, set by Roger Maris, fittingly, in 1961.

attention has been paid to the numbers 56 (longest hitting streak in games) and .406 (the last season BA over .400).

2.6.1   Hitting streaks

Perhaps the most fabled baseball record (and the one which has recently drawn the most attention from statisticians) is Joe Dimaggio's 1941 record of obtaining one or more hits in 56 consecutive games. This record has never been challenged seriously (Pete Rose's 44-game hitting streak in 1978 being the second longest).

Warrack (1995), modeling at-bats as a Bernoulli process, used Feller's formula for the probability $P(n, r)$ of obtaining at least one run of r successes in n trials with probability p in each trial

$$P(n, r) = [1 + (n - r)(1 - p)] p^r$$

to estimate Dimaggio's chance of having a 56-game hitting streak at some point in his career. Based on Dimaggio's career .325 Batting Average over $n = 1736$ games, Warrack estimated the probability $p = 0.777$ of getting at least one hit in a game and $P(1736, 56) = 0.000274$ or only 1 chance in 3700 of having a 56-game hitting streak.

2.6.2   Batting .400

Curiously, the other achievement of great statistical interest also occurred in 1941 when Ted Williams became the last batter to have a .400 Batting Average (.406 to be exact) for a season. In order to determine the probability of a batting champion reaching the .400 plateau in a particular season, Adams (1981) performed an analysis based on Relative Average, a player's BA divided by his league's BA for the same year. He found that the Relative Averages of batting champions from 1901 to 1980 followed a normal distribution with mean 1.361 and standard deviation 0.075. He used this relationship to estimate the probability of a batting champion batting .400 or more based on the League Batting Average (LBA)

$$1 - \Phi\left(\left(\frac{.400}{\text{LBA}} - 1.361\right) \Big/ 0.075\right)$$

where Φ is the standard normal cumulative distribution function. Figure 2.7 indicates that this equation (theoretical probability) provides a reasonable fit to the data points (experimental probability).

Lackritz (1996) took another approach, examining the probability of a selected player to hit .400 in a season at some point in his career based on career Batting Average, at-bats per season, and the length of his career. By his calculation, someone like Tony Gwynn, with a Batting Average of .340 over a 15-year career with 540 at-bats per season, has only about a 3% chance of batting .400.

Gould (1996) felt that the lack of present-day .400 hitters is not a result of a current reduction in batting skills. In fact, from his perspective, *improved*

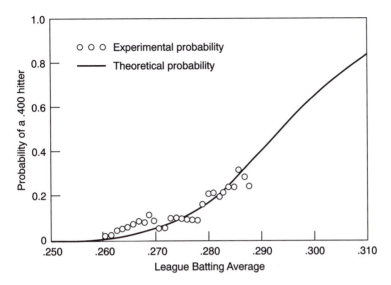

Figure 2.7   Probability of a .400 batting champion as a function of League Batting Average (data from Adams, 1981, p. 83).

playing skills have produced a decrease in variability of batting skill across the player population, making it less likely for such extraordinary batting achievements to be attained. (See Chapter 13's analysis of Ty Cobb's batting .400.)

2.6.3   Career records and superior achievements

Fan interest builds as individual players approach the magic levels of 3000 hits, 500 home runs, and 300 wins. Almost every player who has attained one of these career totals has been elected to the National Baseball Hall of Fame in Cooperstown, NY, the Valhalla of baseball players. James (1982) uses his Favorite Toy technique to estimate the probability that an outstanding player will achieve career records or these levels of superior achievement

$$\text{Probability of reaching goal} = \frac{\text{years remaining} \times \text{annual rate}}{\text{goal} - \text{current total}} - 0.5$$

where years remaining $= 24 - 0.6 \times$ age and annual rate is a weighted average rate from the last three years, the most recent year having weight 3, the previous year weight 2, and the year before that weight 1. James makes no claims for the statistical accuracy of his Toy although he believes it does work. Using probability estimates from James (1983), it is possible to gain some insight into the Toy's accuracy. By summing the probabilities for each player, the expected number of players reaching the goal as predicted by the Favorite Toy in 1982 can be estimated and compared to the number of rated players who actually achieved the goal by the end of the 1997 season. Table 2.16 indicates that the Favorite Toy gives reasonable results.

Table 2.16 Number of players reaching career goals (predicted by Favorite Toy versus actual)

Goal	No. players rated	No. of rated players reaching goal	
		Predicted at end of 1982 season	Actual at end of 1997 season
3000 hits	41	5.82	5
500 HRs	17	3.33	3

2.7   Other topics of enduring interest

Many topics, while not applicable directly to strategy or player evaluation, have drawn the continued attention of researchers. This section provides an overview of these questions and the approaches taken to answer them.

2.7.1   Distribution of runs scored per half-inning

Some researchers have examined the characteristics of the distribution of runs scored and whether it changes during the game. By reducing events to five basic plays (out, walk, single, double, triple, home run) and assuming homogeneous batting ability among players, D'Esopo and Lefkowitz (1977) were able to reduce a Markov model of run scoring to a simpler problem based on sequences of events that score runs. With this model (originally presented in 1960), the authors developed a relatively simple set of equations to estimate the distribution of runs scored in a half-inning. The model provided a reasonable fit to the distribution of runs scored per half-inning in the 1959 National League season. Further, the authors proposed that the model could be used to rate the batting contributions of individual players by estimating the expected number of runs scored per half-inning if the entire batting line-up was composed of that player (similar to James' application of Runs Created). Cover and Keilers (1977) developed a similar model, the Offensive Earned-Run Average, and used it to assert that Ruth and Williams were the greatest hitters of all.

Lindsey (1961) found inhomogeneity in the distributions of runs scored in different innings. The first and third innings have the highest expected number of runs as well as the greatest chance of a big inning and the second inning has the lowest expected number of runs scored. Lindsey concluded this effect was the result of the standard batting line-up being structured with the best batters at the top (appearing in the first and third innings) and the weaker batters at the bottom (batting in the second inning). The negative binomial distribution provided a reasonable fit to the overall distribution of runs scored per half-inning with mean 0.475 runs and standard deviation 1.00 runs. Through several analyses (distribution of total team score, probability of establishing a lead, frequency in overcoming a lead), Lindsey found that baseball scores could be replicated by a model assuming independence of runs scored in different innings.

Rosner *et al.* (1996) modeled the distribution of runs scored per inning as the convolution of a modified negative binomial distribution (for the number of batters faced) and a truncated binomial distribution (for runs scored given the number of batters faced). Their results were used to characterize pitchers by their abilities to get batters out and prevent runners from scoring.

2.7.2 Progress of the score

Baseball games tend to be decided (i.e. leads maintained) at earlier stages than other sports. While teams trailing late in football, basketball and hockey games have about a 20% chance of ultimate victory, the team losing at the end of six innings of a baseball game goes on to win only about 10% of the time. Early leads also hold up more often in baseball (Cooper *et al.*, 1992).

Stern (1994) proposed a general Brownian motion model to estimate the probability of winning given the lead l and the fraction t of the game completed

$$\Pr(l, t) = \Phi\left(\frac{l + (1 - t)\mu}{((1 - t)\sigma^2)^{1/2}}\right)$$

where μ is the average advantage in runs for the home team over the course of a game and σ is the standard deviation of differentials in the final score. Values for $\hat{\mu}$ and $\hat{\sigma}$ are estimated using probit regression of home team wins on home team leads at the end of each inning. These values were estimated as $\hat{\mu} = 0.34$ and $\hat{\sigma} = 4.04$ based on 962 games. Figure 2.8 tests the validity of the model with respect to data compiled by Lindsey (1961) who assumed $\hat{\mu} = 0$. As Lindsey first noted, the runs with the greatest contribution to winning are those scored when the game is close. These runs become more valuable as the game progresses. Stern's model captures these general effects and fits Lindsey's data well in the early innings. However, in the later innings of a close game, the model tends to underestimate the value of the lead. The model may not reflect a team's capability to improve its chances of victory by bringing in a relief pitcher to preserve a lead and close out the game.

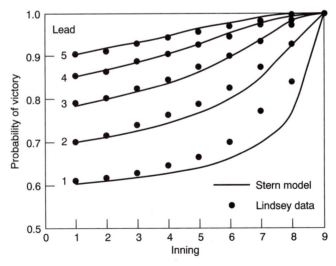

Figure 2.8 Probability of winning given a lead at different stages of a game (data from Lindsey, 1961; model from Stern, 1994).

2.7.3   Luck factor

The winner of the World Series lays claim to being the best team in baseball. But is it really the best? Commentators often obliquely refer to the role of chance in determining champions (e.g. the Philadelphia Phillies won the 1993 National League championship because many players had 'career years'). A recent study indicates that this role may be larger than most fans can imagine. James *et al.* (1993) investigated this luck factor using a simulation of 1000 baseball seasons. The probability of a team win was calculated using the Bradley–Terry choice model. The logarithms of team strength from each season were selected randomly from a normal population with mean 0 and standard deviation 0.19. Based on the MLB structure of the time (two leagues of 14 and 12 teams, each league with two divisions), the study found that the best team (the team with the greatest team strength) won the World Series 25.9% of the time.

Throughout baseball history, the best team in a league wins 73% of its games against the worst team in its league. Thus, even the worst team has about a 10% chance of winning a seven-game series like the World Series (Siwoff *et al.*, 1989, p. 72). The difference between actual games won and games predicted from batting statistics has led to the identification of overachieving teams. James (1983, p. 106) uses this 'Johnson effect' to identify teams that are likely not to maintain this unexpected success into the next season. Chapter 13 presents the theory describing this effect.

2.7.4   Streakiness

Players are prone to experience hot and cold spells, especially in hitting. Sportscasters take note of these hot streaks and slumps and carefully select periods to accentuate the extreme nature of the streak or slump. Casella and Berger (1994) provide an analysis of the effects of this selectivity in presenting Batting Averages.

The question still remains whether these streaks are the result of chance or do they occur because of real changes in performance levels for short periods. (See Chapter 3 for a discussion of a related topic, the 'hot hand' in basketball.) Streakiness has significant strategic implications. If the effect exists, managers should give the edge to using players who are hot now and rest players who are cold.

The results of most studies do not support the existence of streak hitters. Siwoff *et al.* (1987, p. 97) found that the probability of hitting well (or poorly) in a game is independent of whether the batter has been hot or in a slump. Albright (1993) analyzed streakiness in batting by examining 501 sequences of plate appearances for a baseball season. His analysis used logistic regression to incorporate effects on hitting (e.g. day/night, pitcher's ERA, score, baserunners, outs) into his model. Albright concluded that while certain batters did demonstrate streakiness in certain seasons, they did not do so consistently, and the distribution of number of runs (sequences of successful at-bats or unsuccessful at-bats) did not depart significantly from randomness.

Some researchers do believe that the effect exists, but is difficult to detect. Stern and Morris (1993) and Albert (1993) commented on Albright's paper with alternative approaches that could lead to the conclusion that streakiness does exist in batting. Albert (1997) has continued work in this area by analyzing streakiness in Mike Schmidt's home run hitting. Stern (1995) re-examined Albright's data and concluded that evidence of streaks exists if players are analyzed as a group rather than individually.

2.7.5 Clutch hitting

Even more controversial is the existence of clutch hitters, players who perform well in situations with the game on the line. Siwoff *et al.* (1993, p. 85) have even defined a term 'Late Inning Pressure Situation' (LIPS) for such critical states in the game: 'any plate appearance occurring in the seventh inning or later with the score tied or with the batter's team trailing by one, two, or three runs (or four runs if there are two or more runners on base).'

MLB briefly awarded a Game Winning RBI (GWRBI) to the player who batted in the run giving his team a lead which it never relinquished. Players with high GWRBI totals were considered clutch hitters. While GWRBIs are still cited on occasion, MLB dropped its official status because of several criticisms, especially:

(1) GWRBIs favored players in the middle of the line-up who batted with players on base more frequently;

(2) a GWRBI could be awarded early in the game when no clutch situation existed.

In answer to the first criticism, Dewan *et al.* (1990, pp. 120–121) proposed the RBI Percentage

$$\text{RBI Percentage} = \frac{\text{Number of runners batted in from scoring position}}{\text{Number of runners in scoring position}}$$

which roughly normalizes for varying numbers of opportunities. In 1989, RBI Percentages ranged from 16.1% to 35.9% among players with 150 or more opportunities. Siwoff *et al.* (1988, p. 120) countered the second criticism to some degree by replacing the GWRBI with the Go-Ahead RBI, an RBI which puts the team ahead whether or not the lead is later relinquished. Lindsey's results (see Fig. 2.8) support the usefulness of the Go-Ahead RBI.

Power hitters are often subjected to the scorn of fans for not hitting home runs with men on base. Tattersall (1977) summarized the number of home runs hit in each base situation by the greatest home run hitters through the 1975 season. Vic Wertz's 480 RBIs on 266 HRs gave him the highest RBI/ HR ratio (1.80), while Ted Kluzewski's ratio of 1.55 was the lowest. Table 2.17 lists the ratios for hitters in the 600 HR club.

For many, clutch hitting only exists if it can be shown that hitters consistently perform better in clutch situations than in standard situations. Two researchers

Table 2.17 RBI/HR ratios for players with 600 or more home runs (Tattersall, 1977, p. 69)

Player	Number of home runs				
	0 runners	*1 runner*	*2 runners*	*3 runners*	*RBIs/HRs*
Hank Aaron	399	243	97	16	1.64
Babe Ruth	350	249	99	16	1.69
Willie Mays	365	219	68	8	1.57

Table 2.18 Number of good and poor clutch hitters as defined over two years, who improved or declined in the third year (Siwoff *et al.*, 1988, p. 61)

Years	Good in first two years			Poor in first two years		
	In third year			In third year		
	Improved	*Declined*	*% improved*	*Improved*	*Declined*	*% improved*
1985–87	21	22	48.8%	24	40	37.5%
1982–84	24	25	49.0%	22	35	38.6%
1979–81	20	15	57.1%	24	37	39.3%
1976–78	24	25	49.0%	31	33	48.4%
Total	89	87	50.6%	101	145	41.1%

who investigated clutch hitting with respect to this definition have come to opposite conclusions. Cramer (1977) examined the issue through the use of Player Win Average (PWA), the Mills' statistic that incorporates the game state into its measure (see Section 2.2.3). He demonstrated that PWA is correlated with Batter Win Average (similar to James' Runs Created model) across batters. For Cramer, a clutch hitter was a batter whose PWA was consistently higher than expected based on Batter Win Average. He found no significant correlation between prediction residuals for batters from 1969 and 1970. By his definition, clutch hitting does not exist. For four 3-year periods, Siwoff *et al.* (1988, p. 61) identified hitters as 'Good' or 'Bad' clutch hitters in the first two years of the period on the basis of Batting Average in LIPSs as opposed to typical situations. Table 2.18 shows the number of batters who improved or declined in clutch hitting in the third year of each period. Since players tend to bat worse in LIPS than in normal situations, they note that a hitter may be good in the clutch if he is able simply to maintain his BA in LIPSs. Brooks (1989) disputed the statistical significance of their conclusion and criticized their study for ignoring the degree of improvement or decline. Whether clutch hitters exist remains a tantalizingly unresolved question.

2.7.6   Biases in umpiring

In the past, certain umpires have been identified as favoring the home team or favoring the pitcher. Siwoff *et al.* (1993, pp. 37–38) classified 'Pitcher's' and 'Batter's' umpires according to their strikeout-to-walk ratios, which ranged from 1.28 to 2.43 over three seasons from 1990–92. Kitchin (1991) among several analyses of umpire bias concluded that the variation in home game Winning Percentage among different umpires was significant. However, Runquist (1993) found that the distribution of home victories over umpires was not statistically different than chance and that the correlation of umpire records between 1991 and Kitchin's data from 1987–90 was virtually zero. He also found the distribution of runs scored across umpires matched expected statistical variation.

2.8   Closing comments

This chapter has provided a high-level description of statistics and their application in baseball. Its goal has been to provide the reader with a glimpse of the wide variety of questions and techniques that have been applied to baseball statistics. Increasingly, analysts are turning their attention to broader uses of statistics in baseball beyond the standard areas of player performance and strategy. For example, the recent decline in fan interest in baseball often has been attributed to the increasing length of games. Siwoff *et al.* (1993, p. 13) found that this trend may result from the increased attention pitchers give to baserunners, over half a minute more per baserunner in 1992 as opposed to 1982.

Although these analyses have been baseball-oriented, the techniques used may stimulate ideas for applications to other sports as well. Because of limitations in space, many nuances in the statistical analyses discussed have been left to the reader to explore using the references listed in the next section. Interested readers should avail themselves of the vast amount of baseball material available on the World Wide Web. A good starting point is the Major League Baseball web site, currently at www.majorleaguebaseball.com.

Appendix: baseball basics

2.A.1 The baseball season

The Major League Baseball (MLB) regular season starts in early April and ends in late September or early October. Each team plays 162 games. The four best teams in each league enter a knockout tournament of two five-game series succeeded by a seven-game series to determine the league champions. The league champions meet in a seven-game series called the World Series for the championship of Major League Baseball. Separate statistical records are kept for the regular season, league play-offs, and the World Series.

2.A.2 A baseball primer

The official rules of Major League Baseball can be found in many books as well as the MLB web site. This section presents a brief outline to acquaint unfamiliar readers sufficiently to aid in understanding the baseball chapter.

A game of baseball is played by two teams over nine innings. The team with the most runs at the end of nine innings is the winner. In the case of a tie, extra innings are played until one team is ahead at the end of an inning. Each team gets to bat once in each inning, visiting team first, then home team. In its half-inning, the team at bat scores runs until three outs are made, at which point the opposing team comes to bat.

An MLB team is composed of 25 players. At the start of a game, the manager of the team selects nine players and places them in a batting order (or line-up). The batting order can be changed during a game only by replacing a player in the batting order by another player who has not yet played in the game. The first player in the line-up bats first, the second next, and so on. The ninth batter is followed by the first batter. At the start of an inning, the batting sequence continues from where it left off in the previous inning. When the opposing team is at bat, the nine players in the line-up play in the field. An exception is the Designated Hitter (DH) rule (adopted by the American League, but not the National League) in which one player in the line-up does not play in the field; this DH has his field position taken by the pitcher, who does not bat.

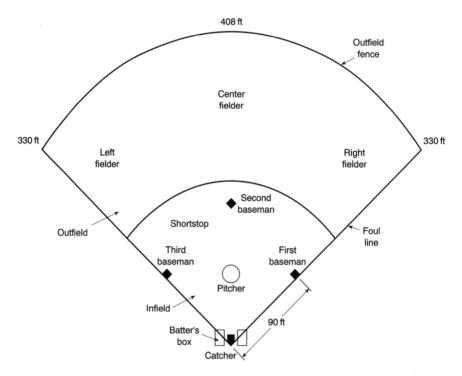

Figure 2.A.1   Baseball playing field (distances are for Veterans Stadium in Philadelphia).

A typical playing field is shown in Fig. 2.A.1, which also shows the standard positions assumed by fielders on the defending team. The three bases and home plate form a diamond, 90 feet on a side. The pitcher's mound is approximately in the center of this diamond. From this position, the pitcher throws the baseball to the batter, who is positioned in the batter's box to the right or left of home plate depending on his handedness.

With each pitch, the batter has the option of swinging or letting the pitch go. If the batter does not swing, the pitch is called a strike by the home plate umpire if it passes over home plate at a height between the shoulders and knees of the batter; it is called a ball otherwise. A strike is also charged if the batter swings and misses the ball. If the batter hits the ball into foul territory, he is charged with a strike if he has fewer than two strikes already. If the batter hits the ball into fair territory, the batter starts to run counter-clockwise around the bases until he is put out, stops safely at a base, or reaches home plate safely.

The batter is out if a batted ball is caught in the air by a fielder or a fielder catches the batted ball and throws it to first base before the batter arrives there. If a batter reaches first base safely, he may advance further around the bases in sequence, but will be called out if a fielder tags the batter with the ball while the batter is not on a base. Runners already on base can also advance on a batted ball as long as they are not tagged out or forced out.

Once the batter and all runners have stopped safely at bases, scored, or have been called out, the batter's plate appearance is concluded and the next batter in the line-up comes to bat. The batter's plate appearance can also end with an out if three strikes are called and can end with a base on balls (or walk, in which the batter is allowed to walk safely to first base) if four balls are called.

Batters continue to come up in a half-inning until three outs are recorded. Each succeeding batter attempts to get safely on base and advance the runners already on base. The team scores a run for each player who has progressed safely to all three bases and arrived safe at home plate during the inning. The opposing team then comes to bat with all bases empty.

2.A.3 Fielding positions

The defensive team in the field has players in nine defensive positions, split into six infielders (pitcher, catcher, first baseman, second baseman, third baseman, shortstop) and three outfielders (left fielder, center fielder, right fielder). In any game, the most important position is the pitcher. The pitcher's task is to start each play by throwing the ball towards home plate in such a way that the batter misses the pitch or cannot hit it well. The pitcher's responsibility is so great that, in each game, one pitcher on the winning side is awarded a Win (W) and one pitcher on the losing side is credited a Loss (L). Pitchers are one of two types, starters and relievers. A starter pitches in the first inning and stays in until he is removed for another pitcher for any of several reasons.

- He is ineffective.
- He has pitched well, but is tiring.
- A pinch hitter has replaced him in his time at bat in order to score runs. Because pitching is a special skill requiring dedicated practice, few pitchers are good hitters.

Starters are generally awarded the win or loss of a game. Starting a game and pitching five or more innings is such a strain on a pitcher's arm that he must rest several days before pitching again. Each team has a pitching staff with four or five regular starters so that a pitcher need only start every four or five days.

Relief pitchers (or relievers) are specialists in replacing starters in the middle of a game, often in the middle of an inning with opposing runners on base. The statistical reward for a relief pitcher is to come into a game with his team ahead and preserve the victory, for which he is awarded a Save. Because they usually pitch no more than one or two innings in any game, relief pitchers can be used in several games in a row.

2.A.4 The box score

During a game, counts are kept of events and summarized in a box score. Box scores are kept on every Major League Baseball game. They are published in newspapers, on websites, and in annual collections by *The Sporting News*. Figure 2.A.2 is a typical box score for a very atypical game, the only perfect game in World Series history. A perfect game occurs if no runner from one team ever reaches base. At the top of the figure is an abbreviated line score which contains two rows of runs scored by each team in each inning. This was a close contest with New York scoring lone runs in the fourth and sixth innings. The 'x' in the ninth inning for New York indicates that the Yankees

Game 5 October 8 at New York

Brooklyn	Pos	AB	R	H	RBI	PO	A	E
Gilliam	2b	3	0	0	0	2	0	0
Reese	ss	3	0	0	0	4	2	0
Snider	cf	3	0	0	0	1	0	0
Robinson	3b	3	0	0	0	2	4	0
Hodges	1b	3	0	0	0	5	1	0
Amoros	lf	3	0	0	0	3	0	0
Furillo	rf	3	0	0	0	0	0	0
Campanella	c	3	0	0	0	7	2	0
Maglie	p	2	0	0	0	0	1	0
a Mitchell		1	0	0	0	0	0	0
Totals		27	0	0	0	24	10	0

a Struck out for Maglie in 9th.

Home Run—Mantle. Sacrifice Hit—Larsen.
Double Plays—Reese to Hodges, Hodges
to Campanella to Robinson to Campanella to
Robinson. Left on Bases—Brooklyn 0,
New York 3. Umpires—Pinelli, Soar,
Boggess, Napp, Gorman, Runge.
Attendance—64,519. Time of Game—2:06.

Bkn.	0 0 0	0 0 0	0 0 0				
N.Y.	0 0 0	1 0 1	0 0 x				

New York	Pos	AB	R	H	RBI	PO	A	E
Bauer	rf	4	0	1	1	4	0	0
Collins	1b	4	0	1	0	7	0	0
Mantle	cf	3	1	1	1	4	0	0
Berra	c	3	0	0	0	7	0	0
Slaughter	lf	2	0	0	0	1	0	0
Martin	2b	3	0	1	0	3	4	0
McDougald	ss	2	0	0	0	0	2	0
Carey	3b	3	1	1	0	1	1	0
Larsen	p	2	0	0	0	0	1	0
Totals		26	2	5	2	27	8	0

Pitching	IP	H	R	ER	BB	SO
Brooklyn						
Maglie (L)	8	5	2	2	2	5
New York						
Larsen (W)	9	0	0	0	0	7

Figure 2.A.2   Box score of Don Larsen's perfect game. Game 5 of 1956 World Series, Brooklyn Dodgers at New York Yankees, 8 October 1956 (data from Cohen *et al.*, 1976).

had no need to bat in the bottom of the ninth inning because the game was already won.

The major portion of the box score is the two tables, side-by-side in the figure, which give the line-ups (batting orders) for both teams. For each player, each table presents offensive and defensive information.

- Offense: the number of at-bats (AB), the number of runs scored (R), the number of hits (H), the number of runs batted in (RBI).
- Defense: the position played (Pos), the number of putouts (PO), the number of assists (A), and the number of errors made (E).

Pitching performances are kept in a separate table (shown below the New York line-up in the figure). This table lists each pitcher along with their counts for innings pitched (IP), hits allowed (H), runs allowed (R), earned runs allowed (ER), bases on balls allowed (BB), and strikeouts achieved (SO). One pitcher on the winning team (Larsen in this case) is awarded the win, marked by a W after his name; similarly, one pitcher on the losing team is marked with an L as the loser of the game.

Other significant events in the game (e.g. home runs, double plays), the attendance, and duration of the game are presented below the tables.

A fan quickly scanning this box score realizes that this was a well-pitched game on both sides, one which 'neither pitcher deserved to lose' to use an old sports announcer's cliché. The perfection of Larsen's performance is evident in the totals of no hits, no walks and no runs (plus no listing of hit batsmen); his performance was given equally flawless support by the Yankees in the field (i.e. no errors). Hitting heroics were few in a game with only five hits, but Mantle's home run is a notable standout.

References

Adair, R. K. (1990) *The Physics of Baseball*. New York: Harper & Row.

Adams, D. (1981) The probability of the league leader batting .400. *Baseball Research Journal*, **10**, 82–84.

Adams, D. (1982) Average pitching and fielding skills. *Baseball Research Journal*, **11**, 104–108.

Albert, J. (1993) Comment. *Journal of the American Statistical Association*, **88**(424), 1184–1188.

Albert, J. H. (1997) The home run hitting of Mike Schmidt. Technical Report, Department of Mathematics and Statistics, Bowling Green State University.

Albert, J. H. (1998) Sabermetrics. In *Encyclopedia of Statistical Sciences* (edited by S. Kotz, C. B. Read and D. L. Banks). New York: John Wiley.

Albright, S. C. (1993) A statistical analysis of hitting streaks in baseball. *Journal of the American Statistical Association*, **88**(424), 1175–1183.

Bennett, J. (1993) Did Shoeless Joe Jackson throw the 1919 World Series? *The American Statistician*, **47**, 241–250.

Bennett, J. (1994) MVP, LVP, and PGP: a statistical analysis of Toronto in the World Series. In *1994 Proceedings of the Section on Statistics in Sports*. American Statistical Association, 66–71.

Bennett, J. M. and Flueck, J. A. (1983) An evaluation of major league baseball offensive performance models. *The American Statistician*, **37**(1), 76–82.

Bennett, J. M. and Flueck, J. A. (1984) Player game percentage. In *1984 Proceedings of the Social Statistics Section*. American Statistical Association, 378–380.

Boyle, D. (1996) Major personnel changes and team performance. *The Baseball Research Journal*, **25**, 80–84.

Boynton, B. (1995) Managers and close games. *The Baseball Research Journal*, **24**, 81–87.

Brooks, H. (1989) The statistical mirage of clutch hitting. *The Baseball Research Journal*, **18**, 63–66.

Browning, R. (1980) These numbers don't lie. *Sports Illustrated*, 7 April, 70–73.

Bukiet, B., Harold, E. R. and Palacios, J. L. (1997) A Markov chain approach to baseball. *Operations Research*, **45**(1), 14–23.

Casella, G. and Berger, R. L. (1994) Estimation with selected binomial information or do you really believe that Dave Winfield is batting .471? *Journal of the American Statistical Association*, **89**(427), 1080–1090.

Cohen, R. M., Neft, D. S., Johnson, R. T. and Deutsch, J. A. (1976) *The World Series*. New York: The Dial Press.

Cook, E. (1977) An analysis of baseball as a game of chance by the Monte Carlo method. In *Optimal Strategies in Sports* (edited by S. P. Ladany and R. E. Machol), pp. 50–54. New York: North-Holland.

Cook, E. and Garner, W. R. (1966) *Percentage Baseball*, Second Edition. Cambridge, MA: The MIT Press.

Cooper, H., DeNeve, K. M. and Mosteller, F. (1992) Predicting professional sports game outcomes from intermediate game scores. *Chance*, **5**(3–4), 18–22.

Cover, T. M. and Keilers, C. W. (1977) An offensive earned-run average for baseball. *Operations Research*, **25**, 729–740.

Cramer, R. D. (1977) Do clutch hitter's exist? *Baseball Research Journal*, **6**, 74–79.

Cramer, R. D. (1980) Average batting skill through major league history. *Baseball Research Journal*, **9**, 167–172.

Deane, B. (1985) Heresy! Players today better than oldtimers. *The Baseball Research Journal*, **14**, 52–54.

Deane, B. (1996) Normalized winning percentage. *The Baseball Research Journal*, **25**, 42–44.

D'Esopo, D. A. and Lefkowitz, B. (1977) The distribution of runs in the game of baseball. In *Optimal Strategies in Sports* (edited by S. P. Ladany and R. E. Machol), pp. 55–62. New York: North-Holland.

Dewan, J., Zminda, D. and STATS, Inc. (1990) *The STATS Baseball Scoreboard*. New York: Ballantine Books.

Dewan, J., Zminda, D. and STATS, Inc. (1991) *The Scouting Report: 1991*. New York: HarperCollins.

Dewan, J., Zminda, D. and STATS, Inc. (1994) *STATS 1994 Baseball Scoreboard*. New York: Ballantine Books.

Dickson, P. (1989) *The Dickson Baseball Dictionary*. New York: Avon Books.

Freeze, R. A. (1974) Monte Carlo analysis of baseball batting order. In *Optimal Strategies in Sports* (edited by S. P. Ladany and R. E. Machol), pp. 63–67. New York: North-Holland.

Gould, S. J. (1996) *Full House*. New York: Harmony.

Hurley, W. (1993) What sort of tournament should the World Series be? *Chance*, **6**(2), 31–33.

James, B. (1976) Big league fielding stats do make sense! *Baseball Digest*, March 1976, 70–75.

James, B. (1977) The relief pitcher's ERA advantage. *Baseball Research Journal*, **6**, 114–116.

James, B. (1982) *The Bill James Baseball Abstract 1982*. New York: Ballantine Books.

James, B. (1983) *The Bill James Baseball Abstract 1983*. New York: Ballantine Books.

James, B. (1984) *The Bill James Baseball Abstract 1984*. New York: Ballantine Books.

James, B. (1985) *The Bill James Baseball Abstract 1985*. New York: Ballantine Books.

James, B. (1986) *The Bill James Historical Baseball Abstract*. New York: Villard Books.

James, B. (1994) *The Politics of Glory*. New York: Macmillan.

James, B. (1997) *The Bill James Guide to Baseball Managers*. New York: Scribners.

James, B., Albert, J. and Stern, H. S. (1993) Answering questions about baseball using statistics. *Chance*, **6**(2), 17–22, 30.

Johnson, L. and Ward, B. (1994) *Who's Who in Baseball History*. New York: Barnes & Noble Books.

Katz, S. M. (1986) Study of 'the count' yields fascinating data. *The Baseball Research Journal*, **15**, 67–72.

Kindel, S. (1983) 'The hardest single act in all of sports.' *Forbes*, 26 September, 180–187.

Kitchin, R. (1991) Do the umps give a level field? *The Baseball Research Journal*, **20**, 2–5.

Klein, J. (1983) Computerball is here. *Sport*, April 1983, 33–37.

Koppett, L. (1991) *The New Thinking Fan's Guide to Baseball*. New York: Fireside Books.

Lackritz, J. R. (1996) Two of baseball's great marks: can they ever be broken? *Chance*, **9**(4), 12–18.

Ladany, S. P. and Machol, R. E. (Eds) (1977) *Optimal Strategies in Sports*. New York: North-Holland.

Lieff, M. (1992) Measuring front office judgments. *The Baseball Research Journal*, **21**, 28–32.

Lindsey, G. R. (1959) Statistical data useful for the operation of a baseball team. *Operations Research*, **7**, 197–207.

Lindsey, G. R. (1961) The progress of the score during a baseball game. *Journal of the American Statistical Association*, **56**(295), 703–728.

Lindsey, G. R. (1963) An investigation of strategies in baseball. *Operations Research*, **11**, 477–501.

Lowry, P. J. (1992) *Green Cathedrals*. New York: Addison-Wesley.

Machol, R. E., Ladany, S. P. and Morrison, D. G. (Eds) (1976) *Management Science in Sports*. New York: North-Holland.

Mills, E. G. and Mills, H. D. (1970) *Player Win Averages*. South Brunswick, NJ: A. S. Barnes.

Mosteller, F. (1952) The World Series competition. *Journal of the American Statistical Association*, **47**, 355–380.

Neft, D. (1986) Is Ozzie Smith worth $2,000,000 a season? *The Baseball Research Journal*, **15**, 43–48.

Neft, D. S., Johnson, R. T., Cohen, R. M. and Deutsch, J. A. (1974) *The Sports Encyclopedia: Baseball*. New York: Grossett & Dunlap.

Palmer, P. (1982) Runs and wins. *The National Pastime*, **1**(1), 78–79.

Pankin, M. D. (1992) Finding better batting orders. *The Baseball Research Journal*, **21**, 102–104.

Petersen, Jr., A. V. (1977) Comparing the run-scoring abilities of two different batting orders: results of a simulation. In *Optimal Strategies in Sports* (edited by S. P. Ladany and R. E. Machol), pp. 36–38. New York: North-Holland.

Rosner, B., Mosteller, F. and Youtz, C. (1996) Modeling pitcher performance and the distribution of runs per inning in major league baseball. *The American Statistician*, **50**, 352–360.

Rubin, E. (1958) An analysis of baseball score by innings. *The American Statistician*, **12**(April), 21–22.

Runquist, W. (1993) How much does the umpire affect the game? *The Baseball Research Journal*, **22**, 3–8.

Schutz, R. W. (1995) The stability of individual performance in baseball: an examination of four 5-year periods, 1928–1932, 1948–1952, 1968–1972, 1988–1992. In *1995 Proceedings of the Section on Statistics in Sports*. American Statistical Association, 39–44.

Seifert, S. (1994) On batting order. *The Baseball Research Journal*, **23**, 101–105.

Siwoff, S., Hirdt, S. and Hirdt, P. (1985) *The 1985 Elias Baseball Analyst*. New York: Collier Books.

Siwoff, S., Hirdt, S. and Hirdt, P. (1987) *The 1987 Elias Baseball Analyst*. New York: Collier Books.

Siwoff, S., Hirdt, S. and Hirdt, P. (1988) *The 1988 Elias Baseball Analyst*. New York: Collier Books.

Siwoff, S., Hirdt, S. and Hirdt, P. (1989) *The 1989 Elias Baseball Analyst*. New York: Collier Books.

Siwoff, S., Hirdt, S., Hirdt, T. and Hirdt, P. (1991) *The 1991 Elias Baseball Analyst*. New York: Fireside Books.

Siwoff, S., Hirdt, S., Hirdt, T. and Hirdt, P. (1993) *The 1993 Elias Baseball Analyst*. New York: Fireside Books.

Smith, J. H. (1956) Adjusting baseball standings for strength of teams played. *The American Statistician*, **10**(June), 23–24.

Stern, H. S. (1994) A Brownian motion model for the progress of sports scores. *Journal of the American Statistical Association*, **89**(427), 1128–1134.

Stern, H. S. (1995) Who's hot and who's not: runs of success and failure in sports. In *1995 Proceedings of the Section on Statistics in Sports*. American Statistical Association, 26–35.

Stern, H. S. (1997) Baseball by the numbers. *Chance*, **10**(1), 38–41.

Stern, H. S. and Morris, C. N. (1993) Comment. *Journal of the American Statistical Association*, **88**(424), 1189–1194.

Tattersall, J. C. (1977) The Henry Aaron home run analysis. *The Baseball Research Journal*, **6**, 66–70.

Thomas, D. C. (1994) Baseball's amateur draft. *The Baseball Research Journal*, **23**, 92–96.

Thomas, D. C. (1996) First round picks: how teams have done. *The Baseball Research Journal*, **25**, 117–120.

Thorn, J. and Palmer, P. (1985) *The Hidden Game of Baseball*. New York: Doubleday.

Thorn, J. and Palmer, P. (Eds) (1989) *Total Baseball*. New York: Warner Books.

Thorn, J. and Palmer, P. (Eds) (1994) *Total Baseball*. CD-ROM, Portland, OR: Creative Multimedia.

Trueman, R. E. (1976) A computer simulation model of baseball: with particular application to strategy analysis. In *Management Science in Sports* (edited by R. E. Machol, S. P. Ladany and D. G. Morrison), pp. 1–14. New York: North-Holland.

Verducci, T. (1997) Bat man. *Sports Illustrated*, 28 July 1997, 40–47.

Warrack, G. (1995) The great streak. *Chance*, **8**(3), 41–43, 60.

Watts, R. G. and Bahill, A. T. (1990) *Keep Your Eye on the Ball*. New York: W. H. Freeman and Company.

Weiskopf, H. (1977) Hitters can be ranked. *Sports Illustrated*, 18 July 1977, 24–25.

Wright, C. R. and House, T. (1989) *The Diamond Appraised*. New York: Simon and Schuster.

Wright, R. O. (1994) A case for the DH. *The Baseball Research Journal*, **23**, 55–61.

3

Basketball

Robert L. Wardrop

University of Wisconsin–Madison, USA

3.1 Introduction

Basketball was invented in 1891 by James Naismith, a physical education instructor at the YMCA Training School in Springfield, Massachusetts, USA. The game achieved almost immediate acceptance and popularity, and the first collegiate game, with five on each side, was played in 1896 in Iowa City, Iowa, USA.

Several important developments took place during the 1930s. The ten- and three-second rules were introduced, the center jump after every made basket was eliminated, and the one-hand shot gained acceptance. The Fédération Internationale de Basketball Amateur was formed in 1932, and basketball became an Olympic sport in 1936.

Professional basketball in the United States dates from the formation of the National Basketball League in 1898. This league merged with the three-year-old Basketball Association of America in 1949 to become the National Basketball Association (NBA). The American Basketball Association began play in 1967, and merged with the NBA in 1976, but only a few of its teams survived the merger.

Currently, there are two fledgling major women's professional basketball leagues in the United States. There are a number of men's and women's professional leagues around the world. Many Europeans currently play in the NBA.

Michael Jordan of the NBA's Chicago Bulls is, arguably, one of the most widely recognized and most popular athletes in the world. While the NBA finals are not nearly as popular as American professional football's Super Bowl, the men's National Collegiate Athletic Association (NCAA) basketball tournament is the crown jewel of college athletics in the United States. (And the women's tournament has experienced a tremendous growth in popularity in recent years.)

Considering its recent surge in popularity, the amount of statistical research in basketball has been relatively small compared with other sports. Most research in refereed journals has focused on modeling shooting (Section 3.4) or modeling the NCAA basketball tournament (Section 3.2). There has been an excellent paper on player types (Section 3.3). The rating of players, however, has not been the subject of refereed research, but has been considered in the popular press (Section 3.3). In addition, basketball has been innovative in its use of statistical graphics. See Chapter 11 of this volume for a clever graphical display of a basketball game (Westfall, 1990) and descriptions of shot charts.

3.2   Modeling the NCAA tournament

The three papers that are discussed in this section analyze men's NCAA basketball tournaments. The methods could be applied to women's tournaments too.

The NCAA basketball tournament consists of four regional tournaments, the winners of which advance to the Final Four tournament to determine the US collegiate champion. Every March across the US, participants in countless 'office pools' vie, under a variety of scoring methods, to be best at predicting the outcomes of tournament games. Each paper mentioned in this section produces either conjectured or estimated probabilities for the outcomes of any game and, thus, could be of interest to pool-participants. None of the papers, however, directly addresses any gaming issues.

Schwertman *et al.* (1991) (henceforth, SMH) introduce and study three models for predicting the winner of a men's NCAA regional basketball tournament.

A regional tournament consists of 16 teams, seeded $1, 2, \ldots, 16$, by a panel of experts, with the number 1 seed going to the team perceived to be the best, the number 2 seed going to the team perceived to be second best, and so on. Let $P(i,j)$ denote the probability that seed i will defeat seed j. The three models of SMH are three specifications of the matrix of values $P(i,j)$. SMH do not attempt to compare their individual values of $P(i,j)$ with data; instead, they use the $P(i,j)$ values to compute the probability that seed i will win the regional tournament.

Model 1 specifies $P(i,j) = j/(i+j)$. SMH describe this model as yielding probabilities (Schwertman *et al.*, 1991, p. 37), 'That heavily favor the number one seed.' For example, $P(1,2) = 2/3$, but $P(12,13) = 13/25$, a considerable difference, despite the fact that in each case the opponents differ by one seed position.

Model 2 specifies $P(i,j) = 0.5 + (j-i)/32$. Probabilities that can be obtained from this formula are equally spaced on the interval $[0,1]$: $1/32, 2/32, 3/32, \ldots$, and $31/32$, which SMH argue corresponds to the teams having (Schwertman *et al.*, 1991, p. 37), '... roughly, a uniform distribution in strengths...'.

Model 3 modifies Model 2 by assuming that the strengths of the 292 Division 1 basketball teams have a normal distribution and that the 64 teams in the overall tournament are the 64 strongest teams. These considerations yield $P(i,j) = 0.5 + \delta(S(i) - S(j))$, where $S(i)$ is the normal score measure of strength of team i. SMH then select $\delta = 0.2814$ so that $P(1,16)$ will equal $31/32$, as it does for model 2.

Figure 3.1 illustrates how the three models perform for six years of data (24 regional tournaments). The expected counts for Models 1 and 2 match the observed counts quite well for the number 2 seeds and number 3 seeds, but perform somewhat poorly for the number 1 seeds and the number 4 or larger

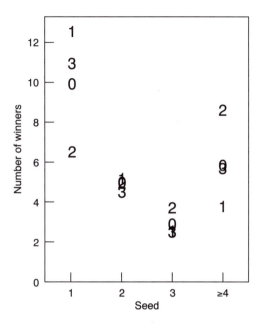

Figure 3.1 A comparison of Schwertman, McReady and Howard's models (1, 2 and 3, respectively). 0 denotes the observed number of regional winners; 1, 2 and 3 denote the expected number of regional winners for Models 1, 2 and 3, respectively.

seeds. All the Model 3 expected counts match the observed counts quite well. The chi-squared goodness-of-fit statistics and P-values for the three models are:

Model 1: $\chi^2 = 1.85$ and $P = 0.603$.
Model 2: $\chi^2 = 2.73$ and $P = 0.435$.
Model 3: $\chi^2 = 0.20$ and $P = 0.978$.

Schwertman *et al.* (1996) (henceforth, SSH) update and expand the work of SMH. The three models of SMH completely specify the values of $P(i,j)$. SSH consider these three models plus eight new regression models that use data from ten years of tournaments to estimate the values of $P(i,j)$. For each regression model, the response equals 1 if the stronger team (lower seed) wins and equals 0 otherwise. The eight regression models are obtained by crossing three dichotomous choices.

(1)   The predictor seed position can enter the regression in a linear ($j - i$ as in Model 2 above), or nonlinear (the normal scores of Model 3 above) manner.
(2)   The regression can be ordinary or logistic.
(3)   The intercept can be specified or estimated. (Specified means that $P(i,i)$ is forced to be 0.5 for all i.)

The 11 resulting models are evaluated in two ways. First, a chi-squared goodness-of-fit statistic is computed to compare the conjectured (for the first three models) or estimated values of $P(i,j)$ to the empirical values obtained from the ten years of data. The researchers perform this comparison twice:

(1)   for the 52 seed-pairs that competed at least once in ten years;
(2)   for the 26 seed-pairs that competed at least five times in ten years.

The researchers believe that the latter comparison is more meaningful. For this second comparison, the researchers discover that among the eight regression models, those with a nonlinear seed predictor are superior to those with a linear one. In addition, the ordinary models perform better than the logistic ones. Finally, whether the intercept is specified or estimated makes little difference in the performance of the models. Interestingly, Model 3 performs quite well; it is surpassed only by the two ordinary regression models (intercept specified or not). (The line in Model 3 is not steep enough.)

Second, the 11 models are compared by using their conjectured (estimated) $P(i,j)$ values to compute (estimate) the probability of seed i winning the tournament. Using the goodness-of-fit approach of SMH, SSH find that the nonlinear seed predictor models are superior to the other models. Unlike the first evaluation, however, the two logistic regression models are superior to the two ordinary regression models. Interestingly, Model 3 is the third best model by this criterion, outperforming its two ordinary regression counterparts.

Carlin (1996) provides an alternative approach to SMH and SSH. The earlier works use seed position as the only predictor of outcome. Carlin suggests that one might get improved models by using a computer ranking of teams, in particular the Sagarin ratings published in *USA Today*, or casino point spreads as predictors.

Carlin's answers have two new features. First, the probability of seed i winning is allowed to vary from region to region, and from year to year. Second, the probability of winning the region need not decrease with seed number. For example, in 1994, Carlin gave eventual winner, number 2-seed Arizona, a higher probability of winning than number 1-seed Missouri. On the other hand, in the Southeast region, Carlin gave eventual winner number 2-seed Duke a much lower probability of winning than both number 3-seed Kentucky and number 1-seed Purdue.

In summary, Carlin's method is interesting, but he does not present enough data to argue convincingly that it is superior to SMH's Model 3.

See Chapter 10 for general analysis of tournament structures and Chapter 12 for other systems of rating teams.

3.3   Player performance

A key element in analyzing the success or failure of basketball teams is the study of its individual players. This section looks at two aspects of player analysis: (1) the categorization of players by type and its usefulness in structuring a team; and (2) the rating of individual players using computations based on standard statistical totals.

3.3.1   Player types

Ghosh and Steckel (1993) (GS) analyze data from the 1973–74 and 1987–88 NBA seasons. Following their lead, I will emphasize the more recent data. After eliminating seldom-used players, GS perform a cluster analysis based on 11 statistics (the statistics can be inferred from the discussion below). GS use a K-means algorithm for clustering and select a six-cluster solution. The names, characteristics, and selected players of each cluster are provided below.

(1) *Scorers.* Scorers take the most shots, score the most points, and are the best free throw shooters. Scorers include Dominique Wilkins, Larry Bird, Michael Jordan, Alex English and Clyde Drexler.

(2) *Dishers.* Dishers lead in assists and steals, and are almost as good as scorers at shooting free throws. They are the worst at shot blocking and rebounding, and commit the fewest number of fouls. Dishers include Magic Johnson, Isaiah Thomas, John Stockton, Mark Price and Mugsy Bogues.

(3) *Bangers.* Bangers lead in offensive and defensive rebounds, and have the second best field goal percentage. They are the worst free throw shooters. Bangers include Dennis Rodman, Joe Kleine, Kevin Willis, Robert Parish and Charles Oakley.

(4) *The inner court.* The inner court has the highest field goal percentage and attempts the most free throws. They are second in scoring, blocks and offensive rebounds, and third in defensive rebounds. Inner court players include Kevin McHale, Hakeem Olajuwon, Kareem Abdul-Jabbar, Charles Barkley, Patrick Ewing and Karl Malone.

(5) *Walls.* The walls lead in blocked shots and fouls committed. They are second in defensive rebounds and third in offensive rebounds. They attempt the fewest field goals and free throws, and are lowest in steals, assists, and points. Walls include Tree Rollins, John Salley, Alton Lister and Mark Eaton.

(6) *Fillers.* The fillers are a heterogeneous group. They are second highest in steals and fouls committed. Fillers include a former star (Jack Sikma), future stars (Scottie Pippen and Reggie Miller) and some marginal players (Brad Sellers and Dennis Hopson).

Table 3.1 presents a cross-classification of player cluster by position, and reveals the following patterns.

• Scorers tend to be small forwards or shooting guards; dishers are guards; the bangers, inner court, and walls are dominated by centers and power forwards; and fillers tend to be forwards or shooting guards.

• Small forwards tend to be scorers or fillers; centers tend to be bangers or inner court; shooting guards are nearly uniformly distributed among

Table 3.1 Cross-classification of player cluster by position, 1987–88 NBA season

Position	Cluster						
	Scorers	*Dishers*	*Bangers*	*Inner court*	*Walls*	*Fillers*	*Total*
Small forward	15	0	3	3	0	12	33
Power forward	1	0	12	8	5	6	32
Center	0	0	8	9	4	0	21
Shooting guard	14	10	0	0	0	9	33
Point guard	0	28	0	0	0	1	29
Total	30	38	23	20	9	28	148

scorers, dishers, and fillers; point guards, with one exception, are dishers; and power forward is the most diverse position, including a number of bangers, inner court, walls and fillers.

For the 1973–74 season the six clusters include scorers, dishers, bangers, and the inner court. But the walls and fillers of the later season are replaced by bangers II, and dishers II. As the names suggest, these clusters were less effective versions of their non-Roman-numeraled brethren. GS summarize the changes over time with the following passage (Ghosh and Steckel, 1993, p. 50).

> The evolution of the walls seems to be part of a pattern of increased specialization in role in recent years. . . . The players in 1973–74 were more multidimensional. Today, they contribute in more specialized, yet varied, ways.

For the 1987–88 season, GS (p. 52), 'Grouped the teams based on the distribution of each team's players across the six role clusters.' This analysis creates four clusters of teams, with teams in the same cluster having the same 'types' of players. Generally, this clustering is ineffective, yet instructive. For example, cluster 4 consists of nine teams, including the Lakers with its NBA high of 62 victories and the Clippers with its NBA low of 17 victories. Interestingly, the Clippers and Lakers have almost identical types of players. Specifically, each has two scorers (Scott and Worthy, for the Lakers, versus Dailey and Woodson), two dishers (Cooper and Johnson versus Drew and Valentine), and two bangers (Green and Thompson versus Cage and Norman). The only difference is in the remaining player; the Lakers had inner court Abdul-Jabbar versus wall Benjamin for the Clippers.

Similarly, cluster 3 includes the very successful Celtics (57 wins) and the relatively weak Spurs (31 wins) that share similar profiles. Each has one scorer and two dishers. The Celtics' remaining two players are a banger and an inner court, while the Spurs remaining three players are all inner court.

Neither clusters 1 nor 2 include any of the top teams (the maximum number of victories among their teams is 44 in an 82-game season). Cluster 2 teams are distinguished by having too many scorers, while cluster 1 teams have too many fillers. Furthermore, teams in clusters 1 or 2 have few effective big men (bangers or inner court).

3.3.2   Rating players

The team with more points wins the game. But how can one measure an individual player's contribution to his team's success? While I found no research papers that address this difficult question, it has been tackled by several popular writers on NBA basketball including Bellotti (1992), Heeren (1988) and Trupin and Couzens (1989) (henceforth, TC), whose work will be reviewed here. These authors have compiled a large amount of data from NBA games, and I recommend these books as sources of data and ideas for future research. Heeren is a sportswriter; Bellotti's book identifies him as a statistical consultant to NBA teams and an award-winning sportswriter; I do not know the backgrounds of Trupin and Couzens.

An obvious limitation in rating players is that one must work with statistics that are readily available. Certain player statistics are naturally viewed as being positively related to a team's success: points, offensive rebounds, defensive rebounds, assists, steals and blocked shots. Others are naturally viewed as being negatively related to a team's success: field goals missed, free throws missed and turnovers. The number of fouls committed is a trickier statistic; some fouls are good (e.g. fouling a player who has an easy shot), and others are bad (e.g. an offensive foul), but only the number of fouls committed is tabulated. In addition, these numerous statistics are correlated, both at the player and team level of the sampling unit. For example, at either the player or team level, steals and rebounds might be negatively correlated, while rebounds and blocked shots might be positively correlated. Users of multiple regression know that a single predictor can, for example, be positively related to the response, but make a negative contribution after adjusting for the effects of other predictors.

In addition, important events occur during a basketball game that do not appear in the statistical summary. These events can be positive (e.g. setting a pick), or negative (e.g. guarding one's man poorly).

Each book presents a score that is a starting point for rating each player. Bellotti and Heeren decide to keep the formula for their score simple. Add together all the good statistics, and subtract all the bad statistics. The only difference between Bellotti and Heeren is that the former subtracts one-half the number of fouls per game and the latter ignores fouls. TC use an occasionally obscure argument to assign different weights to the various good and bad statistics. All three books include rationales for the formulas. But with no clear way to measure response (a team's success), it seems impossible, at this time, to account for the correlation among the statistics. Table 3.2 presents the coefficients assigned to each game-statistic that are needed to compute a player's score.

Heeren takes a player's score and divides by minutes played to obtain what he calls his TENDEX rating. (A more advanced version adjusts for the 'pace' of a game and will not be discussed here.) There are many sensible features to Heeren's analysis. For example, he realizes that his scoring system favors certain positions. Using seven years of data he concludes that while the mean TENDEX per player is about 0.500, the approximate means by position are: centers, 0.550; power forwards, 0.525; small forwards, 0.500; point guards, 0.475; and shooting guards, 0.450. Heeren also suggests analyzing games by comparing the two teams position-by-position (starting centers, power forwards, and so on) using the TENDEX rating. He realizes the limitations of his method, but it is a good start on the difficult problem of understanding how much each player influences the outcome of a particular game.

Bellotti calls the player's score his 'points created average (per game)'. Like Heeren, Bellotti also divides the score by minutes played to obtain the player's

Table 3.2 Coefficients of various statistics for three formulas for computing a player's score

Author	Pts	O Reb	D Reb	Ast	Stl	Blk	FG made	FG miss	FT made	FT miss	TO	Fouls
Bellotti	1	1	1	1	1	1	0	−1	0	−1	−1	−0.5
Heeren	1	1	1	1	1	1	0	−1	0	−1	−1	0
TC	0	0.85	0.5	1	1	1.4	1.4	−0.6	1	0	−0.8	0

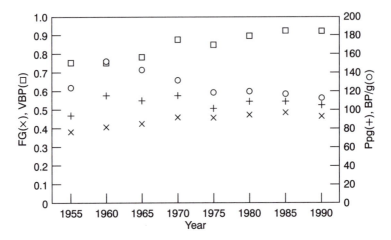

Figure 3.2   NBA offensive efficiency by year. FG is the proportion of field goals made; Ppg is points per team per game; BP/g is ball possessions per team per game; and VBP is value of ball possession and equals Ppg divided by BP/g.

'productivity' and, again like Heeren, seems to feel that it is more meaningful to compare players after adjusting for minutes played.

One purpose of a rating system is to decide which of your favorite star players actually is better. Star players play many minutes per game, and I suspect that within a given season, dividing or not dividing the player's score by minutes played will have only a minor impact on the rankings of star players. (But see the discussion below for comparing players from different eras or looking at career ratings.) Another purpose of rating players is to detect a player who is an emerging star. (This is obviously of interest to NBA executives and coaches.) For this purpose, dividing by minutes played could well make it easier to find players who have been particularly productive in the limited opportunities afforded them.

Bellotti (1992, p. 5) reports some interesting evolutionary trends in the NBA that bear on the issue of rating players. Figure 3.2 presents his findings. While interpreting these data, keep in mind that the 24-second shot clock rule took effect in 1954. Looking at five-year intervals, league-wide field goal percentage was at 38.5% in 1955, grew steadily until 1985 (except for a small decline from 1970 to 1975) when it peaked at 49.1%, and then dropped somewhat to 47.4% in 1990. The number of ball possessions per team per game grew tremendously from 1955 to 1960, then dropped steadily until it reached, in 1975, a level below the 1955 value. The value remained relatively constant over the next decade, and then declined a further five possessions per team per game in 1990.

The mean points per game have shown a more random pattern. Excluding the low value in 1955, the highest value (in 1970) is only 13.7% larger than the lowest value (in 1975). Finally, the value of a ball possession (points per game divided by ball possessions per game) has grown monotonically, except for a small dip in 1975.

Here is a possible explanation for these patterns; the reader is invited to suggest others. The 1955 data are difficult to interpret because of the recent introduction of the 24-second clock. By 1960, teams were overreacting to the shot clock by having very short possessions, which led, of course, to many ball possessions per game. These short possessions often led to poor shot selection and a subsequent low value of a possession. Over time, the teams became accustomed to the shot clock and worked for better shots, which require more time.

As Bellotti correctly points out, any meaningful comparison of players from different eras must adjust for the value of a ball possession. For example, a 1960 rebound is less valuable than its 1990 kin, and a 1990 turnover is more serious than one in 1960. Thus, in order to make comparisons across seasons Bellotti modifies his 'simplified version' of the points created formula (given above). In this 'complex version', a point scored is still a point scored, but now events that 'create possessions' (e.g. rebounds, steals and blocked shots) are multiplied by the average point value of a ball possession (VBP). An assist turns an ordinary ball possession into two points; hence, the number of assists is multiplied by (2−VBP). Missed shots, turnovers, and (one-half of the) fouls lose possessions; hence, they are multiplied by VBP.

Armed with his complex version, Bellotti is able to find the top NBA players of all time, and ranks them by points created per minute (PC/m). Bellotti's ranking is reasonable, but has some flaws. Most notably, Wilt Chamberlain is ranked best ever with 0.880 PC/m, while his old nemesis Bill Russell is ranked only 17th with 0.681 PC/m. Russell and Chamberlain (both centers) squared off against each other in many championship contests that Russell's Boston Celtics team almost always won. For this reason, many sportswriters and knowledgable fans consider Russell to be Chamberlain's superior as a team player. Bellotti's rating system does not capture this player characteristic. It is unfair to blame this on Bellotti, because the available statistics simply do not adequately measure team play. The remaining members of Bellotti's top ten (circa 1992) are:

> 2. Michael Jordan; 3. Magic Johnson; 4. Kareem Abdul-Jabbar; 5. Bob Pettit; 6. Oscar Robertson; 7. Larry Bird; 8. Charles Barkley; 9. Elgin Baylor; and 10. Hakeem Olajuwan.

As Heeren points out, Bellotti's scoring system is tough on shooting guards, and not many are ranked among his top players. After Jordan, Jerry West is ranked 11th, but the next shooting guard I recognize is Clyde Drexler at 30th or George Gervin at 36th. Kareem is the main beneficiary of using PC/m instead of PC/g, most likely because of his long playing career with reduced playing time near its end. The ranking based on PC/g is:

> 1. Wilt Chamberlain; 2. Oscar Robertson; 3. Michael Jordan; 4. Bob Pettit; 5. Elgin Baylor; 6. Larry Bird; 7. Magic Johnson; 8. (Tie) Kareem Abdul-Jabbar and Bill Russell; and 10. Jerry West.

3.4 The hot hand

Ideally, this section would be entitled *Modeling shooting*. I would report that this topic has generated a great deal of interest among researchers, and that I will be reporting on the results of seven research papers. The research shows that on most occasions studied, the simple model of Bernoulli trials is adequate to describe the outcomes of a player's shots, but on many occasions the Bernoulli trials model is inadequate.

I spend an embarrassingly large proportion of my life watching sports on television, and it is indeed a rare event for me to sit through a telecast without hearing something like

> Starks has made his first two shots of the game. He has the hot hand tonight!

My response to such 'inference' by an announcer has progressed from amusement to annoyance to disgust. I have yet to meet a statistician familiar with such broadcast pronouncements who is not, at the very least, dismayed by their occurrences.

While I do not want to cross the line and become the type of historical novelist who can read the minds of his or her subjects, I cautiously infer from the titles and tones of the seven papers that these researchers share my contempt for the reasoning exhibited by the announcers. For example, five of the seven papers include the phrase *hot hand* in their title and the remaining pair of papers mention it in their text.

Note that Chapter 2 discusses a related topic, streak hitting in baseball.

3.4.1   Initial research

Research on the hot hand began with Gilovich *et al.* (1985) (henceforth, GVT). This research reappeared with some changes in Tversky and Gilovich (1989a) (henceforth, TG1). Tversky and Gilovich are psychologists; I do not know the remaining author's training, but I suspect that he is also a psychologist. Thus, it is not surprising that this research includes an investigation of what people *believe to be true* about basketball shooting. Space limitations prevent me from discussing their findings in detail, but an examination of one item is essential.

A convenience sample of 100 'avid basketball fans' was told to consider a hypothetical player who shoots 50% from the field (Tversky and Gilovich, 1989a, p. 20). Each fan was asked two questions about this player.

> 1. (2.) What is your estimate of his field goal percentage for those shots that he takes after having just made (missed) a shot?

The mean of the responses to questions 1 and 2 were 61% and 42%, respectively.

Of course, an important issue is whether these 100 persons are representative of some larger population of fans, but I will ignore this issue because I do not have any basis for commenting on it. My interpretation of this survey is simple; *these fans are really misguided*! A 61% shooter is a dramatically different player than a 42% shooter, and it is ridiculous to believe that such variability in performance is the norm! I am amazed, but not surprised, by these fans' beliefs. TG1 make a positive contribution to the literature by reporting this finding.

GVT and TG1, however, make an egregious error by anointing this 19 percentage point difference the 'gold standard' to which all experimental and observational data are to be compared. It is proper and interesting to note that real data rarely achieves the level of dependence these fans expect. However, it is grossly improper, as these writers do repeatedly, to label data as unimportant simply because they do not match the beliefs of ill-informed fans. Simply put, I believe that departures from Bernoulli trials can be important and interesting without achieving the size or omnipresence suggested by this sample of fans.

To put it bluntly, are these psychologists suggesting a new approach to science? By analogy, if a sample of people believe that smoking three packs of cigarettes a day shortens life by 15 years, but it is shown statistically to shorten life by 10 years, are we to conclude that cigarettes are harmless because they don't match the expectations of the sample?

GVT and TG1 examine data from three sources: a controlled study of college basketball players, game-data of NBA players, and free throw game-data of NBA players. I will begin by examining their data from a controlled study.

Each of 26 members of Cornell University's men's and women's varsity teams attempted 100 shots. The researchers compute the lag one serial correlation for the 26 players. They report (Tversky and Gilovich, 1989a, p. 20) that only one player, '... produced a significant positive correlation (and we might expect one significant result out of 26 just by chance), ...'. They neglect to report, however, that this player's one-sided P-value is 0.00023 and that the probability – on the i.i.d. assumption – of one of more P-values so small is only 0.006! Thus, this one significant result cannot reasonably be attributed to chance. (This player made 72% of his shots after a hit, but only 35% after a miss – a huge difference!)

Next, the researchers asked each Cornell player to predict the outcome of the next shot. (The experimental method was quite clever; see pp. 308–309 of Gilovich *et al.*, 1985, for details.) GVT report that four of 26 players obtain a statistically significant positive relationship between prediction and outcome. Perhaps realizing that four out of 26 are too many to 'expect by chance', they state that the four correlations 'were quite low (0.20 to 0.22)' (Gilovich *et al.*, 1985, p. 309). They fail to mention, however, that success rates of 0.60 and 0.40 yield a correlation of 0.20. Would any reasonable person argue that an estimated change of 20 percentage points in a player's shooting ability is a quite small effect? Interestingly, in the paper by Tversky and Gilovich (1989a), the researchers are misleading in their reporting of the prediction experiment. They do not mention the four significant results, but state instead (Tversky and Gilovich, 1989a, p. 21) that, 'The players were generally unsuccessful in predicting their performance'.

Before turning to the other two data sets, it is important to note an inherent weakness in the methods used in GVT and TG1. The test statistics used in these papers can be effective at detecting non-independence in the shots, but are largely ineffective at detecting non-stationarity in a sequence of shots. For example, consider the following alternative to Bernoulli trials as a model for the Cornell data. Suppose that a player's first ten shots must be successes (that is, $p = 1$), and the remaining 90 shots are Bernoulli trials with $p = 0.5$. In other words, the player has the (extremely) hot hand for the first ten shots, and for the remainder of the sequence the i.i.d. model is correct.

A simulation study of this alternative reveals that GVT's and TG1's tests involving conditioning on the previous one, two or three outcomes, as well as their so-called test of stationarity, have estimated power in the range of 0.13 to 0.20 for a one-sided test with $\alpha = 0.05$. By contrast, if the test statistic is taken to be the longest run of successes, the estimated power is approximately 0.50. More work is needed in this area, especially with an application to real data, but it seems clear that the tests reported in GVT and TG1 are too narrow and that it might be fruitful to examine real data with test statistics like the longest run of successes, or the longest sequence of $n - 1$ successes in n shots.

Suppose that for a particular player on a particular occasion, the hot hand is present in the form of a lag one autocorrelation. It is convenient and accurate to label such non-independence an *omnipresent effect* because its influence is felt by every shot. Every time a player shoots, the chance of making the shot depends on the outcome of the previous shot. By contrast, suppose instead that the hot hand is present in the form of non-stationarity; call this an *occasional effect* for obvious reasons.

This raises the question, 'Why do GVT and TG1 only look for an omnipresent effect?' There is circumstantial evidence that they do not understand the difference between the two departures from i.i.d. In particular and curiously, TG1 do not seem to understand the quote that begins their paper. Professional basketball player Purvis Short states,

> You're in a world all your own. It's hard to describe. But the basket seems to be so wide. No matter what you do, you know the ball is going to go in.

TG1 state that Short is describing the hot hand. Can anyone seriously argue that Short is describing an omnipresent effect? An effect that is present with *every shot*; that kicks in after every success and disappears after every failure? Of course not! Short is describing an occasional effect.

Hooke (1989) provides very entertaining and insightful thoughts on the inherent difficulty of using statistical methods to study phenomena as complicated as a game of basketball or baseball. He also discusses what I have described above as omnipresent versus occasional effects. He notes (Hooke, 1989, p. 36):

> In statistical language, we don't have a well-formulated alternative hypothesis to test against the null hypothesis. Thus, we invent various measures (such as the serial correlations of Tversky and Gilovich).... My intuition tells me that the alternative hypothesis is not that there is a 'hot hand' effect that is the same for everyone, but that the real situation is much more complex.... A measure that is appropriate in detecting the effect for one of these types [of hot hands] may not be very powerful for another.

GVT and TG1 also present and analyze data from the 1980–81 season for the home games of nine players on the Philadelphia 76ers NBA team. The researchers use the same test statistics as in their controlled study. Thus, the earlier comments about these test statistics demonstrated inadequacy with simulated data, and possible inadequacy with real data, still apply.

For eight of the nine players, the success rate of a shot was higher after a miss than after a hit; for seven of nine, the success rate was higher after two misses than after two hits; and for all nine players the success rate was higher after three misses than after three hits. GVT note that these results contradict fans' expectations, but do not explore the meaning of all these negative associations. For example, if you perform Mantel–Haenszel type analyses (see Wardrop, 1995, for a similar analysis in a different setting) to combine information across players, you get z-scores of -2.09 for the lag one autocorrelations, -1.82 for the data after two hits versus the data after two misses, and -3.18 for the data after three hits versus three misses. Clearly, these negative associations should not casually be dismissed as due to chance. Their meaning, however, requires further study. They might be important or they might be an artifact of these data; for example, in the first paper mentioned in the next subsection, no such negative associations are found.

3.4.2 Alternative approaches

Two other papers examine game-data on NBA players. Larkey *et al.* (1989) (henceforth, LSK) code data from 39 NBA telecasts in the Pittsburgh market during the 1987–88 season. LSK examine data for 18 players for whom, by their criteria, there is an ample amount of data for analysis. Notably, their player-set includes Detroit's Vinnie Johnson, who has a reputation for being an extremely streaky shooter, and many of the top stars in the NBA.

LSK do not attempt statistical inference; their work is purely descriptive. They want to see whether they can find any justification for the widespread belief that Vinnie Johnson is a particularly streaky shooter. They conjecture that whatever Johnson is doing, it must be noticeable and memorable, and perhaps unlikely.

Essentially, LSK believe that the TG1 analysis of the Philadelphia data is correct, but does not go far enough. To that end, LSK begin by computing 'Conditional probabilities of a hit and autocorrelation'. LSK's findings are qualitatively similar to those of GVT and TG1, except that LSK do not find so many negative associations.

LSK next examine what they call the 'Ratio of acontextual occurrences to opportunities'. First, they focus on *perfect runs*, such as making four shots in a sequence of four attempts. For example, suppose that a player attempts ten shots during a game and obtains the following sequence of success (S) and failure (F) outcomes: F, F, S, S, S, S, S, F, F and S. This player had seven opportunities to make four shots in a row, namely shots one through four, two through five, and so on. This player achieved four successes in a row twice – shots three through six, and four through seven. For this game, the player's ratio of occurrences to opportunities for four consecutive successes is $2/7$. LSK present these ratios for their entire set of data for each player for runs of three through eight successes. By this measure, Dennis Rodman achieves the highest ratio for each run length, usually much larger than the runner-up's ratio. By this measure, Vinnie Johnson is not atypical. Second, LSK look at imperfect runs – making n out of $(n + 1)$ with n ranging from 2 to 7. The co-winners for this measure are Michael Jordan and Kevin McHale, with Vinnie Johnson slightly behind the co-leaders. Thus, the ratio of acontextual occurrences to opportunities is *not* the measure (if, indeed, one exists) that separates Johnson from other players.

Next, LSK switch from the ratio of 'occurrences to opportunities' to the ratio of 'occurrences to expectations'. The idea is that Rodman ranks highest on perfect runs not because he is a streak shooter, but because he has the highest shooting percentage in the group of players studied. Nobody would seriously suggest Rodman is a great shooter – quite the contrary, he has a high shooting percentage because he rarely attempts difficult shots. 'Occurrences to expectations' essentially adjusts the earlier ratios for different shooting percentages. Vinnie Johnson is a 46% shooter, so his number of occurrences, for example, of an imperfect streak of 5 out of 6, is divided by the number to be expected, under the assumption of i.i.d., for a 46% shooter.

With the adjustment, Rodman is best at 8 out of 8, but Robert Parish is the best for all other perfect sequence run lengths. For the imperfect sequence, Vinnie Johnson is best for runs of 4 out of 5 up to 7 out of 8, second best (to Michael Jordan) for runs of 3 out of 4, and ordinary for runs of 2 out of 3 (Jordan is best). Thus, adjusting for expectations instead of opportunities makes Johnson more noticeable.

LSK next introduce the notion of context. Although their argument contains some flaws (discussed below) it is a very clever idea and is worthy of further study. Suppose that player A makes 5 shots out of 5 during a game, and that the shots are uniformly distributed over the length of the game. By contrast, player V makes five consecutive shots without any intervening shots in the game. LSK argue that V's performance, compared to A's, is more noticeable and memorable to a fan (these claims seem obvious), and also more unlikely than A's performance. They argue that V's performance is less likely than A's because it never occurred in their data set. In fact, there was no such 4-shot streak in their data set, and only two 3-shot streaks without intervening shots. By contrast, in approximately 100 instances, a player made five consecutive shots with intervening shots by other players. (LSK appear to overlook a rather obvious flaw in their example. Every time V makes a basket, the other team would get the ball! Thus, for V to achieve the sequence they propose, both of the following must be true:

- The opponent's offense is so inept, or V's team's defense is so outstanding that V's team steals the ball four times in a row.
- V is such a blatant ball-hog or his team-mates are so selfless that he is allowed to shoot after every steal.

And, of course, V must make the shots.)

LSK next compute the ratio of occurrences to expectations in the context of 20 field goal attempts. For example, the numerator of one such ratio is the number of times Vinnie Johnson made 5 shots out of 5 during a period of 20 overall field goal attempts in a game, instead of during an entire game. The denominator is trickier. LSK compute the number of times Johnson would be *expected* to achieve the given sequence. The expectation is computed by adjusting for a player's field goal percentage and by estimating the probability (γ) that any given shot will be taken by him. With this approach, Vinnie Johnson emerges as dramatically different from other players. For example, he is the only player to make 7 out of 8 shots in context and his ratio is 10 050. Only eight players made 5 out of 5 in context, and Johnson's ratio of 81.69 is almost three times the next largest, Joe Dumars' 28.37. LSK summarize by stating (Larkey *et al.*, 1989, p. 30):

> Vinnie Johnson's reputation as a streak shooter is apparently well deserved; he is different from other players in the data in terms of noticeable, memorable field goal shooting accomplishments.

The issue of *Chance* that contains the LSK paper follows it with a 'rebuttal' by Tversky and Gilovich (1989b) (henceforth, TG2). They have three main criticisms of LSK.

First, TG2 point out that the estimate of γ is flawed. LSK estimate γ for Johnson by multiplying 0.50 (the proportion of a game's shots taken by the Pistons) by 0.13 (the proportion of Pistons' shots taken by Johnson) to obtain 0.065. But Johnson cannot shoot when he is on the bench! Given that Johnson's average playing time was about one-half of a game, the estimate of γ is too small by a factor of two. The effect of having γ too small is to deflate LSK's denominator and, thus, inflate their ratio. Of course, the estimate for γ is too small for all players, but the distortion should be strongest for Johnson because he played the fewest minutes, on average, of the 18 players studied. (TG2 do not mention the other obvious flaw in the model; namely, after a made shot, the other team gets the ball. Thus, after a made shot, γ is too large because the most likely event by far is that the opponents will shoot next. If this error in the model could be corrected, it should help LSK's argument because variation in γ will lead to less variation in the number of shots a player takes out of 20, and this should decrease the denominator in the ratio.)

Second, TG2 claim that the entire 'context' argument of LSK gets its validity from a *single* sequence of 7 out of 7 by Johnson. Third and finally, TG2 claim that the 7 out of 7 sequence by Johnson was, in fact, a 6 out of 7 sequence that was miscoded. TG2 end with a claim, essentially, that the flaws noted above completely invalidate the positive results of LSK. The second and third criticisms are serious; TG2 should have substantiated them. As they stand, I suspect that the criticisms are overblown

I wonder what results would be obtained if LSK had computed the ratio of contextual occurrences to opportunities, because computing expectations appears to be hopelessly complicated. But I have seen no further work on this problem.

The final paper to look at NBA game-data is Forthofer (1991). Forthofer uses newspaper box scores to examine all NBA players in all NBA games during the 1989–90 season. He pares his database to 123 men who averaged more than nine shots per game and played in 50 or more games.

Forthofer has three categories of streak shooters: those with 'hot and cold streaks', those with 'a hot hand', and those with 'cold hands'. Of the 123 players in his database, 17 players fall into one or more of these categories. (Forthofer's narrative of who falls into which categories does not quite match his table; for example, from his Table 1, Rickie Pierce is cold, and hot and cold, but not hot, while in the narrative the opposite is posited.) I will use Karl Malone, one of Forthofer's streak shooters, to illustrate his method of classification.

For the season, Malone made 56.2% of his shots. In a game against Charlotte, Malone made 22 of 28 shots. Now, $22/28 = 0.786$, a good shooting game for Malone. Given that X has a binomial distribution with $n = 28$ and $p = 0.562$, it can be shown that $P(X \geqslant 22) = 0.0120$. This game is classified as 'hot' because the upper tail probability is 0.05 or smaller and it is labeled 'hot and cold' because either the upper or lower tail probability (in this case, of course, it is the upper) is 0.025 or smaller. In another game, Malone made 6 of 18 against Sacramento. The lower tail probability for these data is 0.0433. Thus, this game counts as cold, but not hot or cold (and obviously not hot).

Overall, Malone played 82 games and had eight cold games, four hot games, and three hot and cold games. Next, 5% of 82 equals 4.1; thus, Malone is classified as cold because $8 > 4.1$, but is neither hot because $4 \leqslant 4.1$, nor hot and cold because $3 \leqslant 4.1$.

I applaud Forthofer's goal of looking for the most extreme players, but his criterion for labeling players is somewhat flawed. In particular, the easiest way to be labeled a streak shooter by him is to have played in 59, 77, 78 or 79 games. For example, Pierce played in 59 games and had three hot games. Because 3 is more than 5% of 59 (but exactly 5% of 60 games) he is labeled hot; if Pierce had played one more game, chances are he would have lost his label. Similarly, four players achieved distinction by having four notable games out of 79, two by having four out of 78, and one by having four out of 77. If we eliminate these cases, we are left with only ten players, not 17, who exhibit some form of streakiness.

In any event, Forthofer's main idea – to identify the streakiest shooters – is a good one. It would be interesting to see his analysis repeated over several years to discover whether the set of players exhibiting streakiness is fairly constant.

3.4.3 Free throws

TG1 introduce their third set of data with a question they asked their sample of basketball fans (Tversky and Gilovich, 1989a, p. 20):

> When shooting free throws, does a player have a better chance of making his second shot after making his first shot than after missing his first shot?

Sixty-eight percent answered yes. GVT and TG1 analyze data on nine members of the 1980–81 and 1981–82 Boston Celtics teams and conclude that the correct answer is no, and, hence, that a majority of the fans are incorrect.

Wardrop (1995) suggests a different interpretation of these free throw data. He shows that if the data are aggregated over players, then the correct answer to the question becomes yes. GVT correctly mentioned in a footnote (Gilovich *et al.*, 1985, p. 304) that such aggregation is inappropriate for answering their question. Wardrop points out, however, that aggregation might be appropriate for the purpose of trying to understand why the fans believe what they do. In order to apply GVT's analysis of the Celtics to a fan's experiences, one must assume that the fan has a separate two by two contingency table for each of the hundreds, if not thousands, of players he or she has seen play.

This is a big assumption to make! Wardrop suggests that it might be more reasonable to assume that the fan has a table only for the aggregated data. Thus, instead of concluding that (Tversky and Gilovich, 1989a, p. 21), 'People ..."detect" patterns even where none exist,' it would be more productive to teach people the dangers of aggregation. After all, Simpson's Paradox would not be called a paradox if its conclusion were natural!

Wardrop also shows that the Celtics players, as a group, were statistically significantly better free throw shooters on their second attempt than on their first attempt. This rather obvious feature of the data was overlooked by GVT and TG1, perhaps revealing again their habit of focusing on dependence and ignoring non-stationarity.

3.5 Suggestions for future research

As a broad general comment, future research should include the women's game. A good starting point would be to apply the work on the NCAA men's tournaments to the women's. More work is needed on the interesting topic of player types. Player ratings are very important to fans and management, and it is time that sophisticated statistical researchers helped the sportswriters!

Regarding the hot hand, research should focus on modeling processes, not on what we label the results. Test statistics should not be limited to those that are well-known to have approximate chi-squared distributions. Instead, more real data need to be examined and the statistical procedures should be designed to fit the data rather than the other way around.

If you are looking for a topic for research, read the three books mentioned in Section 3.3 and remember the following words that I heard attributed to Herb Robbins: 'A mathematical statistician studies what uneducated people do and shows that it is optimal'.

Acknowledgments

Many thanks to Jay Bennett for his many helpful suggestions that greatly improved the quality of this chapter. This chapter is dedicated to my late thesis advisor, Professor Norman Starr of the Department of Statistics at the University of Michigan.

References

Bellotti, B. (1992) *The Points Created Basketball Book, 1991–92*. New Brunswick, New Jersey: Night Work Publishing.

Carlin, B. (1996) Improved NCAA basketball tournament modeling via point spread and team strength information. *The American Statistician*, **50**(1), 39–43.

Forthofer, R. (1991) Streak shooter – the sequel. *Chance*, **4**(2), 46–48.

Ghosh, A. and Steckel, J. (1993) Roles in the NBA: there's always room for a big man, but his role has changed. *Interfaces*, **23**(4), 43–55.

Gilovich, T., Vallone, R. and Tversky, A. (1985) The hot hand in basketball: on the misperception of random sequences. *Cognitive Psychology*, **17**, 295–314.

Heeren, D. (1988) *The Basketball Abstract*. Englewood Cliffs, New Jersey: Prentice Hall.

Hooke, R. (1989) Basketball, baseball, and the null hypothesis. *Chance*, **2**(4), 35–37.

Larkey, P., Smith, R. and Kadane, J. (1989) It's okay to believe in the 'hot hand.' *Chance*, **2**(4), 22–30.

Schwertman, N., McReady, T. and Howard, L. (1991) Probability models for the NCAA regional basketball tournaments. *The American Statistician*, **45**(1), 35–38.

Schwertman, N., Schenk, K. and Holbrook, B. (1996) More probability models for the NCAA regional basketball tournaments. *The American Statistician*, **50**(1), 34–38.

Trupin, J. and Couzens, G. S. (1989) *HoopStats: The Basketball Abstract*. New York: Bantam Books.

Tversky, A. and Gilovich, T. (1989a) The cold facts about the 'hot hand' in basketball. *Chance*, **2**(1), 16–21.

Tversky, A. and Gilovich, T. (1989b) The 'hot hand': statistical reality or cognitive illusion? *Chance*, **2**(4), 31–34.

Wardrop, R. (1995) Simpson's paradox and the hot hand in basketball. *The American Statistician*, **49**(1), 24–28.

Westfall, P. (1990) Graphical presentation of a basketball game. *The American Statistician*, **44**(4), 305–307.

4

Test Statistics

Stephen R. Clarke

Swinburne University of Technology, Australia

4.1 Introduction

With origins that can be traced back to the 13th century, the first set of rules for cricket were written in 1744. One hundred years later, on 24–25 September 1844, Canada played the USA at St George's Cricket Club Ground, Manhattan, New York, USA. The game had spread from England throughout the British Empire and beyond. Cricket is now administered by the International Cricket Council, and matches between countries of a suitable standard are called Test Matches and are usually scheduled as five-day games of six hours of play each (30 hours of play over five days). About 250 Tests were played between 1877 and 1935, but with the expansion in the number of Test playing countries a similar number were played in the 1980s. Currently, nine Test-match playing countries (England, Australia, West Indies, India, Pakistan, Sri Lanka, New Zealand, South Africa and Zimbabwe) play irregular series against each other consisting of between one and six Tests. While the major interest for statisticians is Test cricket, many official matches scheduled to be played over at least three days are deemed First Class and these constitute the majority of cricket records. The domestic competitions of countries with Test status, of which the English County Cricket Championship is the best known, are in this category.

Cricket is played between two teams of 11 players on large bounded oval-shaped grounds of various sizes. The main action takes place in the centre of the ground on a grass pitch 22 yards long and about six feet wide. A wicket consisting of three stumps supporting two bails forms a target for the bowlers at each end of the pitch. Unlike pitchers in baseball, bowlers are not allowed to throw but use a stiff arm action to deliver the ball on the run, and usually bounce the ball off the pitch before it reaches the batsman. Batsmen play in pairs, one on strike facing the bowler and one at the bowler's end, and score a run each time they run the length of the pitch, thus changing ends. A line on the pitch about three feet in front of each wicket is known as the crease, and a batsman is dismissed if he fails to ground his bat or part of his body over the crease before the fielding side hit the corresponding wicket with the ball. There is no foul area, and the batsmen do not have to run when they hit the ball. A long hit may give batsmen time for up to four runs, while hitting the ball to the boundary automatically scores four, and over the boundary on the full (without hitting the ground) scores six. The many means of dismissal include being bowled (the ball hitting the wicket and dislodging one or both bails), Leg Before Wicket or LBW (the ball hitting the batsman's legs when it would otherwise have hit the wicket), caught (a fielder catching the ball off the bat), and run out (failing to make an attempted run). Bowlers bowl balls in sets of six, called overs, with alternate overs bowled by different bowlers from opposite ends. A team's innings ends when ten wickets have fallen (leaving one batsmen not dismissed or not out) or the captain 'declares' the innings closed (leaving two batsmen not out). A good source of information on cricket is the CricInfo web site which includes a section devoted to explaining cricket to initiates at www-usa. cricket.org/link_to_database/ABOUT_CRICKET/EXPLANATION.

In the traditional form of the game, each team has two innings, and a match is played over a fixed maximum period of two to five days. In most domestic round robin competitions, first innings points would be given for the team leading after each team has batted once, but outright victory is the major goal. This is achieved by obtaining a higher total score than the opponent obtains in two completed innings. In Test cricket, since only outright victory counts as a win, the losing team generally must be dismissed twice to get a result. This means games lasting five days often end in a draw (i.e. are unfinished). The famous 'Timeless Test' (England versus South Africa, Durban, 1939) was scheduled to be played to the finish with no time limit, but was abandoned and declared a draw on the tenth day of play because the England team had to board its ship home. To overcome the high frequency of draws, and generally improve spectator interest, a one-day form of the game was introduced in the 1960s. Each team has only one innings lasting about three hours, in which they each face a maximum number of balls (usually 300). While less tactically subtle, failing severe interruption by the weather, a decision is always achieved.

The difficulty of scoring runs and of being dismissed depends very much on the quality of the pitch, which varies greatly from match to match and generally deteriorates during a match. Thus, while 300 may be a respectable score in a Test innings, the highest team score in a Test match was 952 for 8 wickets by Sri Lanka against India in 1997, while the lowest was 26 by New Zealand against England in 1955.

A typical scorecard for one innings, as usually published in the daily press or cricketing almanacks, contains a list for each batsman in batting order with his total score, the method of dismissal and the bowler responsible. The bowler is still credited with taking the wicket even if the batsman is caught. For each bowler, the number of overs bowled, the number of maidens (overs in which no score was made), the total runs scored off their bowling and the number of wickets taken are given. Also included is the team score when each batsman was dismissed. This allows the calculation of the partnership – the total score made while each pair of batsmen were batting. The batting order is usually determined by ability, with the first six being recognized batsmen and the last four players selected for their bowling. Milestones for batsmen are the multiples of 50; in particular, a century. For bowlers, five wickets in an innings or ten in a match are comparably difficult achievements.

Table 4.1 Test career record of Ian Botham

	Matches	I	NO	Runs	HS	Ave	100	50	Ct	St
Batting	102	161	6	5200	208	33.54	14	22	120	–

	Balls	Maidens	R	W	Ave	Best	5	10	SR	Econ
Bowling	21 815	788	10 878	383	28.40	8-34	27	4	56.9	2.99

While the traditional scoresheet used by officials has more information, such as a batsman's individual scoring shots, it is a non-trivial task to reconstruct a ball-by-ball account of the match from the score sheets. Since official score sheets are usually only available from the particular association running the match, and these may not be archived, almost all analysis is done from published scorecards. (Officials actually burned the official score book used for the historic 1960 first ever tied Test.) Some televised matches use computer systems that keep ball-by-ball data, including where on the ground the ball was hit, but such data are not freely available.

An example of a Test career record, that of English all-rounder Ian Botham (1955–), is shown in Table 4.1. A lack of symmetry in the scorecard information is reflected in the career records. For a bowler, we know how many balls he bowled (Balls), runs allowed (R) and wickets taken (W). For a batsman, we are told runs scored (Runs) and whether he lost his wicket or not (I – NO and NO respectively), but are not told how many balls he faced. So, for bowlers, we can calculate a 'strike rate' equal to the number of balls bowled per wicket taken (SR = Balls/W), but cannot calculate the run rate (Runs/Balls) for batsmen. Only in recent years has the number of balls faced by the batsman been collected, but it is rarely published in the newspapers. Even if balls faced by each batsman is published, it is impossible to reconstruct balls faced by a particular partnership.

Like all sports, changes to rules require some care when statistics from different eras are compared. While an over is now standardized as six balls, Australia used eight-ball overs for 60 years prior to 1978. The rules for LBW have undergone several modifications, and the definitions of a no ball and a wide may alter from series to series, or from Test to one-day cricket. The treatment of sundries (no balls, wides, leg byes and byes[1]) has also changed over the years. For example, no balls are now recorded against a bowler's figures, but this was not always the case. In some competitions, a no ball has been credited as two runs, whereas traditionally it has been worth one run. It has recently been suggested to discontinue counting a wide as a ball faced by batsmen. This seems reasonable, since a wide, by definition, is a ball impossible to hit.

While earlier publications (Haygarth, 1862–85, 1925) detail the history of cricket for each season from 1746, *Wisden Cricketers' Almanack* (Engel, 1997) has been the traditional Bible of cricket statistics since its first publication in 1864. Each annual edition contains statistics from the previous English domestic season, plus full scorecards of every Test match played around the world

---

[1] See the glossary for definitions of these terms.

during the year, and coverage of cricket in 40 different countries. Over 40 000 First Class matches have been played, and the maintenance and publication of their records is a major aim of the Association of Cricket Statisticians and Historians (ACS). The association publishes a quarterly journal, *The Cricket Statistician*, containing non-technical articles on cricket statistics. All ACS scorecards are gradually being transferred to CricInfo, a fan-based organization that aims to provide cricket scores and records via the World Wide Web (currently at www.cricket.org). Further details on both these organizations and several similar ones can be obtained from the web. While such organizations provide cricket statistics, the capabilities of their memberships to perform analyses of a technical nature are limited. Thus, while a huge number of statistics and records are collected and published, there is little attempt at any serious statistical analysis.

Cricket has the distinction of being the first sport used for the illustration of statistics. In *Primer of Statistics*, Elderton and Elderton (1909) used individual scores of batsmen to illustrate frequency distributions and elementary statistics. Elderton (1927) used scores of batsmen to illustrate the exponential distribution, and Wood (1941) investigated the idea of consistency. These efforts resulted in Wood (1945) and Elderton (1945) reading separate papers at the same meeting of the Royal Statistical Society. These papers have some claim as the first full quantitative papers applying statistics to sport. The papers are accompanied by 17 pages of discussion, demonstrating the great interest generally created by papers in sport. Yet in spite of this interest, the topics raised were ignored in the professional statistical literature for over 30 years. In contrast to baseball, few papers in the professional literature analyse cricket, and two rarely examine the same topic. This allows us the luxury of looking in some detail at virtually all the published material using statistics in cricket.

4.2   Distribution of scores

The distribution of batting scores was first discussed by Elderton and Elderton (1909). Graphs of a batsman's scores were used to illustrate skewed distributions. While no formula was given, a theoretical graph (see Fig. 4.1) which could only be that of the negative exponential distribution is shown. Elderton (1927) formalized this by using the same scores to illustrate the Pearson Type

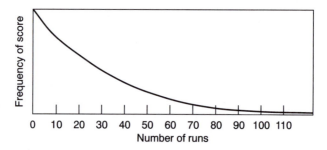

Figure 4.1   Curve for cricket scores, showing the number of times each score is made, the score made most frequently being 0 (redrawn from Elderton and Elderton, 1909, p. 45).

X or negative exponential distribution. If a batsman scores only singles and his probability of dismissal is constant, his scores should follow a geometric distribution, the discrete equivalent to the negative exponential. Elderton (1945) obtained a reasonable fit of the geometric distribution to the individual scores over three years of four early cricketers.

Wood (1945) took this further and compared the scores and several statistics of several groups and many individual batsmen with that expected using the geometric distribution. Although no significance tests were used, the fit was generally fair. However, the number of 'ducks' (a score of zero) and scores less than five were fewer than predicted, while the number of centuries was greater than expected, suggesting that the probability of dismissal is not constant. Cricket folklore says batsmen are more prone to dismissal early in their innings, perhaps get nervous or careful when their score reaches 90, and tire or hit out later in their innings. In fact there are several reasons why scores should not be exactly geometric. The score does not increment by one, but advances by jumps of usually 1, 2, 3, 4 or 6. As well as changing throughout an innings, one would certainly expect the probability of dismissal to change from innings to innings, depending on the quality of the pitch, or the opposition. Wood raised the possibility that the discrepancies are the result of combining two or more geometric distributions, and investigated this by looking at batsmen's scores over several years both individually and in combination. He found that while combining several geometric progressions understated the expected number of zeros and centuries, the effect is small and does not explain the discrepancy with the observed scores.

Other authors argue that because the chance of dismissal varies throughout an innings, and from one innings to another, the negative binomial distribution may describe the number of runs scored in completed innings. Using data from Wood for three batsmen, and data for three contemporary batsmen, Reep *et al.* (1971) found only one batsman's scores in each set were approximately negative binomial. Pollard (1977) applied a chi-square goodness-of-fit test to Elderton's data and found the geometric distribution performed slightly better than the negative binomial. However, both fitted distributions had higher variances than the empirical data. Pollard *et al.* (1977) claimed the excess of high scores will not be present if partnerships are investigated, as several partners may be dismissed while a player is making a high score. They obtained a good fit of the negative binomial to all partnerships for a team in the English County championships. On the other hand, Croucher (1979) found a negative binomial failed to fit the total number of runs for each completed Australian partnership in 82 England–Australia Tests.

In the papers of Elderton and Wood, we first come across a continuing problem in cricket – how to handle the not out scores (i.e. scores by batsmen who are not dismissed). While Elderton and Elderton (1909) added the next score to a previous not out score, early papers generally treated not outs as completed innings. Kimber and Hansford (1993) looked at the empirical discrete hazard and smoothed empirical hazards of the Test, first class and one day international scores of several batsmen, and compared them with the constant hazard expected for the geometric distribution (see Fig. 4.2 for Don Bradman's hazard). For most batsmen, the region from 0 to 5 runs is higher than expected. However, they found no compelling evidence that the hazard is not otherwise

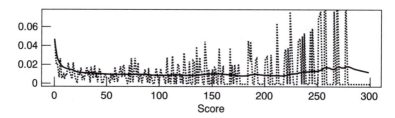

Figure 4.2 Empirical hazards (···) and smoothed empirical hazards (—) for the First Class scores of Sir Donald Bradman (1908–) (redrawn from Kimber and Hansford, 1993, p. 447, Fig. 2d).

constant and concluded that the tail of the score distribution for a batsman is roughly geometric.

Many arguments against the geometric distribution for scores do not apply to the number of balls a batsman faces. This count certainly increments by one. A batsman can alter the degree of risk he takes, in order to play each ball with the same chance of dismissal. Early in the innings he is just content to survive, whereas later when he is settled he will take risks in order to score. Similarly, on a bad pitch or against good opposition he may play more carefully, and adjust his scoring rate to keep roughly the same risk of dismissal. For these reasons, Clarke (1991) suggested the distribution of the number of balls faced by a batsman in an innings may be geometric. However, using ball by ball data for Australian batsmen from all matches in the 1989–90 one-day series involving Australia, Pakistan and Sri Lanka, he failed to obtain a better fit of the geometric distribution to balls faced than to scores. Strangely, the results for scores tended to be the reverse of those found in Test cricket; the number of very low values and very high values was less than expected. The limited nature of one-day cricket obviously produces a distribution of scores different from the traditional game.

Because teams generally play under similar conditions throughout a match, a strong positive correlation between performances in a match may be expected. However, various investigations have failed to produce much evidence of this correlation. Elderton (1945), by visual inspection, found no correlation between the scores of two pairs of county championship opening batsmen. Croucher (1982b) calculated the correlation for each team between 25 pairs of completed innings in Tests between Australia and England. He found zero correlation for Australia, and a non-significant *negative* correlation for England. Testing a belief held by fans that long partnerships are generally followed by short ones, Croucher (1979) found some evidence to the contrary, that partnerships following century partnerships tend to be *longer* than usual.

The distribution of scores is important as it affects the appropriateness of other statistics. Kimber and Hansford (1993) questioned the use of the batting average. This is defined as the total number of runs scored divided by the number of times the batsman has been dismissed. Thus, runs scored in innings in which the batter has not been dismissed are included in the numerator, but the denominator does not count those innings. Because of the handling of not outs, many cricket fans have a certain unease with the current statistic, as it appears to give inflated averages – a player may have an average larger than his greatest score. Since more than 10% of scores are not outs, it is an important statistical issue.

Kimber and Hansford demonstrate the batting average B has the desirable property of consistency irrespective of the censoring mechanism for not out scores only if the distribution of scores is geometric. Since this is not the case exactly, they claimed the batting average does not estimate the population mean. Applying the methods of survival analysis, they defined a non-parametric alternative A which effectively distributes the not out scores using the empirical distribution of any higher scores. A parametric adjustment is needed when the batsman's highest score is not out, as there is no empirical information on his chances of dismissal above that score. The method has the disadvantage of needing recalculation from scratch with each new score. Kimber and Hansford calculated the new average for several batsmen and also estimated other statistics describing the underlying distribution of scores (e.g. centiles (quantiles) and bin probabilities). Table 4.2 gives some examples for test and first class cricket. Doubtful that such a radical change would be accepted by players and administrators, they finally recommended a slight alteration to the career statistics of batsmen. The number of innings in which the batsman scored 100 runs or more (100) and the number of innings in which the batsman scored between 50 and 99 runs (50) are standard career statistics. Table 4.3 gives several examples of how Kimber and Hansford supplemented these totals with the percentage of innings that are 50s and 100s after estimating inflation for not outs.

Elderton and Elderton (1909) used the standard deviation of scores as a measure of consistency, with zero implying perfect consistency. By contrast, because the mean and standard deviation of the geometric distribution are (roughly) equal for a large mean, Wood (1945) suggested using the coefficient of variation (CV) multiplied by 100 as a measure of consistency, with the closer to 100, the more consistent a batsman. In his analysis, Wood found CVs as low as 96 and as high as 139, but mainly in the range 100–109. Pollard et al. (1977) claimed a high CV indicates a batsman has problems early, but scores runs more easily later in the innings. Using a model where the number of balls faced was geometric, while the score off each ball had any unspecified alternative distribution, Clarke (1991, 1994) showed that perfectly consistent batsmen will have CVs greater than 100, and that perfectly consistent batsmen with different scoring distributions off each ball will have different CVs. Thus, it is not possible to have a single measure (CV closeness to 100 as proposed by Wood) which indicates perfect consistency for all batsmen. Still, the regular publication of the standard deviation or CV of scores would assist fans in judging the consistency of batsmen. Analogous questions on the distribution and consistency of runs and wickets for bowlers have not been analysed in the literature.

4.3 Tactics

Some quantitative work has been done in the area of tactics. The limited nature of the innings in one-day cricket creates a trade-off between run rate and loss of wickets. Traditional tactics suggest an innings starts cautiously, with teams scoring slowly and preserving wickets. The run rate steadily increases, until near the end of the innings there is often a frenzy of runs scored and wickets

Table 4.2 Summary statistics for various players in Tests and First Class cricket (from Kimber and Hansford, 1993, p. 450, Table 3)

Player	Type of cricket	Average		Centile					Probability mass (%) for the following scores			
		Standard B	New A	10th	25th	50th	75th	90th	0–9	10–49	50–99	100+
Armarnath	Tests	42.50	41.45	0	8	35	63	100	27.56	39.13	22.84	10.47
Bradman	Tests	99.94	98.98	1	16	66	169	244	17.50	27.88	14.97	39.65
Gower	Tests	44.31	44.30	5	11	28	61	102	21.00	47.91	20.04	11.05
Bradman	First Class	95.14	94.76	3	22	66	139	232	15.39	26.89	19.33	38.39
Trumper	First Class	46.58	47.36	2	10	28	67	110	23.87	41.70	22.13	12.30

Table 4.3 Summary of some sets of scores from Tests and First Class cricket (from Kimber and Hansford, 1993, p. 450, Table 4)

Player	Type of cricket	I	NO	H (High)	Runs	A	100	(%)	50	(%)
Armarnath	Tests	113	10	138	4378	41.45	11	(10.5)	24	(22.8)
Bradman	Tests	80	10	334	6996	98.98	29	(39.7)	13	(15.0)
Gower	Tests	200	16	215	8154	44.30	18	(11.1)	39	(20.0)
Bradman	First Class	338	43	452[++]	28067	94.76	117	(38.4)	69	(19.3)
Trumper	First Class	395	21	300[++]	17420	47.36	45	(12.3)	86	(22.1)

[++] Not out score.

fallen. Clarke (1988) analysed this by setting up a dynamic programming model with the number n of balls remaining in the innings marking 'time'. The model assumed all batsmen were of the same ability with the probability of dismissal on each ball p_d depending only on the run rate R runs per six-ball over. For the innings of the team batting first, the states were the number of wickets remaining, i, and the objective function was the expected number of runs scored in the remainder of the innings, $f_n(i)$. The functional equation

$$f_n(i) = \max_R \{ p_d f_{n-1}(i-1) + R/6 + (1 - p_d) f_{n-1}(i) \}$$

with suitable boundary conditions could be solved iteratively. Table 4.4 gives part of the output for the first-innings model.

In the innings of the team batting second, the state was the number of wickets remaining, i, and the number of runs remaining to pass the opponent's score, s, and the objective function was the probability of exceeding the opponent's score. Given the batsman's scoring profile per ball (the probability p_x of scoring x runs, $x = 0$–6), the functional equation becomes

$$P_n(s, i) = \max_R \left\{ p_d P_{n-1}(s, i-1) + \sum_{x=0}^{6} p_x P_{n-1}(s - x, i) \right\}$$

Again, this can be solved iteratively, and Table 4.5 gives part of the output.

Under the assumed relationship between run rate and dismissal probabilities (necessitated by a lack of data), Clarke's results suggested that teams should

Table 4.4 Optimal expected score in the remainder of the innings under optimal policy (from Clarke, 1988, Table 2)

	Expected score in remaining balls $f_n(i)$							
Overs to go ($n/6$)	Wickets in hand i							
	2	4	5	6	7	8	9	10
1	9	12	12	12	12	12	12	12
5	27	39	42	45	47	49	51	53
10	38	59	67	73	77	82	85	88
20	49	83	96	107	117	126	134	141
25	53	91	106	119	131	142	152	160
30	56	97	114	129	142	155	166	176
40	60	106	126	144	160	175	189	202
50	63	113	135	155	174	191	207	222

Table 4.5 Probability of scoring a further *s* runs with 300 balls to go (from Clarke, 1988, Table 3)

s	Wickets in hand							
	2	4	5	6	7	8	9	10
100	0.200	0.638	0.805	0.907	0.961	0.985	0.995	0.998
125	0.094	0.432	0.622	0.776	0.881	0.942	0.975	0.990
150	0.042	0.262	0.430	0.600	0.743	0.848	0.917	0.958
175	0.016	0.137	0.257	0.403	0.555	0.691	0.799	0.878
200	0.006	0.065	0.137	0.241	0.367	0.502	0.629	0.739
220	0.002	0.031	0.073	0.140	0.234	0.348	0.471	0.591
225	0.002	0.025	0.061	0.120	0.206	0.312	0.431	0.551
250	0.000	0.009	0.024	0.053	0.101	0.170	0.258	0.359
275	0.000	0.003	0.008	0.020	0.043	0.079	0.132	0.200
300	0.000	0.001	0.003	0.007	0.017	0.034	0.061	0.100

score more quickly early in the innings – in fact, at any stage they should score at a slightly greater rate than the expected rate for the remainder of the innings. Such tactics have become more accepted recently and, in particular, have been used by the current World champions, Sri Lanka. Johnston (1992) later showed the recommendations were valid under a range of relationships between run rate and dismissal probabilities. Clarke and Norman (1995, 1997, 1998) have extended the models to allow for batsmen of unequal ability. Near the end of an innings in Test or one-day cricket, two batsmen of widely different ability will often refuse a possible run early in the over to protect the weaker batsman from taking strike. They investigated the point in the over and in the innings when this tactic is optimal with respect to different objective functions.

The way batsmen are dismissed has received some attention. For example, Croucher (1982a) analysed dismissals in the 96 Australia–England Test matches in the period 1946–80. He investigated the various types of dismissal with respect to in the period position and location (England or Australia). Table 4.6 shows that about 20% of batsmen 1 to 7 are bowled, but this increases steadily to nearly 38% for number 11. The LBW dismissal rate is about 14%

Table 4.6 Australia and England combined dismissals for all tests 1946–80 expressed as a percentage for each batsman (from Croucher, 1982a, p.181, Table 3)

Batsman No.	Type of dismissal				
	Bowled	Caught (by wicket keeper)	Caught (by fieldsman)	LBW	Other
1	19.6	16.3	45.6	13.6	4.8
2	21.1	17.6	41.4	14.3	5.7
3	22.9	12.7	48.8	11.4	4.2
4	20.1	16.2	44.6	15.6	3.5
5	20.5	17.1	45.3	14.1	3.0
6	20.0	20.0	44.2	11.2	4.6
7	20.9	18.3	40.7	13.1	7.1
8	23.7	16.3	44.0	12.5	3.5
9	30.5	15.4	42.7	6.9	4.5
10	35.0	14.7	37.8	7.4	5.1
11	37.9	13.1	36.3	8.1	5.6

for batsmen 1–8 but much less for batsmen 9–11. In contrast, the percentage of batsmen caught is reasonably constant. A thorough investigation of modes of dismissal would be interesting and could find application to vary tactics in different countries.

4.4 Rating players

The major system of rating players is the Deloitte ratings which were created by Deloitte, Haskins and Sells in 1987. After several mergers, they are now called the Coopers & Lybrand ratings. The system rates the current Test form in both batting and bowling for all international players, with ratings updated after each Test match. The algorithm takes into account not only the results of the match, but also the difficulty of the pitch and the strength of the opposition. However, the latter are not estimated subjectively, but calculated from the details in a typical summary score card using weighted averages. The ratings range from 0 to 1000 and have been backdated to the late 1970s. A top Test player will have a rating around 700. Table 4.7 gives the top ten batsmen as of 30 September 1997, and Fig. 4.3 shows a graph of the bowling ratings for Courtney Walsh over several years.

Because of its proprietary nature, it is difficult to obtain details of the algorithm, and even more difficult to get details of the statistical work behind its derivation. Berkmann (1990) contains a few pages of description of the algorithm, along with ratings for all players from the 1980s. A weighted average is worked out for each performance (defined for the purposes of calculating the ratings as an innings for a batsman, and 30 runs conceded for a bowler), combined to give a player rating for each Test, which is then smoothed with past ratings to produce a new rating. The statistics used to calculate a batsman's rating are runs scored, whether not out or out, the team score, wickets fallen, and the match result. The opposition bowling strength is also used but this is estimated from the statistics. For bowlers, the statistics used are runs conceded, wickets taken (including which batsmen), balls bowled per wicket (strike rate), runs conceded per ball, team scores and match result. The ratings take a player's basic statistics (e.g. runs scored or wickets taken) and multiplies it by a series of factors depending on whether the performance is above or below average with respect to playing conditions. For example, a batsman's runs are increased if the

Table 4.7 Coopers & Lybrand top ten batsmen ratings on 30 September 1997

Rank	Player	Country	Rating	Career average
1	S. R. Waugh	Aus	870	49.66
2	G. P. Thorpe	Eng	803	42.34
3	S. R. Tendulkar	Ind	792	52.09
4	S. Chanderpaul	WI	770	53.85
5	S. T. Jayasuriya	Sri	769	49.84
6	B. C. Lara	WI	763	54.10
7	B. M. McMillan	SA	736	42.55
8	R. S. Dravid	Ind	717	48.52
9	Saeed Anwar	Pak	711	48.30
10	D. J. Cullinan	SA	709	40.40

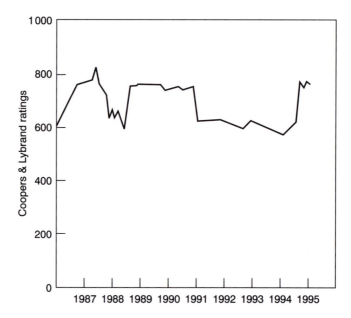

Figure 4.3 The Coopers & Lybrand bowling rating for Courtney Walsh for 1987–95 (from WWW Coopers & Lybrand ratings – www.uk.coopers.com/cricketratings/ratingsguide/interpret/interpret.html).

pitch is 'difficult', as measured by the ratio of the number of runs per wicket in the match to the average of 31 found in previous Test matches. Similarly, an average of the bowlers' ratings, weighted by the number of overs each bowled in the innings, is used to determine the strength of the bowling. A batsman's score is then increased if this is higher than the average for Test bowlers. For a bowler, the number of wickets he is credited with depends on the ratings of the batsmen he dismisses.

A major point of contention is that the algorithm adjusts ratings depending on the result of the match. It rewards good performances of players on the winning side in a match more than good performances by players on the losing side, and penalizes poor performances by players in the losing side more than poor performances by players in the winning side. Thus, above average players on the winning side are given a bonus, below average players on the losing side are penalized, and others unadjusted. A player's rating is thus clearly affected by the ability of his team. Ted Dexter, one of the initiators of the scheme, said 'the pleasant surprises include ... the considerable accuracy of the ratings when used to compare the relative strengths of the Test playing countries' (Berkmann, 1990, p. iii). This is hardly surprising for an algorithm that gives a greater rating to a good performance in a winning team. In addition, the rating is not symmetrical with respect to batsmen and bowlers. A bowler's strike rate is included in the calculation, but not a batsman's scoring rate. Similarly, the overall opposition bowling strength is taken into account for batsmen, but only the strength of the batsmen he dismisses for bowlers. While the ratings have gained some acceptance by cricket followers, the need for some proper analysis of the scheme is evident. For example, a simulation could be used to investigate the effect playing in good and poor teams has on the ratings of players of similar calibre. The effect of batting against good and poor bowling sides should also be analysed. Surprisingly, alternatives based on models of how Test cricket is played have not been proposed.

The Coopers & Lybrand ratings are only applied to Test cricket. Johnston (1992) and Johnston *et al.* (1992, 1993) looked at a method of rating players in one-day cricket, in which traditional statistics are not as relevant. Because the match is limited to a set number of overs, a quick score of 30 may be more valuable to the batting side than a slow century, three wickets for two runs in the last over is of less value to the fielding side than a wicketless maiden over. By comparing the actual number of runs scored each ball with the optimal number given by a dynamic programming formulation, both a batsman's and a bowler's contribution could be measured in a radically different way to normal methods. The context in which events occurred becomes important. For example, Table 4.4 shows the expected score in the remainder of the innings with 60 balls to go and five wickets in hand is $f_{60}(5) = 67.0$. If the expected score in the remainder of the innings with 59 balls to go and five wickets in hand is $f_{59}(5) = 66.5$, then the batsman on strike when 60 balls remain in the innings must score 0.5 runs for the side to maintain the same expected innings score. If the batsman scores two runs in this situation, then he has advanced his team's expected innings score by 1.5 runs. In this case, the batsman's performance measure would increase by 1.5 and the bowler's measure would decrease by 1.5. This concept is similar to the calculation of Player Win Average in baseball which replaces runs with probability of winning (see Chapter 2).

Johnston *et al.* (1993) gave ratings for each player in a one-day series. One difficulty confronting the acceptance of the system is its need for ball-by-ball data. However, with the growing use of computer-assisted scoring, such methods are becoming more viable.

4.5 Umpiring decisions

An area that has received some attention is the contentious one of umpire bias. Only recently has the traditional practice of the home country providing both umpires for Test matches been modified. Because the rule is quite complicated, LBW is the most subjective type of dismissal and often creates controversy. For the period 1877 to 1980, Sumner and Mobley (1981) found significantly fewer LBW dismissals against home teams than visiting teams in India, Pakistan and South Africa, but not in Australia. Croucher (1982a) found that, while on average about 12% of batsmen are dismissed LBW, this percentage varied from 10.6% in Australia to 14.0% in England. This could be due to the different behaviour of pitches in England and Australia. Table 4.8 shows, however, that when subdivided by team, LBW rates for England were fairly constant at 11.6% and 12.7% in Australia and England, but for Australia varied from 9.5% in Australia to 15.4% in England. This could be due to umpire bias, or Australian batsmen not adjusting to conditions. However, breaking batsmen into two

Table 4.8   Percentage of frequency of dismissals: each team in each country (from Croucher, 1982a, p. 184, Table 9)

Team in Country	Bowled	Caught (by wicket keeper)	Caught (by fieldsman)	LBW	Other
Australia in Australia	23.8	15.9	45.2	9.5	5.6
Australia in England	23.3	12.8	44.3	15.4	4.2
England in Australia	22.7	17.0	43.9	11.6	4.8
England in England	24.5	19.3	39.9	12.7	3.6

categories, 1–5 and 6–11, the frequency of LBW decisions showed a dependence of location and category for Australian batsmen but not for English. One interpretation of this is that umpires give decisions against top batsmen but square the account against lower batsmen.

Crowe and Middeldorp (1996) used logistic regression to compare LBW rates for visitors and their Australian opponents for Tests in Australia for the period 1977–94. The odds of an LBW dismissal are defined as the ratio of LBW dismissals to all dismissals by other means. A logistic model fitted the logarithm of the odds for a series of matches to a linear expression using indicator variables for the various countries. Separate models were fitted using only the top six batsmen and all batsmen. An initial model found a significant difference for LBW rates for visiting teams and Australia. Since there was no evidence of the odds for countries changing over time, nor of Australia's odds changing depending on opponent, the overall odds for each country were calculated and compared with the common odds for Australia. For the top six batsmen and all batsmen, only three of the seven countries that had visited Australia during the period (England, Sri Lanka and South Africa) had significantly proportionally more LBW dismissals than Australia (see Fig. 4.4). For example, the odds ratio for England was 1.7, with a 95% confidence interval of 1.2 to 2.5. Interestingly, India, whose captain had complained of umpire bias, had an odds ratio of only 1.059, barely greater than the expected value of 1 if there was no bias. Of course, as Crowe and Middeldorp point out, other factors such as a difference in playing style, more experience on home wickets or different tactics could account for the results.

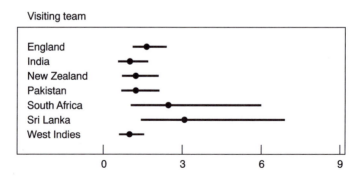

Figure 4.4   Odds ratios for visiting teams versus Australia (top six batsmen) (from Crowe and Middledorp, 1996).

4.6 Rain interruption in one-day cricket

Because of the limited number of balls in one-day cricket and the requirement that a result has to be achieved in a given time period if at all possible, allowance has to be made for rain interruption. For example, if the innings of the team batting second is shortened because of rain, the target score for victory has to be reduced to compensate for the reduced number of overs. Various formulae to adjust the target score have been tried with varying degrees of success. These rules appear to be developed ad hoc by administrators and are rarely based on a proper quantitative study. They are used until a particular situation makes a mockery of the rule, which is then replaced. For example, in one World Cup semi-final, South Africa, who had a reasonable chance of achieving their target when rain interrupted, were then required to face *one* ball and score 22 runs to win when play resumed.

Clarke (1988) suggested that the results of his dynamic programming model could be used to evaluate the effectiveness of rain interruption rules, so that teams would have the same chance of winning after the interruption that they had prior to the rain. For example, Table 4.4 shows a team batting first could expect to score 222 runs in 50 overs. Table 4.5 shows the team batting second has a slight advantage, in that they will exceed this target with probability 0.551. He used a similar argument to show that the rain interruption rule current at the time, where the second team had to score at the same rate for the shortened number of overs, was highly advantageous to the team batting second.

Duckworth and Lewis (1996) and Duckworth (1997) produced a method using tables similar to Clarke's first innings table, but which have been derived from past statistics. They treated balls to go and wickets remaining as run-scoring resources, and fitted an exponential decay model for the average number of runs scored to past data. If $Z(u, w)$ is the average number of runs scored in the remainder of the innings when there are u overs left and w wickets lost, then

$$Z(u, w) = Z_0 F(w)[1 - \exp\{-bu/F(w)\}]$$

where $F(w)$ is the proportion of the asymptotic total score obtained when w wickets are already lost. Using data from over 250 international games, they estimated $b = 0.0315$ and $Z_0 = 283.69$. To estimate $F(w)$ required data on the score at various stages, which proved difficult to obtain. The values currently used are such that, with unlimited overs, the first five wickets each account for 12% of the asymptotic average, while the last five wickets account for 10%, 8%, 8%, 7% and 6%. Figure 4.5 shows the generated family of curves.

The resulting function was used to give a measure of the proportion of these combined resources available at any stage of the innings. Table 4.9 gives an extract. Targets are adjusted depending on the proportion of overs lost due to rain. For example, suppose a team chasing 250 has lost two wickets after ten overs when a rain interruption reduces their available overs from 50 to 40. The table shows at the time of interruption they had 77.6% of resources left, but on resumption had 68.2% left. They lost 9.4% so their target is reduced by 9.4% of 250 or 24 runs. The method can be applied to interruptions at any stage of a match. This system was first used in international competitions in Zimbabwe in 1996, and is being used for domestic competitions in England in 1997.

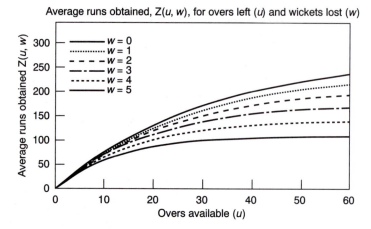

Figure 4.5 Average runs obtained in *u* overs with *w* wickets lost (from Duckworth and Lewis, 1996, p. 57, Fig. 1).

A by-product of the system could be a better method of declaring the winning margin in one day matches. Margins in one day matches are still given using traditional measures of runs if won by the team batting first or wickets if won by the team batting second. These can be quite misleading. For example, a team batting second that scores the winning run on the last ball of the innings may be credited with a six wicket victory. This sounds comfortable, when in fact the team had used virtually all their resources. The method could also be investigated to provide alternative tie-breaking procedures in one-day round-robin tournaments.

4.7 Sundries

In 1981, in response to a request for data for a simulation study of batting order in one-day cricket, a student of the author received a firm refusal from a high ranking Australian cricket administrator which included 'Any analysis that you suggest would be wholly hypothetical, and of no value...the analysis would only be greeted by scorn by those with a proper understanding of cricket. The inherent essence of cricket is its unpredictability; and an attempt to reverse

Table 4.9 Proportion (%) of resources remaining at various stages of a one-day match (from Duckworth, 1997, p. 13)

Overs left	Wickets lost			
	0	*2*	*5*	*8*
50	100	83.8	49.5	16.4
40	90.3	77.6	48.3	16.4
30	77.1	68.2	45.7	16.4
25	68.7	61.8	43.4	16.4
20	58.9	54.0	40.0	16.3
10	34.1	32.5	27.5	14.9

this . . . is something I would not personally encourage.' With the increasing use of science in sport, one hopes this view is not widely held today. Cricket administrators now clearly seek assistance from academics to solve management problems that are not peculiar to cricket (Willis and Armstrong, 1993; Willis, 1994; Wright, 1991, 1992) .

The history of rain interruption rules suggests administrators are less reluctant to seek statistical help with on-field and other problems. Few First Class competitions have the luxury of allowing five days play as in test matches. Consequently, in a high proportion of these matches, neither team achieves an outright victory. In domestic round-robin tournaments, the relative allocation of points for first innings and outright victory varies, and administrators have experimented with bonus points for fast batting or penetrative bowling. Rarely are these experiments based on, or their effectiveness judged by, statistical studies. Bosi (1976) investigated the effect of the introduction of bonus points in county cricket. He claimed that a significant change in the correlation in ladder position using the traditional and new methods shows that the rule alteration affected the way cricket was played.

However, statisticians should play a major role in developing rules, not just evaluating their effect after the fact. The problem of allocating points for unfinished matches should be investigated, with possibly some of the methods used for one-day matches applied. There is currently discussion in cricket about the publication of a world ranking of Test teams. Wisden publish their own table (Engel, 1997, p. 19), based on each country scoring two points for winning a series and one for drawing, but the system accounts for neither margin of victory nor home advantage. With countries playing intermittent series of different lengths, statisticians should investigate and recommend suitable ranking systems before cricket administrators decide on something inappropriate. Similarly, player evaluations should not be considered resolved, and rivals to the Coopers & Lybrand rating should be developed.

Cricket is highly variable. Clarke (1994) showed that with roughly geometric score distributions purely random variations give rise to team scores ranging from 100 to 500 in Test cricket. Johnston (1992) simulated one-day cricket using optimal batting rate policies and obtained scores ranging from 75 to 322. A game with so much variation provides ample scope for statisticians to assist participants, administrators and supporters to separate real effects from random noise.

However, topics that have proved fruitful areas of research in other sports have been largely ignored in cricket. While the difficulty of winning a Test series on foreign soil is recognized, home ground advantage has not been thoroughly studied. Pollard (1986) quoted home advantage in cricket county championships to be 56.1% excluding drawn games. In domestic competitions teams play on pitches of varying quality. Some pitches are more difficult to bat on and hence produce more outright results. Since outright wins are rewarded more than first innings wins, Clarke (1986) showed that some teams not only win a greater proportion of the points awarded on their home ground, but compete for a greater number of points than other teams. Such a system would never be tolerated in other sports.

Surprisingly few alternative statistics have been suggested. Cricket fans seem to be satisfied with dividing the traditional ones into categories. So, a batsman is

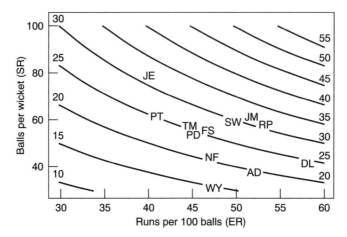

Figure 4.6 Augmented scatter plot of the 12 highest wicket-takers in English First Class cricket in 1991 (WY = Waqar Younis; AD = A. A. Donald; D. L. = D. V. Lawrence; NF = N. A. Foster; PD = P. A. J. DeFreitas; FS = F. D. Stephenson; TM = T. A. Munton; RP = R. A. Pick; SW = S. L. Watkin; JM = J. N. Maguire; PT = P. C. R. Tufnell; JE = J. E. Emburey) (redrawn from Kimber, 1993, p. 83, Fig. 1).

still judged by his average, although this may also be given against a particular country or at a particular ground. With the introduction of one-day cricket, strike rate, economy rate, and run rate have also become popular, and the articles mentioned here have made several suggestions for alternative statistics. With these statistics currently being measured over the life of a player, the use of moving averages could be used to measure current form. There may be advantages in combining the statistics in various ways. For example, Ganesalingam *et al.* (1994) applied multivariate analysis techniques to classify players as batsmen, all-rounders or bowlers. Kimber (1993) compared bowling statistics using scatterplots of strike rate (balls bowled/wickets taken) against economy rate (runs conceded/100 balls) augmented with bowling average (runs conceded/wickets taken) contours. He found some bowlers better than others in all measures, and showed different types of bowlers (spinners and fast bowlers) appeared in different regions of the plot (see Fig. 4.6). Such studies could suggest further statistics or indices to be used in evaluation of players and selection of teams.

4.8 Conclusions

In spite of a huge collection of statistics on cricket dating back over 200 years, little attempt at serious analysis has appeared in the professional literature. Many questions in cricket could be investigated using relatively elementary statistical techniques. Is one team better than another? Is one batsman better than another? Does the rate of dismissal vary? Of all the sports in this text, cricket has the distinction of having statistics that stretch back the longest, the first use of sport in a statistics text to illustrate statistical principles, the first full quantitative paper, and yet probably the fewest serious papers analysing the statistics in the professional literature. It is surprising that more statistical analysis has not been undertaken.

Chapter 2 of this book shows the large number of papers written on baseball – best batting order, value of player, measurement of hot streaks, etc. Similar research could be done in cricket. One of the neglected areas calling for study is bowling. When thanking Elderton, Wood (1945) said 'At last a great statistician has discovered what is, I believe, the richest field of statistical material left untilled. I have scratched over its surface, but other statisticians will find in it materials for all sorts of statistical experiments, particularly in the bowling analyses'. This statement is still valid today.

Acknowledgments

I wish to thank Greg McKay, a member of the Association of Cricket Statisticians from Mont Albert, and David Richards, School of Mathematical Sciences, Swinburne University, who provided editorial comments and suggestions.

References

Berkmann, M. (1990) *Deloitte Ratings: the Complete Guide to Test Cricket in the Eighties*. London: Transworld.

Bosi, A. (1976) The effect of bonus points on the Cricket County Championships. *Mathematical Spectrum*, **9**(3), 75–82.

Clarke, S. R. (1986) Another look at the 1985/86 Sheffield Shield competition cricket results. *Sports Coach*, **10**(3), 16–19.

Clarke, S. R. (1988) Dynamic programming in one-day cricket – optimal scoring rates. *Journal of the Operational Research Society*, **39**, 331–337.

Clarke, S. R. (1991) Consistency in sport – with particular reference to cricket. *Proceedings 27th NZ Operational Research Conference*, 30–35.

Clarke, S. R. (1994) Variability of scores and consistency in sport. In *Mathematics and Computers in Sport* (edited by N. de Mestre) Bond University, Gold Coast, Queensland, pp. 69–81.

Clarke, S. R. and Norman, J. M. (1995) Protecting a weak batsman from taking strike at the start of a new over – a simple dynamic programming model in cricket. In *Network Conferencing, Proceedings of ASOR: The 13th National Conference* (edited by D. L. Hoffman) Canberra, pp. 45–55.

Clarke, S. R. and Norman, J. M. (1997) To run or not? Some dynamic programming models in cricket. In *Proceedings of the Twenty-Sixth Annual Meeting of the Western Decision Sciences Institute* (edited by R. L. Jenson and I. R. Johnson), pp. 744–746.

Clarke, S. R. and Norman, J. M. (1998) Dynamic programming in cricket – protecting the weaker batsman. *Asia-Pacific Journal of Operational Research*, in press.

Croucher, J. S. (1979) The battle for the ashes – a statistical analysis of post-war Australia–England cricket. In *4th National Conference of the Australian Society of Operations Research*, pp. 48–58.

Croucher, J. S. (1982a) Anglo-Australian test cricket dismissals 1946–1980. *Bulletin in Applied Statistics*, **9**, 179–193.

Croucher, J. S. (1982b) Australia–England test cricket. *Teaching Statistics*, **4**(1), 21–23.

Crowe, S. M. and Middeldorp (1996) A comparison of leg before wicket rates between Australians and their visiting teams for Test cricket series played in Australia, 1977–94. *The Statistician*, **45**, 255–262.

Duckworth, F. (1997) Dealing satisfactorily with interruptions in one-day matches. *The Cricket Statistician*, No. 98, 10–15.

Duckworth, F. and Lewis, T. (1996) A fair method for resetting the target in interrupted one-day cricket matches. In *Mathematics and Computers in Sport* (edited by N. de Mestre) Bond University, Gold Coast, Queensland, pp. 51–68.

Elderton, W. E. (1927) *Frequency Curves and Correlation.* London: Layton.

Elderton, W. E. (1945) Cricket scores and some skew correlation distributions. *Journal of the Royal Statistical Society (Series A)*, **108**, 1–11.

Elderton, W. P. and Elderton, E. M. (1909) *Primer of Statistics.* London: Black.

Engel, M. (Ed) (1997) *Wisden Cricketers' Almanack.* Guildford: John Wisden.

Ganesalingam, S., Kumar, K. and Ganeshanandam, S. (1994) A statistical look at cricket data. In *Mathematics and Computers in Sport* (edited by N. de Mestre) Bond University, Gold Coast, Queensland, pp. 89–104.

Haygarth, A. (1862–1885, 1925) *Cricket Scores and Biographies of Celebrated Cricketers*, Vols I–IV, Lilywhite; Vols V–XV, Longman.

Johnston, M. I. (1992) Assessing team and player performance in one-day cricket using dynamic programming. M.App.Sc Thesis, Swinburne University.

Johnston, M. I., Clarke, S. R. and Noble, D. H. (1992) An analysis of scoring policies in one-day cricket. In *Mathematics and Computers in Sport* (edited by N. de Mestre) Bond University, Gold Coast, Queensland, pp. 71–80.

Johnston, M. I., Clarke, S. R. and Noble, D. H. (1993) Assessing player performance in one-day cricket using dynamic programming. *Asia-Pacific Journal of Operational Research*, **10**, 45–55.

Kimber, A. (1993) A graphical display for comparing bowlers in cricket. *Teaching Statistics*, **15**(3), 84–86.

Kimber, A. C. and Hansford, A. R. (1993) A statistical analysis of batting in cricket. *Journal of the Royal Statistical Society A*, **156**, 443–445.

Pollard, R. (1977) Cricket and statistics. In *Optimal Strategies in Sport* (edited by S. P. Ladany and R. E. Machol) New York: North-Holland, pp. 129–130.

Pollard, R. (1986) Home advantage in soccer: a retrospective analysis. *Journal of Sport Sciences*, **4**, 237–248.

Pollard, R., Benjamin, B. and Reep, C. (1977) Sport and the negative binomial distribution. In *Optimal Strategies in Sport* (edited by S. P. Ladany and R. E. Machol) New York: North-Holland, pp. 188–195.

Reep, C., Pollard, R. and Benjamin, B. (1971) Skill and chance in ball games. *Journal of the Royal Statistical Society A*, **134**, 623–629.

Sumner, J. and Mobley, M. (1981) Are cricket umpires biased? *New Scientist*, 2 July, 29–31.

Willis, R. J. (1994) Scheduling the Australian state cricket season using simulated annealing. *Journal of the Operational Research Society*, **45**, 276–280.

Willis, R. J. and Armstrong, J. (1993) Scheduling the World Cup of cricket – a case study. *Journal of the Operational Research Society*, **44**(11), 1067–1072.

Wood, G. H. (1941) What do we mean by consistency? *The Cricketer Annual*, 22–28.

Wood, G. H. (1945) Cricket scores and geometrical progression. *Journal of the Royal Statistical Society (Series A)*, **108**, 12–22.

Wright, M. B. (1991) Scheduling English cricket umpires. *Journal of the Operational Research Society*, **42**, 447–452.

Wright, M. B. (1992) A fair allocation of county cricket opponents. *Journal of the Operational Research Society*, **43**, 195–201.

5

Soccer

John M. Norman

University of Sheffield, UK

5.1 Introduction

Soccer, as Association Football is commonly known, is a game lasting 90 minutes, played between two teams, each containing 11 players. The game is played with an inflated spherical ball and the teams try to score goals; that is, to get the ball over the opposing team's goal line, between the goal posts and under the crossbar. The main rules of the game are known to most schoolchildren throughout the world. It is surprising that a game familiar to so many has a history of little more than 100 years.

Perhaps the main reason is that it was not until the middle of the 19th century that the rules of football became codified. Football in various forms has been played for centuries. Some forms were played at schools such as Eton (the Wall Game), in towns (such as Ashbourne in Derbyshire) and at universities. These games seem mostly to have been played without rules and were often fierce affairs: indeed, the town-based game (still held annually at Ashbourne) remains a game for young men not afraid of getting hurt. Marples (1954) notes the death of a student in Oxford in 1303 'whilst playing the ball in the High Street'.

Marples dates the beginning of soccer at 8 December 1863, when a group representing various clubs in England adopted a code of rules, which from then on became the rules of the Football Association. Marples quotes two decisive rules:

- 'no player shall run with the ball';
- 'neither tripping nor hacking [kicking an opponent on the shin] shall be allowed and no player shall use his hands to hold or push his adversary'.

These rules regarding handling and running with the ball defined the essential difference between soccer and rugby and their adoption marked the beginning not just of the Association code, but of the Rugby code also, and its descendant, American football.

In just over 100 years, the rules of the game have remained essentially unchanged. There have been minor changes, such as those allowing for substitutes, the determination of a result in some games by penalty shoot outs, and developments in the offside rule, but a game of soccer played today would be instantly recognized by a football Rip van Winkle of 1863 who returned to the game over 100 years later.

5.2 Sources of information

UK football is well covered in the many yearbooks published by Rothmans (Rollins, 1995) and others. The Association of Football Statisticians, with 1400 members in 38 countries, publishes regular reports and digests, but the most convenient source in present times is the Internet, where a little surfing will often produce the information required. Deloitte and Touche publish an *Annual Review of Football Finance* (Boone, 1997), which is more wide ranging than its title suggests, covering the major European clubs as well as those in the UK. Academics interested in psychology and physiology in soccer will find material of interest in the *Proceedings of the World Congress in Science and Football*, held every four years (Reilly *et al.*, 1988, 1993), with abstracts of papers published in the *Journal of Sports Science* (various, 1995). Those interested in the sociology of soccer may turn to the publications of the Sir Norman Chester Centre for Football Research at the University of Leicester (see, for example, Williams *et al.*, 1989). Seddon's *A Football Compendium* (1995) deserves special mention. It covers a vast range of topics associated with football and every entry is fully referenced; many references also carry a short abstract of their contents. This book is an invaluable source.

To the statistician, the game presents many features of interest. In this chapter we can deal with only some of them.

This chapter will cover research in modelling scores, home ground advantage, simulated games, the effect of a red card, betting on results, factors making for team success and econometric models. Other interesting research areas such as the effect of three points instead of two for a win and other competition rules are beyond the scope of this chapter.

5.3 Modelling soccer scores

Possession of the ball is important in soccer, since whenever a team has the ball it has an opportunity to attack the opposing team's goal and perhaps score. The probability p that any attack will result in a goal is small, but there are many such attacks in a game. If attacks are independent, then the distribution of the number of goals scored will be binomial and, for small p, the Poisson approximation will be a good fit. However, as Pollard *et al.* (1977) point out, this is not a realistic model of a real-life game of soccer. In particular, the Poisson assumption may need to be modified in two ways. First, teams have different rates of scoring, varying from one match to another. Second, within a match, a team's goal scoring rate may change as the match develops: a team that is losing may concentrate on attack in the hope of at least levelling the score, even though this tactic may leave it liable to concede further goals. Pollard *et al.* therefore suggest that a negative binomial

Table 5.1   Number of goals scored per game by individual teams

Number of goals (r)	English First Division Football League 1967–68		World Cup 1974	
	Observed	Expected	Observed	Expected
0	225	226.6	28	31.4
1	293	296.4	24	20.0
2	224	213.9	13	11.4
3	114	112.6	5	6.2
4	41	48.3	4	3.3
5	15	17.9	0	1.8
6	9	5.9	0	0.9
7 +	3	2.5	2	1.0
Total	924	924.1	76	76.0
Mean	1.51		1.28	
Variance	1.75		2.54	
$P(\chi^2)$	0.57		0.42	

Source: Pollard *et al.* (1977)

distribution would be more appropriate and obtain a good fit to actual data (Table 5.1).

Some years earlier, in the first analysis of the distribution of the total number of goals scored in a match, Moroney (1951) had come to the same conclusion. Pollard *et al.*, however, suggest a way in which the negative binomial distribution of goals might arise, as a result of the way in which the game is played. In soccer, the ball is passed from player to player on the same team until possession is lost, an infringement of the rules occurs, the ball goes out of play, or a goal is scored. Pollard *et al.* study the distribution of the length of pass-moves and again find that a negative binomial distribution fits the data well (Table 5.2).

Pollard *et al.* apply the negative binomial distribution to other ball games, again finding a good fit to the data, suggesting, as Maher (1982) points out,

Table 5.2   Passing movements. Forty-two English First Division Football League matches, 1957–58

Number of passes	Observed	Expected
0	10 187	10 143
1	6 923	7 022
2	3 611	3 553
3	1 592	1 578
4	608	651
5	280	257
6	107	98
7	33	37
8	9	13
9 +	11	8
Total	23 361	23 360
Mean	1.02	
Variance	1.50	
$P(\chi^2)$	0.13	

Source: Pollard *et al.* (1977)

that a single negative binomial distribution applies to the number of goals scored in a match by a team, irrespective of the quality of that team or the quality of the opposing team. Indeed, in an earlier paper, Reep and Benjamin (1968) remark that 'chance does dominate the game'.

Hill (1974) observes that skill as well as chance must be involved in a game of soccer. He compares the final league tables of the four English and two Scottish Football Leagues with expert forecasts of the final positions made before the start of the season. All six rank correlations are positive, which, even ignoring their values, gives a probability of 1/64 using a sign test. As Maher remarks, over a whole season, skill rather than chance dominates the game. In any one match, chance plays an important part (injuries to players, the ball hitting the woodwork, incorrect decisions by the referee and so on), but over many matches, the effect of chance must be reduced.

Maher's paper has been the basis of several others. He points out that Pollard *et al.* take no account of the quality of each of the two teams playing in a match, and argues that teams are not identical and that a good team playing against a poor team is more likely to score goals and to win. He uses data from many matches to infer the quality of individual teams by maximum likelihood estimation. The mean of the distribution of the number of goals scored in a match will vary from team to team so that the distribution of goals scored by all teams would be a mixture of Poisson distributions with different means, which could well result in the negative binomial distribution found by Moroney and Pollard *et al.*

Maher considers a game between team i, playing at home against team j, in which the score is (x_{ij}, y_{ij}), and X_{ij} is Poisson with mean $\alpha_i \beta_j$ and Y_{ij} is Poisson with mean $\gamma_i \delta_j$, with X_{ij} and Y_{ij} independent. The parameters represent the strength of the home team's attack (α), the weakness of the away team's defence (β), the weakness of the home team's defence (γ), and the strength of the away team's attack (δ). Maher finds that a reduced model in which $\delta_i = k\alpha_i$, $\gamma_i = k\beta_i$ for all i is the most appropriate of several models he investigates. Thus, the quality of a team's attack and a team's defence depends on whether it is playing at home or away. Home ground advantage ($1/k$) applies with equal effect to all teams. Although a soccer match does not, in fact, comprise two independent games taking place at opposite ends of the pitch, Maher shows that this relatively simple model gives a reasonably good fit to the data.

5.4   Home ground advantage

The existence of a home advantage in most sports is now well documented. Courneya and Carron (1992) give a summary of the work done. For a league in which each team plays the same number of matches at home and away, Pollard (1986) measures home advantage as the percentage of games won by home teams with respect to all games played. Thus, 50% would indicate an absence of home advantage. This measure is clearly inappropriate as a measure of the home advantage enjoyed by an individual team, as it does not take into account the quality of teams: a good team playing away may be expected to beat a much weaker team playing at home. In an unbalanced competition (in which teams do not play the same number of games at home and away), mathematical models need to be fitted to estimate team ability and home advantage as in Kuk (1995). However, in a balanced competition, the analysis can be much simpler. Clarke and Norman (1995) give a method which, although equivalent to fitting a model to the individual match results by least squares, can be applied using simple arithmetic on the final league table. Clarke and Norman work on the English Football Association league tables for the ten years 1981–90, and Clarke (1996) extends the work to 1991–96.

The model is similar to that used by Stefani (1983, 1987), Stefani and Clarke (1992) and Clarke (1993), which has been successful in predicting match results. The winning margin in a match between team i and team j played at the home ground of team i is modelled as

$$w_{ij} = u_i - u_j + h_i + \varepsilon_{ij}$$

where u_i is a measure of team i's ability, h_i is a measure of team i's home advantage and ε_{ij} is a zero-mean random error. The arithmetic used to estimate the parameters for team i in an N-team league is described by Clarke (1996):

H = total home goal difference of all teams/$(N - 1)$

$$h_i = \frac{(\text{home goal difference for team } i - \text{away goal difference for team } i - H)}{(N - 2)}$$

u_i = (home goal difference for team $i - (N - 1)h_i)/N$

The results for the Premier Division in 1995–96 are shown in Table 5.3.

Counting a draw as half a win, we see that, at home, Leeds have won exactly 50% of games and 50% of goals, so that, under the definition of percentage of games or goals scored at home, the team has zero home advantage. Leeds won an equivalent of 9.5 games at home and six games away for a 3.5 game home advantage. In terms of goals, Leeds had a zero home goal difference and a (-17) away goal difference. It might be said that Leeds had a $0 - (-17) = 17$ goal home advantage. However, virtually all teams, irrespective of ability, have a home advantage measured in this way. The source of this spurious home advantage may be seen in the calculation of h_i. The difference in a team's home and away performance is made up of one component due to that team's home advantage and another due to the total home advantage of all the teams. Thus, although a team may do better at home than away, this has to be adjusted relative to the collective advantage enjoyed by all teams. Table 5.3 shows three clubs with a negative home advantage: for all these clubs the difference between their home and away performance is less than nine goals, the total home advantage of all teams. For Leeds,

$$h = (0 - (-17) - 9.05)/(20 - 2) = 0.44 \text{ and}$$

$$u = (0 - (20 - 1)(0.44))/20 = -0.42$$

Clarke and Norman (1995) and Clarke (1996) show that a team's home advantage is variable from year to year. On average, the home ground advantage is worth just over 0.5 of a goal per game, a value constant over all divisions, a result which seems to negate crowd size as a cause of home advantage. However, there is a significant year effect: home advantage was higher than average in 1982, 1983 and 1985 and lower in 1981, 1987 and 1989. The degree of this

Table 5.3   Results for 1995–96 Premier Division

Team	Home					Away					Pts	u	h
	Wins	Draws	Losses	Goals for	Goals against	Wins	Draws	Losses	Goals for	Goals against			
Manchester U	15	4	0	36	9	10	3	6	37	26	82	0.98	0.39
Newcastle U	17	1	1	38	9	7	5	7	28	28	78	0.40	1.11
Liverpool	14	4	1	46	13	6	7	6	24	21	71	0.54	1.16
Aston Villa	11	5	3	32	15	7	4	8	20	20	63	0.43	0.44
Arsenal	10	7	2	30	16	7	5	7	19	16	63	0.60	0.11
Everton	10	5	4	35	19	7	5	7	29	25	61	0.64	0.16
Blackburn R	14	2	3	44	19	4	5	10	17	28	61	−0.17	1.50
Tottenham H	9	5	5	26	19	7	8	4	24	19	61	0.72	−0.39
Nottingham F	11	6	2	29	17	4	7	8	21	37	58	−0.40	1.05
West Ham U	9	5	5	25	21	5	4	10	18	31	51	−0.22	0.44
Chelsea	7	7	5	30	22	5	7	7	16	22	50	0.14	0.27
Middlesboro	8	3	8	27	27	3	7	9	8	23	43	−0.31	0.33
Leeds United	8	3	8	21	21	4	4	11	19	36	43	−0.42	0.44
Wimbledon	5	6	8	27	33	5	5	9	28	37	41	0.02	−0.34
Sheffield Wed	7	5	7	30	31	3	5	11	18	30	40	−0.15	0.11
Coventry City	6	7	6	21	23	2	7	10	21	37	38	−0.36	0.27
Southampton	7	7	5	21	18	2	4	13	13	34	38	−0.64	0.83
Manchester C	7	7	5	21	19	2	4	13	12	39	38	−0.95	1.11
QPR	6	5	8	25	26	3	1	15	13	31	33	−0.47	0.44
Bolton Wand	5	4	10	16	31	3	1	15	23	40	29	−0.38	−0.39
Totals	186	98	96	580	408	96	98	186	408	580	1042	0.00	9.05

Source: Clarke (1996)

effect varies too much to be a chance fluctuation. Weather conditions were studied, but could not be demonstrated to be a significant explanatory variable. There is some evidence for a significant club effect but this is not conclusive. (Among other things, club effect might result from an unusual pitch. For example, Yeovil's former ground had a marked slope that often allowed the home team to defeat stronger teams.)

Barnett and Hilditch (1993) look specifically at the effect of an artificial pitch on home advantage. Queen's Park Rangers (QPR) from 1981–82 to 1987–88, Luton from 1985–86 to 1989–90, Oldham from 1986–87 to 1990–91 and Preston from 1986–87 to 1990–91 all had artificial pitches. Barnett and Hilditch find a significant advantage in home match performance for teams employing an artificial pitch, which they estimate might yield, over the whole season, an improved goal difference of about six goals for such a team. Clarke and Norman obtain a similar result: in their data, the 22 seasons played on an artificial pitch produced a significantly higher home advantage than the other 898 seasons.

Stefani and Clarke (1992) state that home advantage may be thought of as h_{ij} and, for each pair of teams, the difference between home and away matches gives an estimate of this for each year. The data of Clarke and Norman gave 10 153 match pairs, played between 2865 club pairs. To see whether distance had any effect, the straight line distance between each pair of clubs was calculated and the pairs of clubs were separated into groups in multiples of 50 miles apart. The results are shown in Fig. 5.1, which clearly shows increasing home advantage with increasing distance between clubs. Dowie (1982) suggests that the fatigue resulting from travel may reduce the level of a team's performance when playing away, although he suggests that the major factors in home advantage may be the tempo of the sport and the proximity (and enthusiasm) of the supporters. Chapter 12 of this book provides a more general description of predictive models and their incorporation of home field advantage.

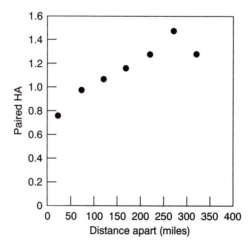

Figure 5.1 Average paired home advantage HA (goals) versus distance apart of clubs (redrawn from Clarke and Norman, 1995).

5.5   Simulated games

There are now several 'fantasy leagues' operated by national newspapers in the UK. In these leagues, points are scored according to the simulated performance of 'virtual' players managed by individual readers. The virtual players have some of the characteristics of their real-life namesakes, such as transfer values and playing ability. Lee (1997) simulates the Premier Division of the Football League in the 1995–96 season, whose final results were presented earlier in Table 5.3. The question Lee asks is whether Manchester United deserved to win the League. In one sense, the answer must be yes, because they gained the most points (three for a win and one for a draw). Yet some of the games may have been very close, with the outcome being largely due to chance.

Lee's approach is similar to that of Maher (1982), in that he assumes a common home advantage for all teams, and a Poisson distribution of goals scored in a match by each of the two teams. However, he fits parameters for both attack and defence capabilities for both home and away teams, using maximum likelihood. Using these parameters, he estimates the probability of each possible result for each of the possible match pairs: for example, he estimates the probabilities of Manchester United winning, drawing or losing at home to Liverpool as 0.48, 0.25 and 0.26 respectively. The 1995–96 season was simulated 1000 times. Table 5.4 shows the results. It seems that Manchester United were deservedly top of the table but that Liverpool may be considered somewhat unlucky. The large differences between the actual points and the simulated mean points for some teams, such as Manchester United and Newcastle at

Table 5.4   Results from simulating the 1995–96 season

Team	Actual points 95/96	Poisson model expected points	Simulated mean points	Simulated std. dev. points	Proportion at top of table
Manchester U	82	75.7	75.5	7.1	0.38
Newcastle U	78	70.7	70.5	7.8	0.16
Liverpool	71	74.9	74.9	7.5	0.33
Arsenal	63	63.8	63.6	7.7	0.03
Aston Villa	63	63.7	63.6	7.4	0.03
Blackburn R	61	61.2	61.4	7.4	0.03
Everton	61	64.9	65.0	7.5	0.04
Tottenham H	61	60.2	60.8	7.5	0.01
Nottingham F	58	50.0	49.5	7.4	0.00
West Ham U	51	46.3	46.1	7.7	0.00
Chelsea	50	53.4	53.5	7.4	0.00
Leeds U	43	41.4	41.4	7.4	0.00
Middlesboro	43	41.5	41.8	7.4	0.00
Wimbledon	41	44.7	44.7	7.6	0.00
Sheffield Wed	40	44.8	44.9	7.2	0.00
Coventry City	38	41.2	41.4	7.6	0.00
Manchester C	38	35.7	35.4	6.9	0.00
Southampton	38	39.6	39.5	7.0	0.00
QPR	33	39.9	40.1	7.3	0.00
Bolton Wand	29	33.9	34.0	7.2	0.00

Source: Lee (1997)

the top and QPR near the bottom, indicate that extreme records are often the result of chance.

5.6 The effect of a red card

Ridder *et al.* (1994) investigate the effect of a player's expulsion on the result of the match. For a serious infringement, a referee may show a player a card, either yellow or red, depending on the severity of the offence. Once a player has been shown a yellow card, a further offence which would warrant a yellow card results in a red card being shown. A red card is a sign of immediate expulsion from the match.

Ridder *et al.* consider a model in which the two teams score goals according to two independent Poisson processes, but as the scoring intensities of each team are allowed to change throughout the match, the Poisson processes are non-homogeneous. The ratio of the scoring intensities of the two teams when full is constant, but, after a player has been shown a red card, his team's scoring intensity diminishes. The authors estimate the extent of this diminution by a conditional maximum likelihood estimator, using data from the period 1989–92 relating to the Dutch professional league.

Among their results, they show the critical time during the progress of a match, after which it is worthwhile for an unethical defender to commit a 'professional foul'. Table 5.5 shows how this time relates to the relative strength γ of the teams (the ratio of defender to attacker) and the probability of the attacker scoring on the play. It is assumed that the objective of the defending team is to minimize the probability of losing the match.

5.7 Betting on match results

Betting on sports, covered in more detail in Chapter 12, has long been popular in the UK, although until recently, most betting in soccer was through the form of the football 'pools', in which the principal outlay was on matches thought likely to result in draws. Jackson (1994) remarks on the continued steady growth in sports betting and on the many different ways to bet on sports. He investigates index betting in particular, in which (in the case of soccer) the total goals scored in a match may be bought or sold as if it were a commodity, or an object of speculation like the future price of some commodity. He quotes

Table 5.5 Time (minute of game) after which a defender should stop a breaking-away player, by probability of score and relative strength of the defender's team

	Probability of score		
Relative strength of teams, γ	*0.3*	*0.6*	*1*
0.5	70	42	0
1	71	48	16
2	72	52	30

Source: Ridder *et al.* (1994)

an example of the offer by the firm Spread Bet International (SBI) in relation to the England v. Republic of Ireland match in the World Cup finals of 1990. Prior to the match, SBI offered to sell 2.4 goals scored or buy 2.1 goals scored. In the event, the score was England 1, Republic of Ireland 1 for a total of two goals scored. A punter who sold SBI 2.1 goals scored at £10 per goal would have sold £21 worth of goals prior to the match. In buying back the actual result of two goals, the bettor had to pay only £20 = £10 × 2, realizing a profit of £1. On the other hand, a bettor who bought 2.4 goals at £10 per goal would have lost £4 = £10 × (2.4 − 2). Thus, if SBI's offer had been able to balance bets bought and sold, SBI would have realized a profit of £3 for every £10 per goal bought/sold.

Fixed odds betting is also increasing in popularity. Here, bookmakers offer odds on the various outcomes of a match. As Dixon and Coles (1997) explain, in the simplest case, this comprises just the outcome of the match; a home win, an away win, or a draw, although more complicated bets can be made on the actual result or even the score at half-time. The challenge is to find 'good bets' in which the punter deems the probability of occurrence to be more than the probability determined by the bookmaker's odds, giving a positive expected return. Dixon and Coles note that odds are fixed one week before the matches are played so that any inefficiencies in the bookmaker's odds can be exploited. They base their analysis on Maher's (1982) model, with independent Poisson distributions of goals scored by either side.

The probabilities derived from a bookmaker's odds offered for a match typically sum to more than unity; for example, odds offered on a home win, draw and away win might be 8:13, 12:5 and 4:1, with corresponding probabilities 0.62, 0.29 and 0.20 which sum to 1.11, representing an expected profit to the bookmaker of 11%. If, say, b_h is the bookmaker's probability of a home win (scaled so that all three probabilities sum to unity) and p_h is the punter's estimated probability of a home win (which Dixon and Coles estimate using maximum likelihood), then it will be worthwhile betting on a home win if $p_h > b_h$, neglecting the bookmaker's 'take'. A sensible strategy would be to bet on a home win if $p_h/b_h > r$, where r takes account of the bookmaker's take. Typically, with a bookmaker's take of 10%, we might require $r > 1.1$. Figure 5.2 shows home win probabilities estimated by Dixon and Coles against

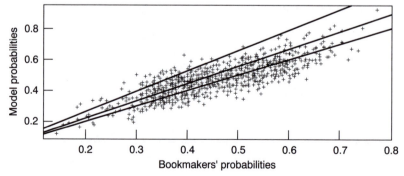

Figure 5.2 Model probability estimates plotted against odds for all matches where odds are available. This plot is for home wins (redrawn from Dixon and Coles, 1997).

scaled bookmakers' probabilities for 1548 soccer matches played in the 1993–94 and 1994–95 seasons. The three straight lines correspond to values of *r* of 1.0, 1.1 and 1.3 respectively. There are clearly opportunities for profitable betting, although these are obtained only by extensive computation. During a television programme, Dixon was invited by a sceptical bookmaker to invest £500 which the bookmaker would provide, with any profits going to a charity of Dixon's choice. Dixon was able to use his strategy to generate a small, but very satisfying profit.

5.8   What makes a successful team?

dell'Osso and Szymanski (1991) study the performance of leading English clubs in the 20-year period 1970–89, looking at three teams in particular: Nottingham Forest, Manchester United and Liverpool. Figure 5.3 shows the relationship between expenditure on players' wages and transfer fees (relative to the average) and league position for 12 leading clubs. The calibration of the vertical axis is

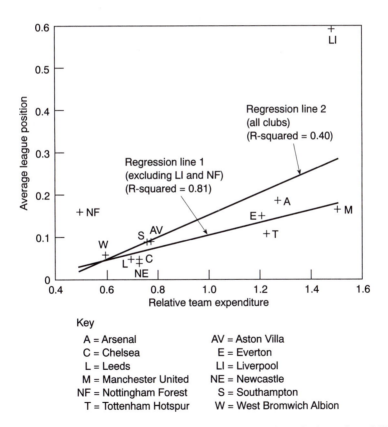

Figure 5.3   Average league position versus relative team expenditure (redrawn from dell'Osso and Szymanski, 1991).

not clear from the text but, essentially, Liverpool performed more successfully than any other team (they were top of the League 11 out of the 20 seasons). Liverpool spent as much money as any other club, while Nottingham Forest spent less. These two clubs stand apart from the rest. The fit of regressing position on expenditure is much improved when these two clubs are excluded.

del'Osso and Szymanski argue that to be profitable, firms (and clubs in the Football League are indeed firms) must have a distinctive capability. This capability might be a technological advantage, a monopoly of some factor, a reputation, or what the authors term 'architecture'. Architecture is what binds an organization together; it is linked with teamwork and synergy, the whole being greater than the sum of its parts. It is hard to think of a possible technological advantage in soccer, but in yachting, for example, technology seems to play a major part in determining success in the America's Cup. del'Osso and Szymanski link the success of each of the three clubs to a single different factor: Nottingham Forest's to their possession of the unique management capability of Brian Clough, Manchester United's to their reputation and Liverpool's to their architecture.

Table 5.6 shows Clough's career, with his teams' average positions in the League's four divisions over various time periods. The periods in which Clough was manager have been set back with a one year lag to represent the time taken for a manager's work to take effect. Position 1 represents the leadership of the (then) First Division, Position 2 the second place and so on. The bottom position of the First Division is followed immediately by the top position of Division 2, and so on. Clough represents a distinctive capability and could raise a club's performance to a high level, a level which could not be sustained after his departure.

Manchester United has been the most famous club in England over many years and has enjoyed consistently high attendances. del'Osso and Szymanski suggest that the club's reputation has enabled it to purchase superstars, but that the club's performance has been below what might have been expected. It would be interesting to learn if the authors would maintain this view after the club's performance in the years following their study.

Table 5.6 Performances of Brian Clough's teams (overall, before, during with a one year lag, and after Clough was manager)

	Years	Position
Hartlepool	1958–89	88.4
Before	1958–65	92.9
(manager 1965–67)	*1966–68*	*74.7*
After	1969–89	88.7
Derby County	1946–89	26.2
Before	1946–67	29.5
(manager 1967–73)	*1968–74*	*6.9*
After	1975–89	25.4
Nottingham Forest	1946–89	16.5
Before	1946–75	21.3
(manager 1974–93)	*1976–89*	*7.0*

Source: dell'Osso and Szymanski (1991)

Liverpool's success has not been due to any one individual, though some individuals have made important contributions, both on and off the field. The club has had excellent managers and excellent players, yet the only explanation of their continued success seems to be what the authors call their cooperative behaviour, shown in many ways, but on the pitch by their distinctive passing game.

5.9 Econometric models

Professional soccer can be a test bed for econometric models. Success may be measured by a club's position in the League and its operating profits and the accounts of every club can be inspected at Companies House. A club's competitors are known and the little movement into and out of the League follows strict rules. Thus Szymanski and Smith (1997) are able to write about football as an industry:

> The English Football League...is facing declining demand and competition from more technologically advanced leisure products. Most firms make losses, its plant is antiquated and grossly under-utilised and heavy investment is required both to meet government mandated safety standards and to improve the quality of the product sufficiently to compete in modern markets. Firms have little control over their main input cost, players, which are traded on a competitive market. Of course, many mature industries are profitable and there is considerable evidence, particularly from international comparisons, that a restructured industry could earn money.

Szymanski and Smith (1997) develop an econometric model of the English Football League for the period 1974–89. In their model, football skills are bought on a competitive market for players and the amount of skill purchased by a club determines its position in the League. A club's position in the League determines the revenue it earns. Other things being equal, a club can trade profit for an improvement in its League position. A club's objective is to get as high up in the League as possible subject to a limit on its financial loss.

The model performs reasonably well: a few of the top clubs make profits, the rest suffer losses. There is a high degree of concentration with a few clubs accounting for a high proportion of total income. The model estimates the money required to move a team up the League, an estimate which compares well with the £12 million paid for players by the owner of Blackburn Rovers, who shifted the team from near the bottom of one division to the top of the division above in only two seasons. Other big spenders who have tried to improve the standing of their clubs have not always been so successful.

Dobson and Gerrard (1997) have used the team performance–club profit framework introduced by Szymanski and Smith (1997) to produce a model of player transfer fees in the English League. A transfer fee is the sum paid by one club to another to obtain the services of a player. They use their model to confirm that the selling club is able to profit by exploiting the difference between the value placed on the transferred player by the purchasing club and the reserve price for the player (the least sum the selling club is willing to accept).

Dobson and Goddard (1995) examine the determinants of the demand for professional league football in England in the period 1925–92. They find evidence of a significant effect of the admission price on the demand for football in the UK. Clubs that have been in the League a long time are found generally to enjoy higher attendances. They also, find, however, that clubs from towns with a high proportion of professional and managerial employees are not much affected by price changes, or indeed by the club's form. Finally, they note the presence of a 'loyalty' factor, which seems to affect all clubs, regardless of position or status.

5.10 Conclusions

It is not long ago that an academic interest in sport would have been regarded with suspicion; today, this is less often the case. Sport is worth studying for its own sake, but there are other reasons why it merits academic study. Soccer, in particular, as we have seen, has furnished a model for economic competition. It has also been useful as a source of examples and exercises in teaching. Wright (1997), for example, uses the Football League table tie-breaking rules to illustrate the difference between interval and ratio levels and Bland (1995) must be one of many hundreds of schoolboy projects concerned with the statistical analysis of soccer results. There is still much work to be done.

Acknowledgments

I would like to thank Stephen Clarke for his collaboration in research over many years and Mandy Robertson for her aid in the production of the chapter.

References

Barnett, V. and Hilditch, S. (1993) The effect of an artificial pitch surface on home team performance in football (soccer). *Journal of the Royal Statistical Society Series A*, **156**, 39–50.

Bland, N. D. (1995) A mathematical analysis of football. *Mathematics Project*, Pimlico School, London.

Boone, G. (ed.) (1997) *Deloitte & Touche Annual Review of Football Finance*. London: Deloitte & Touche.

Clarke, S. R. (1993) Computer forecasting of Australian rules football for a daily newspaper. *Journal of the Operational Research Society*, **44**, 753–759.

Clarke, S. R. (1996) Home advantage in balanced competitions – English soccer 1990–1996. In de Mestre, N. (ed.) *Mathematics and Computers in Sport* Bond University, Gold Coast, Queensland, 111–116.

Clarke, S. R. and Norman, J. M. (1995) Home ground advantage of individual clubs in English soccer. *The Statistician*, **44**, 509–521.

Courneya, K. S. and Carron, A. V. (1992) The home advantage in sport competitions: a literature review. *Journal of Sport and Exercise Psychology*, **14**, 13–27.

dell'Osso, F. and Szymanski, S. (1991) Who are the champions? (an analysis of football and architecture). *Business Strategy Review*, **2**, 113–130.

Dixon, M. J. and Coles, S. G. (1997) Modelling association football scores and inefficiencies in the UK football betting market. *Journal of the Royal Statistical Society Series C*, **46**, 265–280.

Dobson, S. M. and Goddard, J. A. (1995) The demand for professional league football in England and Wales, 1925–92. *The Statistician*, **44**, 259–277.

Dobson, S. M. and Gerrard, W. (1997) Testing for rent-sharing in Football Transfer fees: Evidence from the English Football League. Leeds University Business School Working Paper E97/03.

Dowie, J. (1982) Why Spain should win the World Cup. *New Scientist*, 693–695.

Hill, I. D. (1974) Association Football and statistical inference. *Applied Statistics*, **23**, 203–208.

Jackson, D. A. (1994) Index betting on sports. *The Statistician*, **43**, 309–315.

Kuk, A. Y. C. (1995) Modelling paired comparison data with large numbers of draws and large variability of draw percentages among players. *The Statistician*, **44**, 523–528.

Lee, A. J. (1997) Modelling scores in the Premier League: is Manchester United really the best? *Chance*, **10**, 15–19.

Maher, M. J. (1982) Modelling association football scores. *Statistica Neerlandica*, **36**, 109–118.

Marples, M. (1954) *A History of Football*. London: Secker and Warburg.

Moroney, M. J. (1951) *Facts from Figures*. Harmondsworth: Penguin.

Pollard, R. (1986) Home advantage in soccer: a retrospective analysis. *Journal of Sports Sciences*, **4**, 237–248.

Pollard, R., Benjamin, P. and Reep, C. (1977) Sport and the negative binomial distribution. In S. P. Ladany and R. E. Machol (eds) *Optimal Strategies in Sports*. New York: North-Holland, 188–195.

Reep, C. and Benjamin, P. (1968) Skill and chance in association football. *Journal of the Royal Statistical Society Series A*, **131**, 581–585.

Reilly, T., Lees, A., Davids, K. and Murphy, W. J. (eds) (1988) *Science and Football*. London: E. and F. N. Spon.

Reilly, T., Clarys, J. and Stibbe, A. (eds) (1993) *Science and Football II*. London: E. and F. N. Spon.

Ridder, G., Cramer, J. S. and Hopstaken, P. (1994) Down to ten: estimating the effect of a red card in soccer. *Journal of the American Statistical Association*, **89**, 1124–1127.

Rollins, J. (ed) (1995) *Rothman's Football Yearbook 1995–96*. London: Headline.

Seddon, P. J. (1995) *A Football Compendium*. Boston Spa: The British Library.

Stefani, R. T. (1983) Observed betting tendencies and suggested betting strategies for European football pools. *The Statistician*, **32**, 319–329.

Stefani, R. T. (1987) Applications of statistical methods to American football. *Journal of Applied Statistics*, **14**, 61–73.

Stefani, R. T. and Clarke, S. R. (1992) Prediction and home advantage for Australian rules football. *Journal of Applied Statistics*, **19**, 251–261.

Szymanski, S. and Smith, R. (1997) The English football industry: profit, performance and industrial structure. *International Review of Applied Economics*, **11**, 135–154.

Various (1995) Communications to the Third World Congress of Science and Football. *Journal of Sports Sciences*, **13**, 499–522.

Williams, J., Dunning, E. and Murphy, P. (1989) *Hooligans Abroad: the Behaviour and Control of English Fans in Continental Europe*. London: Routledge and Kegan Paul.

Wright, D. B. (1997) Football standings and measurement levels. *The Statistician*, **46**, 105–110.

6

Statistics in Golf

Patrick D. Larkey

Carnegie Mellon University, USA

6.1 Introduction

Statistical applications in golf address four main questions.

(1) How can a player learn from his or her past performance or from the performance of others to improve future performance? Millions of amateur and professional golfers around the world are intensely interested in the answer to this question.

(2) What should the basis be for a handicap that will yield fair competitions in several formats, e.g. stroke play, match play, better ball, alternate shot, scramble, etc., among players of differing ability? The amateur game is predominantly a 'net' game where net score = (gross score – handicap). Curiously, the winners of handicap events usually believe that the handicaps were fair while losers usually believe that they are getting too few strokes and/or giving too many. Whatever system is used, it is constantly under attack and a firm analytic foundation is a necessary but not sufficient basis for securing broad acceptance of the handicap system.

(3) Which professional player has been 'best' in some historical period (e.g. a year or an era) for some set of competitions (e.g. Majors or all sponsored events on one of the official 'tours')? There is great popular interest in comparing players, current to current, past to past, and current to past. Also a host of multi-tournament prizes and honors and qualifications for future events are determined through ranking systems.

(4) What will professional players' performances be in the next competition or some larger set of upcoming competitions? There is a healthy, worldwide gambling industry built on professional golf, legal and illegal, and the burgeoning Fantasy Golf and Pick-Your-Pro games calling for predictions.

This chapter briefly reviews some of the work that has been done with respect to these questions and subsidiary questions, and a few works that have used golf as a domain for examining other issues. Before proceeding with the review, however, it is important to understand the game and its numbers. In golf, as in many other application domains, insufficient attention is given to how measurements are made and what they mean, which is a necessary, if not sufficient, step in doing useful analyses.

6.2   Golf and its quantitative measures

The objective in golf, generally, is to minimize the number of times you strike the ball in moving it from the teeing area to the cup on the green. In Match Play, where your score is compared with opponent(s) hole-by-hole, the total round score is much less important – you can have a few very bad holes and still win the match – than in Stroke Play, where low total score (gross or net) wins.

6.2.1   Round scores

Golf naturally yields a number of quantitative measures. Players count, perhaps with some help from their opponents, the number of times they strike the ball on each hole and sum this for a round of 18 holes. The basic performance observation in golf is the reported round score, the number of strokes taken for 18 holes. For the 15 776 rounds played by 195 players who played at least 48 rounds on the US PGA TOUR in 1997, the mean score was 71.36 (median = 71; mode = 71; maximum = 87; minimum = 60; and standard deviation = 3.15). The distribution of scores is very close to normal with slight skewness to the right. The mean raw 18-hole score for US amateurs is about 90.

There are two excellent studies on the distribution of scores. Mosteller and Youtz (1993) studied the scores in the last two rounds in 33 professional tournaments. They fit a reduced variance Poisson model to the data with good results and found independence of scores between rounds. Scheid (1990) studied the round and hole-by-hole scores of 3000 amateurs. He found the normal approximation to be good except that the tail of bad scores was heavier than expected. Both studies show this long tail of bad scores. It is much easier for golfers to let scores deteriorate than to lower them. There have only been three scores of 59 for a round on the PGA TOUR in official events over its 60+ year history. The equivalent region above the mean contains innumerable scores of 80 and higher; even the best players have occasionally produced one of these. Scheid also found independence violated for both round-to-round and hole-by-hole in this data set.

Round score is a random variable. The choice of distribution depends, as always, on the needs of the analysis as well as on fit. While quantitative measures are a 'natural' product of golf, their interpretation is anything but simple. Reported scores are a complex function of player ability, equipment, course difficulty (length, speed of greens, difficulty of pin placements, hazards, etc), scoring conditions (wind speed and direction, temperature, humidity, etc) and the competitive format (stroke play, match play, or various team events).

Golf performance requires complex and disparate physical and cognitive skills. Tommy Bolt, a prominent golf professional in the 1950s, once noted that, '(i)n golf, driving is a game of free swinging muscle control, while putting is something like performing eye surgery using a bread knife for the scalpel.' The head of a professional's driver routinely exceeds 100 miles per hour (mph) at the point of impact in making a drive; Tiger Woods and John Daly achieve 120 to 130 mph. At those speeds, a degree or two difference in the angle of the clubface or swingpath leads to wildly different ball flights in terms of distance, direction, and ball movement. Putting requires dramatically different skills than driving. There is the cognitive task of judging the correct line and speed. Then there is the physical task of striking the ball with the putter to execute the plan, perhaps with nothing, a two-dollar Nassau bet, your club championship or hundreds of thousands of dollars and a Major title riding on the outcome.

Scoring depends on much more than player ability. Courses differ significantly in their scoring difficulty. Some of these differences are constants reflecting fundamental differences in course architecture and terrain; USGA (United States Golf Association) research indicates that length is the most important difference among courses for determining their scoring difficulty. Other differences depend on more ephemeral conditions. It matters how the course is 'set up'. Pins may be cut behind hazards in a sloped area, making it more difficult for both approach shots and putting. The lawnmower settings matter a lot; if the greens are cut to the nub to make putts faster and the rough is allowed to grow several inches, a course can be several strokes more difficult on average for all players. Playing in wind, wet, cold or extreme heat generally makes scoring harder.

The nature of the competition also matters for scoring. Round scores are accomplished in a variety of competitive formats, including some formats where minimizing the overall number of strokes is not the primary objective. For example, the round score may result from Better Ball competition (a team game in which the team score for each hole is the lower score by an individual on the two-player team), where risk-taking may minimize the team's expected score but not the individual's score. If my partner has a par for the team secured, I may attempt a shot where I expect to make birdie (one less than par) 10% of the time, par (the norm for the hole – three to five strokes) 10% of the time, bogey (one over par) 40% of the time, and double-bogey (two over par) or worse 40% of the time, rather than a shot that would protect my individual score with a 50% chance of par and 50% chance of bogey but have no chance of helping the team. If the hole is a par 4, my expected individual score if I select the shot that minimizes the expected joint product for the team is 5.1 ($0.1 \times 3 + 0.1 \times 4 + 0.4 \times 5 + 0.4 \times 6$); the expected result selecting the shot that minimizes my personal score is 4.5 ($0.5 \times 4 + 0.5 \times 5$). The team expectation if I take the selfish shot is 4, my partner's score; if I take the selfless shot, it is 3.9 ($0.9 \times 4 + 0.1 \times 3$). The implications for successful play in different formats and for interpreting scores from team competitions as evidence of individual ability in individual competitions are obvious.

A further complication of the observed round score from such a format is that many putts are conceded when the individual's score cannot improve the team score; in individual stroke play, some of these putts would be missed.[1] Finally, many players perform better or worse with, respectively, the security or strain

---

[1] There is a difficult tension between speed of play and accurate scoring. It speeds play to have individual players who can no longer contribute to their team 'pick up' but are required to estimate 'the score they probably would have made' in recording their individual score.

of a partner than they would in individual stroke play. The upshot of these complications in interpreting reported scores for the main amateur question, devising a handicap system that provides a basis for fair matches between those of unequal ability, is that scores from one competitive format may not adequately predict performance in other formats. While professionals do not have the handicapping problem, the format differences for performance were never clearer than in the 1997 Ryder Cup competition where the US players did well in individual competition but poorly in team competitions, and lost the overall event. Team members are selected on the basis of their ability to perform individually. Since the US team usually dominates the individual portion of the competition with players who are, on the whole, more successful in individual competitions prior to the Ryder Cup, and loses the team portions – winning turns on the size of victories in each portion – there must be some flaw in the US player selection for team membership or in their pairing or playing strategy for team competitions.

As if interpreting scores were not complicated enough, there are also issues of player integrity, particularly in the amateur game where reported scores are the basis of handicapping. As a wag once noted, 'Nothing increases golf scores so much as a careful observer.' Some amateurs report scores in which they have unknowingly or willfully violated the rules of golf. They improve lies without penalty. They fail to take 'stroke and distance' as the penalty when they hit a shot out of bounds. They give themselves putts 'in the leather'. They 'count funny'. Most of these rule violations result in lower scores and lower handicaps than should be the case. These practices and selective reporting, where better scores are reported and poorer scores are not, contribute to 'ego' handicaps where the players' abilities are overstated in low handicaps that give them no chance to succeed in formal competitions, but assuage their egos in presenting their level of 'golfing ability' to the larger world. On the other side are players who build 'betting' handicaps. Popularly known as 'sandbaggers', these players selectively report a subset of their best scores and intentionally play poorly when the stakes are low in order to have a handicap that understates their ability for use when the stakes are higher. Given the natural variation in scoring, distinguishing the cheaters from honest golfers who are just having 'a good day' is a very interesting statistical problem.[2]

6.2.2 Tournament measures

The basic observations from a professional golf tournament appear in the format of Table 6.1.

The modal tournament in professional golf is four rounds of stroke play with a starting field of 156 players which is cut to the low 70 scorers and ties after the second round of play. For each player making the cut (1) place, (2) scores in each of four rounds, (3) total score, and (4) prize money are reported. Players not making the cut receive no prize money and have only their two round scores recorded.

[2] Knuth *et al.* (1994) address this problem directly. Heiny and Crosswhite (1986) address it in passing.

Table 6.1 Leaders in the PLAYERS Championship, Ending: 30 March 1997

Player	Round scores	Total	Prize money ($)
Steve Elkington	66-69-68-69	272	630 000.00
Scott Hoch	69-71-65-74	279	378 000.00
Loren Roberts	70-74-67-69	280	238 000.00
Brad Faxon	70-69-72-70	281	168 000.00
Billy Andrade	68-72-68-74	282	140 000.00
Tom Lehman	67-71-73-72	283	126 000.00
Mark Brooks	72-68-70-74	284	109 083.34
Colin Montgomerie	70-70-71-73	284	109 083.33[3]

For understanding a single event, these observations provide an adequate story.[4] For analyzing the results of two or more tournaments, the observations are inadequate. The first difficulty with the observations is the variety of tournament formats. There are bigger and smaller fields. Two tournaments, the World Series and the Tour Championship, have no cut. One tournament, the International, uses a point system, the modified Stableford (Double Eagle $= +8$, Eagle $= +5$, Birdie $= +2$, Par $= 0$, Bogey $= -1$, and Double Bogey or worse $= -3$), instead of strokes. For historical analysis, in some official match play events (most notably the PGA Championship until 1957), after medal rounds using stroke play, players moved to match play (hole-by-hole between pairs of players where stroke scores are not kept); the results were given in place, prize money, and match result (e.g. '4 and 3' meant the winner was four holes ahead with three holes remaining to play).

Several tournaments, notably Major Championships administered by organizations other than the PGA TOUR, have deviated at times from the modal payoff schedule, where the winner gets 18% of the purse, second place gets 10.8%, third place gets 6.8% and so forth. The PGA TOUR's current standard payoff schedule is shown in Fig. 6.1.

In contrast to most sports, earnings is a major measure of player performance. Unfortunately, the prize money for performances in particular tournaments is a poor surrogate for the value received. Within each year, the size of the tournament purse does not correlate well with the significance of the tournament in terms of its prestige or the strength of the field; the TOUR Championship always has the largest purse but it is less significant than the four Majors (the Masters, the US Open, the British Open, and the PGA Championship), the PLAYERS Championship, and probably others such as the Memorial. There are also non-monetary rewards that vary across tournaments. Winning one of the four Majors, the PLAYERS Championship, or the World Series of Golf earns a ten-year exemption[5] on the PGA TOUR. Winning a regular

---

[3] Current year data of this form and much more detail are available in magazines such as *Golfweek* and *GolfWorld* and from www.pgatour.com.

[4] There may be significant within-tournament problems. Tee times, when players start play, are not randomly assigned for the first two rounds but set for television coverage to ensure that the marquee players rather than unknowns are on the telecast. There are claimed advantages to playing at certain times – greens tend to be smoother for putting for the first groups because they are not yet 'spiked-up' – and, with certain players – for any player who likes large galleries, Tiger Woods is the dream pairing, as was Arnold Palmer in an earlier era.

[5] Exempt players have the right to enter any regularly sponsored PGA TOUR event.

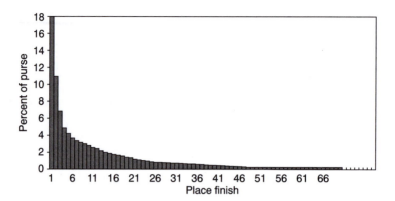

Figure 6.1   Current PGA TOUR payoff schedule.

tour event is worth a two-year exemption and qualification for the Majors and the Tournament of Champions. A second to fifth place finish in the US Open earns an exemption for the Open next year and to other Majors. Tournament victories, particularly Major tournament victories, are reported to lead to lucrative product endorsement contracts. And so on.

Even if these 'extras' for performance were not available, prize money would still be a conceptually troubled measure. There have been dramatic increases in the size of tournament purses available and, of course, the dollar in 1997 is worth substantially less in terms of purchasing power than it was in 1935, the first year of a truly organized US professional golf tour. Gene Sarazen won $1500 from a $5000 purse for winning the 1935 Masters tournament; Tiger Woods won $486 000 from a purse of $2 700 000 for winning the 1997 Masters. All the records and 'statistical accomplishments' based on nominal dollars will become even more meaningless because the PGA TOUR has just negotiated record increases in revenues from television coverage of competitions, which will increase purses significantly over the next few years.

An important problem with all outcome measures on the performance of professional golfers – money, place, strokes, etc – is that the tournaments are flawed natural experiments. Not all players play in all tournaments and participation is not random; there is a complicated set of rules for determining eligibility, and all of the better players play a subset of the tournaments for which they are eligible. Indeed, the greatest players have often played many fewer tournaments than the average. Jack Nicklaus, for example, played no more than 17 or 18 tournaments while the other good players were averaging about 25. Ben Hogan, for physical reasons, played only six tournaments in 1953, winning five of his starts including three Majors. The best players also play disproportionately well in events where the scoring conditions are most difficult and their competitors are the best.

Ranking professional golfers across tournaments is a special case of the following general rating problem: rate each player's performance relative to that of the other $N - 1$ players when N players have participated in varying subsets of M tournaments and there is no useful exogenous information about the relative skills of players or about the relative difficulty of the tournaments.

Larkey (1991d) proposed and successfully applied an algorithmic solution to this problem.[6]

6.3 Existing work

The statistical work on most aspects of golf is very rudimentary. There are lots of numbers and very little credible analysis. Only a handful of peer-reviewed scholarly articles and a few other published works possess a serious analytic bent. The numerous popular statistical analyses of golf are primarily the creations of golf journalists and functionaries working for organizations such as the US PGA TOUR or its European counterpart. Most of this popular work has serious conceptual flaws in which the conclusions drawn do not follow from the analyses.

As stated at the outset, there are four main questions for statistical applications in golf. The following subsections will examine research into these questions and several subsidiary questions and applications that, while not about golf, use golf to study other issues.

6.3.1 How can a player learn from his or her past performance to improve future performance?

Play improvement systems are schemes for players to gather information about their own playing performance and guidelines on how to use this information to improve their play through such mechanisms as reallocating practice time (e.g. spend more time practicing chipping, putting and bunker play than the full swing) or altering playing strategy (e.g. hit fewer drivers). Dozens, if not hundreds, of these systems are available on the World-Wide Web (search on 'golf statistics'). The analytics in these systems tend to be very weak. No theoretical basis is given or made obvious from the data or from the rules. Alternative systems have not been carefully evaluated.

Many golfers are, to put it mildly, obsessive about the game and improving their abilities. They spend a lot of time and money on it, creating a tremendous market for a careful system for measuring aspects of play and using resulting models to guide improvements.

While the bulk of the activity in this area pertains to amateurs, the PGA TOUR introduced statistics on components of play in 1980 including:[7]

Driving distance (yardage on the first shot as measured on two holes in each round for each player);
Driving accuracy (percentage of the time that players' drives finish on the fairway);
Greens-in-regulation (percentage of the time that players put the ball on the green of par 3s in one, par 4s in two or less, and par 5s in three or less);

---

[6] This problem occurs in a number of other application areas including ranking airlines in terms of their on-time performance (Caulkins *et al.*, 1993) and students in terms of their grade performance (Caulkins *et al.*, 1996).
[7] Larkey and Smith (1998) provide an extensive diagnosis of the problems with these measures and recommend improvements.

Sand saves (percentage of the time that a player succeeds in getting the ball in the cup from a greenside bunker in two or less);
Putting (the average number of putts taken on greens-in-regulation);
Eagles (the number of holes on average between eagles – two under par – namely, the number of holes played divided by the number of eagles); and
Birdies (percentage of birdies in holes played).

There has been some interest among professional golfers in using their component performance data to diagnose their game. The component data are heavily used by the media to explain why this player or that has either risen from relative obscurity to prominence (e.g. switched from a wooden to a titanium driver and gained 20 yards per drive) or fallen from prominence to ignominy (e.g. he was the 14th ranked putter last year and 120th this year).

Several authors have attempted to examine the statistical relationship between these measures of component skills and overall performance. Hale and Hale (1990) found that the component statistics did not explain more than about a quarter of the variation in earnings for a restricted sample of players. Jones (1990) confirmed this low explanatory power of component statistics with respect to money earned on a slightly larger sample. Davidson and Templin (1986) and Belkin *et al.* (1994) using larger samples and stepwise regression with stroke average as the dependent variable found an $R^2 \approx 0.85$ with the highest standardized beta coefficient on greens-in-regulation. Wiseman *et al.* (1994) regressed stroke average on four component statistics (greens-in-regulation, putting, driving accuracy, and driving distance) for all three US tours in 1992. The model explained more variance for the seniors and women ($R^2 = 0.88$ and 0.93, respectively) than for the regular tour ($R^2 = 0.67$). They found greens-in-regulation and putting as the best explanations for stroke average on all three tours; the driving components were only significant for the regular tour. Hale *et al.* (1994) used differences in the component statistics between the Europeans and Americans in an attempt to explain differences in Ryder Cup success from 1980 to 1993. They find that the component statistics, with the exception of scoring average, do not explain much. The edge does, however, appear to go to the team with the lowest scoring average among its players. Riccio (1994), with detailed data on Tom Watson's 3719 strokes in the US Open from 1980 through 1993, provided an interesting analysis of why Watson's play deteriorated over that period. He concluded that it was not putting, as conventional wisdom would have it, but tee-to-green play that contributed most to his play deterioration.

On the basis of the consistency of the findings from their explanatory study, Belkin *et al.* (1994) concluded that '(t)he findings of these studies [essentially theirs and that of Davidson and Templin (1986)] yield consensus regarding the predictive validity of PGA statistics and thereby make possible research with these measurements.' There have been, of course, no predictive study and no persuasive analysis of what components of golf performance contribute to overall success.

Several flaws in the design and use of the component statistics are evident. First, there is no adjustment for the relative difficulty among tournaments and fields. So the Driving Accuracy results for the US Open and PGA with narrow fairways are averaged with Driving Accuracy results for the desert tournaments or the Masters where it is much easier to hit the fairway. Second, while the statistics are a lot better in terms of measures than when they were first introduced, there are still a few consistency problems. The Eagles measure began in 1980 as a simple count of the number of eagles regardless of how many and in which events players had participated. The current measure in 1997, the number of holes between eagles, ranges from 104.1 for Tiger Woods to 528 for Kirk Triplett in 194th place to ∞ (or 'undefined' because of a zero divisor) for eagleless Dicky Pride who is dropped from reporting on this category. Third, there has been no attempt to analyze and construct orthogonal measures.

The statistical mavens at the PGA TOUR have unaccountably ignored the use of normalization. Their solution for combining these disparate measures into an All-Around statistic is to discard the interval information on each measure and sum the integer ranks for a new integer measure that is then ordered. Normalizing to preserve the interval information gives a different All-Around ranking because the intervals are, of course, not uniformly distributed (Larkey and Smith, 1998).

Much more rigorous statistical applications are in the work of scientists and engineers who have studied detailed aspects of the golf swing, the physiology of golfers, the impact of golf equipment (particularly balls and club shafts), and course characteristics (e.g. hardness of greens) with respect to player performance. This surge of scientific interest in aspects of golf is directly attributable to the publication of the classic, *The Search for the Perfect Swing*, by Cochran and Stobbs (1968). Cochran and Farrally (1994) is a collection of 92 papers by mathematicians, physicists, kinesiologists, chemists, biomechanical engineers, operations researchers, physicians, psychologists and members of a variety of other disciplines. These works provide references to hundreds of other papers, many published in peer-reviewed technical journals.[8]

Most of the statistical applications in the research by physical scientists and engineers are descriptive statistics characterizing data from experiments in terms of means and standard deviations with a modicum of hypothesis testing and estimation. Many of the experiments are clever and use diverse apparatus in laboratories at the USGA and at leading universities and golf equipment manufacturers around the world. The results are a cornucopia of information about many physical and mental aspects of the game. The detailed descriptive knowledge is impressive. Using this knowledge to improve play is quite another matter. There is a profound chasm between intellectual understanding of what one should do in the golf swing and doing it. The search for the perfect swing and, perhaps more important, how any significant proportion of the human race might learn it and acquire the capacity to use it when they want to is surely ongoing.

6.3.2 What should the basis be for a handicap that will yield fair competitions in several formats among players of differing ability?

The most mature area of statistical analysis in golf is found in the work on handicapping. The handicap system administered by the United States Golf Association (USGA) is a serious analytic endeavor. The primary purpose of

---

[8] An earlier, comparable compendium is Cochran (1990).

this system is to provide a basis for fair matches among golfers of varying ability. In 1971, a Professor of Mathematics at Boston University and avid golfer, Francis Scheid, published a landmark paper in *Golf Digest*, 'You're not getting enough strokes,' taking the USGA to task for flaws in their handicap system that seriously disadvantaged less capable players in competition. The handicap system and the analytic work behind it are much too complex to describe fully in this review. Scheid's initial work modeled 50 players from their 20 most recent rounds and used these models to compute handicaps and to simulate (using an IBM-360!) some 500 000 matches. He also addressed the problem of courses of differing difficulty with data on 6000 rounds on five different courses. The biases in the handicap system of the time against high handicappers and the importance of course differences in terms of scoring difficulty were obvious.

Scheid's conclusions and prescriptions launched a sophisticated and complex research program on handicapping that led to a succession of significant changes in USGA procedures beginning in 1976. Gordon Ewen and Dean Knuth were key collaborators in this work, much of which is in unpublished technical reports to the USGA.

Currently, the individual handicap is based on a 'handicap differential' computed for the best ten of each player's last 20 Adjusted Gross Scores:[9]

Handicap differential

$$= \text{(Adjusted gross score} - \text{USGA course rating)}(113/\text{USGA slope rating)}$$

where:

'Adjusted gross score' is the reported round score. The 'adjustment' is to reduce very high scores on particular holes; the limiting score that one can take depends on one's handicap.

'USGA course rating' is the evaluation of the playing difficulty of a course for scratch golfers under normal course and weather conditions. It is expressed as strokes taken to one decimal place, and is based on yardage and other obstacles to the extent that they affect the scoring ability of a scratch golfer.

'USGA slope rating' is the USGA's mark that indicates the measurement of the relative playing difficulty of a course for players who are not scratch golfers. A golf course of standard playing difficulty would have a USGA slope rating of 113. The lowest USGA slope rating is 55 and the highest is 155.[10]

To derive the actual handicap, one then (1) averages the ten handicap differentials; (2) multiplies this average by 0.96; and (3) deletes all numbers after the tenths digit (no round-off).

[9] The following description is taken from the USGA Handicap Manual, Section 10 which can be found at (www.usga.org/handicap/manual/c2-13.html).

[10] The Men's USGA slope ratings = 5.381 × (Bogey course rating − USGA course rating), where the bogey rating is the evaluation of the playing difficulty of a course for the bogey golfer under normal course and weather conditions. It is based on yardage, effective playing length and other obstacles to the extent that they affect the scoring ability of the bogey golfer. Bogey rating is equivalent to the average of the better-half of a bogey golfer's scores under normal playing conditions.

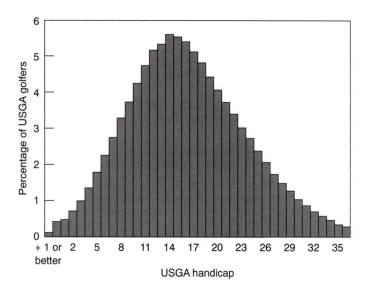

Figure 6.2   Distribution of USGA handicaps. Source: USGA, 1997 (www.usga.com).

The resulting distribution of handicaps from this procedure for all golfers with a USGA handicap is shown in Fig. 6.2.

In case you were wondering, the professional golfers on the PGA TOUR, when handicaps have been calculated for them,[11] range from 0 to +7; they represent something less than a tenth of one percent of all players in terms of scoring ability. Those in the 0 to +3 range don't make much money and don't stay on tour very long.

There are other adjustments, such as those for players whose tournament scores are better than their casual rounds, and for different forms of competition where players get some percentage of their full handicap index or where handicap strokes are given to one or another member of a competing team. Ultimately, the handicap determination for each player and the application of those handicaps to various competitions is left to a Handicap Committee associated with each Golf Club.

While the work on handicapping has made great strides, it remains a work in progress. The business of creating fair matches in an endless variety of possible competitive formats is very complicated. Further modifications to the system are proposed for 1998.

6.3.3   Which professional player has been 'best' in some historical period for some set of competitions?

Most of the work on the myriad possible forms of this question has been done by journalists and the minions of the various golf organizations. They are, almost without exception, statistically challenged and, frequently, arithmetically

---

[11] The top echelon professionals are difficult to handicap precisely because their performances are on courses generally set up to be longer and more difficult than those amateurs experience.

Table 6.2   Career Money Leader list of the PGA TOUR at the end of the 1997 season

Rank	Player	Career money ($)
1	Greg Norman	11 910 518
2	Tom Kite	10 286 177
3	Fred Couples	8 885 487
4	Nick Price	8 794 431
5	Mark O'Meara	8 506 774
6	Davis Love III	8 470 982
7	Payne Stewart	8 465 062
8	Tom Watson	8 307 277
9	Corey Pavin	8 130 356
22	Hale Irwin	5 902 306
23	Bruce Lietzke	5 880 083
24	Brad Faxon	5 842 619
25	Tom Lehman	5 642 999
26	Loren Roberts	5 624 104
27	Jack Nicklaus	5 563 516
65	Lee Trevino	3 478 328
83	Tiger Woods	2 857 427
124	Arnold Palmer	1 904 668
142	Billy Casper	1 691 583
259	Sam Snead	620 126

challenged. Most of the attempts to define 'best' for any period, but particularly for the long historical haul, have taken the form of rankings with both subjective and objective components.

Probably the most foolish and offensive statistic in all of golf is the Career Money Leader list of the PGA TOUR. Excerpts of the list at the conclusion of the 1997 season are given in Table 6.2.

This is the only list, aside from the simple count of tournament victories, that the PGA TOUR offers as a summary of career accomplishments. The victory counts favor the older players who played in eras when fields were less strong and deep and wins were easier to achieve. The career earnings favor the excellent player nearing the end of his career. The future belongs to the young who will play for purses growing geometrically.[12] If Tiger Woods follows the modal professional career profile of peaking at age 30 to 35 (Berry and Larkey, 1998), he will reach the top of this career earnings list in about five to seven years (depending on the performances of others and the rate of growth in purses) and head it for 10 to 15 years before being superseded by a younger player who has played for a lot more money on average over his career.

The PGA TOUR badly needs a statistic of career accomplishment that has the names of Nicklaus, Hogan,[13] Snead, Palmer, Player, Casper, Trevino and Watson in some plausible order at the top of the list. Let us argue statistically about Nicklaus versus Hogan as the best of all time, not Nicklaus versus Roberts who surpassed Nicklaus in career earnings this year but trails Jack

---

[12] The Commissioner of the PGA TOUR, Tim Finchem, recently announced the goal of getting all tournament purses to the $3 500 000 minimum level over the next few years. This is possible, in part, because of a 50% increase in television revenues in the new five-year contract.

[13] Ben Hogan is excluded from the Career Money Leader list because he retired before 1965 – the PGA TOUR's arbitrary cut for inclusion although they have the earlier data in hand.

by 18 Major professional victories (18 to 0) and 66 tour victories (70 to 4). The problem is not just that great players are denied their rightful place in history but that the PGA TOUR actually uses this measure to determine eligibility for the Senior PGA TOUR. In doing so it does a great injustice to the players whose careers did not happen to coincide with inflated tournament purses.

Sam Snead, one of the greatest all-time players, has nominal dollar Career Earnings of \$620 126. In 1998 or 1999 a rookie who has the week of his life for four rounds in a big event may win more than Snead did in his career and pass him on the Career Earnings list. David Duval accomplished this feat in 1997 with his victory at the TOUR CHAMPIONSHIP; his \$720 000 payday there, his third consecutive victory, vaulted him into 58th place on the all-time Career Earnings list with \$3 815 010. No one of sound mind and even the slightest information about the history of professional golf believes that Duval (or virtually all of the 258 players ahead of Snead on the Career Earnings list) has been a better player than Snead over a career; Snead won 81 official PGA TOUR events including six Majors and 135 titles worldwide, and was at or near the top of the annual money list for over 20 years. The mystery is not who is the better player but why the PGA TOUR persists in accumulating and reporting this measure and using it to determine eligibility for both the regular PGA TOUR and the Senior PGA TOUR. The PGA TOUR statisticians also carefully track and report such measures as the 'Most money won in a single season' (also 'by a rookie', 'by a second-year player', 'in first two seasons'), 'Most consecutive years \$100 000 (also \$200 000, \$500 000, \$1 million) or more'. It will come as no surprise that the holders of these records turn over with great rapidity, often annually.

A simple solution to these difficulties is to compute 'Quality points', the ratio of a player's prize to the winner's prize (Larkey, 1990a). Table 6.3 applies this to the data from Table 6.1.

This base outcome measure has several desirable characteristics. It captures the within-tournament incentive schedule for players. It is normalized on the 0–1 interval and transportable across tournaments and time; indeed, it makes it possible to capture the performances of amateurs (what they would have won) and professionals in all formats including matchplay, where stroke information is never available. It has a natural interpretation with respect to place finishes in tournaments.

The most prominent and credible ranking system is the Official World Golf Ranking. This system has been endorsed by all the major organizations in

Table 6.3 Quality points for leaders in THE PLAYERS championship (ending: 30 March 1997)

Player	Prize money ($)	Quality points
Steve Elkington	630 000.00	1.00
Scott Hoch	378 000.00	0.60
Loren Roberts	238 000.00	0.38
Brad Faxon	168 000.00	0.27
Billy Andrade	140 000.00	0.22
Tom Lehman	126 000.00	0.20
Mark Brooks	109 083.34	0.17
Colin Montgomerie	109 083.33	0.17

Table 6.4   Official world golf rankings at the end of the 1997 season

Rank	Player	Country	Pts Avg.	Tot. Pts	No. of Evts
1	Greg Norman	Aus	11.78	542	46
2	Tiger Woods	USA	10.58	455	43
3	Ernie Els	SA	9.53	553	58
4	Nick Price	Zim	9.30	409	44
5	'Jumbo' Ozaki	Jpn	8.86	390	44
6	Colin Montgomerie	Sco	8.66	511	59
7	Davis Love III	USA	8.45	431	51
8	Mark O'Meara	USA	8.20	377	46
9	Phil Mickelson	USA	8.08	404	50
10	Tom Lehman	USA	7.83	423	54

golf as the basis for deciding player eligibility to enter particular tournaments. The Official World Golf Ranking is issued weekly, updated for results from tours in the US, Europe, Australasia and Japan, and Asia and South Africa.[14] It is a point system, with the number of points given according to players' finishing positions in events around the world.[15] The points available for an event depend on a calculation of relative field strength based on the number and ranking of the players entered. The system treats four Majors and the US PLAYERS Championship separately, giving them more weight.[16] There are, however, point minimums for winners (6 for Asia and South Africa, 8 for Australasia and Japan, and 10 for Europe and the United States); selected other tournaments have higher minimums (32 for the Volvo PGA Championship and 16 for the Australian and Japan Opens) 'to reflect their status'.

To get the World Ranking:

> points for each player are accumulated over a two year 'rolling' period with the points awarded in the most recent 52-week period doubled. Each player is then ranked according to his average points per tournament, which is determined by dividing his total number of points by the tournaments he has played over that two-year period. There is a minimum requirement of 20 tournaments for each 52-week period. Points are reduced proportionately for tournaments curtailed to 36 or 54 holes because of inclement weather or other reasons.

The list at the conclusion of the 1997 US season is given in Table 6.4.

The criteria for evaluating such ranking systems are fairly vague. There is, of course, no unimpeachable source of correct rankings against which one might

---

[14] The description of the Official World Ranking is based on the information at www.golfweb.com/ga97/pga/ranking/.

[15] Point systems are very popular in golf. The eligibility of US players to play in the two big biannual international team events, the Ryder Cup and the President's Cup, is determined by a point system. Point assignments are, to a large extent, arbitrary.

[16] 'The winners of the Masters Tournament, United States Open, Open Championship and PGA Championship are awarded 50 points (30 points for second place, 20 for third, 15 for fourth down to a single point for a player completing the final round), and the winner of the PLAYERS Championship is awarded 40 points' (www.golfweb.com/ga97/pga/ranking/).

directly test. *Face validity* is important. The resultant rankings from a system must be plausible to knowledgeable observers. If a historical ranking system produces rankings that are obviously wrong in the sense that a significant majority of knowledgeable observers would strongly disagree with the ordering, they are not credible. *Logical process* in ranking is important. Is the procedure used to construct the ranking logically consistent and non-arbitrary? *Predictive validity* is important for ranking systems where contests among the ranked follow. If the rankings from any system do not predictively correspond more closely to the contest orderings than rankings by other systems, including naive models, then the ranking system is not credible.

Implicit in the use of these rankings, to determine eligibility for tournaments and teams, is an objective of picking out the 'best' players in terms of their expected performance. Exemption from past performances rather than expected performance occurs too often. Ian Baker-Finch's smooth 92 in the first round of the 1997 British Open, playing on his 10-year exemption from having won the 1991 version, and Doug Ford's sporadic appearances on the PGA TOUR, based on his lifetime exemption from Major wins in an earlier era, are only two examples.

The predictive power of the Official World Ranking (and various other less ambitious and less official systems) has never been tested. Face validity is a problem. Jumbo Ozaki is not the fifth best player in the world; he is not even in the top 25 and maybe not even in the top 50.[17] This anomaly reflects the politics of getting worldwide official endorsement for a ranking system and the mechanics of the minimums. The logical process is also a problem. There are too many wholly arbitrary factors in weighting performances, most notably the minimums that give excessive weight to success in events with very weak fields. The significance of the points, pretentiously taken to two significant digits, is not clear. Do they mean that Tom Lehman is 66.4% as good as Greg Norman? What is the maximum possible on this scale? Was Hogan in 1953 with five victories, including three Majors, in six starts about a 45 on a World Ranking scale not eligible because he entered too few tournaments?

Many other ranking systems in professional golf are questionable. For example, the Top Ten Finishes ranking by the PGA TOUR ranks players by a simple count of the number of times they finished in the top ten (or tied for tenth place) in any tournament. All ten places including ties for tenth count 1, everything else counts 0, and there is no regard for how many tournaments were entered. The winner of the modal professional tournament wins 18% of the total purse; tenth place wins 2.7% and eleventh place wins 2.5%. The persistence of Top Ten Finishes as one of the most popular statistics about player performance surely corroborates our worst fears about what innumerate media will attempt to simplify things and about how gullible innumerate fans are. Systems with arbitrary point assignments are used to pick participants for the Ryder and President's Cup teams. There is, in short, a lot of room for improvement in golf's ranking systems.

---

[17] Should anyone associated with the design or operation of this ranking system believe strongly that the ranking of Ozaki is correct and that he has, therefore, an even or better than even chance to beat the five players that follow him on the list in their next common Major, please raise a lot of money and call Pittsburgh; your wagers will be covered.

6.3.4   What will professional players' performances be in the next competition or some larger set of upcoming competitions?

Predicting the performance of professional golfers has not received any serious analytic attention. There are a number of web and fax tip services to which one can subscribe for $29.95 a year and up. To my knowledge, none of these prediction services has much of an analytic basis nor have they been rigorously evaluated for accuracy. Since the advice takes the form of odds rather than point predictions, all of the usual difficulties in defining 'accuracy' are present. Bookmakers and golf magazines also provide odds on individual players before Major tournaments. Many of these 'predictions' have the curious property of implicit 'fair bet probabilities' that sum to considerably more than one (Larkey, 1990b).

6.4   Motivations and opportunities

It is not always easy to get our less athletically-minded academic colleagues (or our less academically-minded athletic colleagues) to take analytic work on sports seriously. This issue should be addressed directly. One compelling reason for studying golf is that it is a relatively simple domain containing theoretical phenomena and methodological issues that occur in other, ostensibly more serious, domains such as business, education, medicine or government. For example, comparing golfers who play in non-random subsets of N tournaments is an isomorph for the problem of comparing students who choose non-random subsets of N classes. The use of the unadjusted Grade Point Average as the numeraire performance measure for students, treating all nominally equivalent grades as equally significant performance accomplishments, has been the source of much mischief in higher and, to a lesser extent, in secondary education, where students have less latitude in choosing courses. It gives incentive to students to avoid difficult instructors and hard courses, contributing to a gradual erosion of standards (e.g. difficult electives have trouble attracting students) and to grade inflation. The golfer isomorph is more easily perceived and the golf data are more easily obtained.

In one interesting use of golf competitions to analyze an important substantive issue, Ronald G. Ehrenberg and Michael L. Bognanno (1990) studied the incentive effects of purse levels in golf tournaments on player performance. From an econometric analysis (Probit and regression) of selected 1984 tournament data,[18] they find that higher prize levels induce more effort/concentration and lower scores from competitors. This is consistent with a common-sense, albeit highly formalized, postulate in the field of Labor Economics that individuals will respond to the prospects of higher compensation with greater effort, and that performance is strongly, positively correlated with levels of effort. Ironically, this result is almost certainly wrong because, as all serious

---

[18] Their data were taken from sources that only have results for players making the cut. They also eliminated the British Open on the pretext that many top players, notably from the US, do not enter. A few US players, including a couple of top players, chose not to enter in 1984, but the field was probably stronger than most, if not all, of the US tournaments because of the presence of better foreign players.

golfers know, trying harder at golf is usually counterproductive. Relaxation is essential to excellent performance and most players, even most professionals, find an inverse relationship between the size of the stakes and their ability to relax and perform well. The players who have regularly 'risen to the occasion' (Jack Nicklaus comes immediately to mind) are few and far between (and rich and famous). Many, if not most, golfers, including professionals in the top echelon, have great difficulty in controlling their concentration and nervous systems in executing shots under extreme pressure – when the stakes are greatest – that would be fully automatic for them in normal circumstances. The relationship between performance and the magnitude of the stakes needs careful re-examination. Golf is not the most promising domain for confirming the conventional wisdom in labor economics.

The data are rich and the issues plentiful in both amateur and professional golf. Some of these theoretical and methodological issues may turn out to be at least as important as the nth marginal contribution to heavily studied, more traditional areas.

Bibliography

Barkow, A. (1989) *The History of the PGA TOUR*. PGA TOUR, New York: Doubleday.

Belkin, D. S., Gansneder, B., Pickens, M., Rotella, R. J. and Striegel, D. (1994) Predictability and stability of Professional Golf Association Tour statistics. *Perceptual and Motor Skills*, **78**, 1275–1280.

Berry, S. M. and Larkey, P. D. (1995) Picking a better ball partner, 7 pp., unpublished.

Berry, S. M. and Larkey, P. D. (1998) The effects of age on the performance of professional golfers. In Cochran, A. J. and Farrally, M. R. (eds.) *Science and Golf III*. London: E. and F. N. Spon.

Caulkins, J., Barnett, A., Larkey, P., Yuan, Y. and Goranson, J. (1993) The on-time machines: some analyses of airline punctuality. *Operations Research*, **41**, 710–720.

Caulkins, J., Larkey, P. D. and Wei, J. (1996) Adjusting GPA to reflect course difficulty, 19 pp., Working Paper, H. J. Heinz III School of Public Policy and Management, Carnegie Mellon University.

Clarke, S. R. and Rice, J. R. (1995) How well do golf courses measure golf ability? An application of test reliability procedures to golf tournament scores. *Australian Society for Operations Research*, **14**(4), 2–11.

Cochran, A. J. (ed.) (1990) *Science and Golf: Proceedings of the First World Scientific Congress of Golf*. London: E. and F. N. Spon.

Cochran, A. J. and Farrally, M. R. (eds.) (1994) *Science and Golf II*. London: E. and F. N. Spon.

Cochran, A. J. and Farrally, M. R. (eds.) (1998) *Science and Golf III*. London: E. and F. N. Spon.

Cochran, A. J. and Stobbs, J. (1968) *The Search for the Perfect Swing*. New York: J. B. Lippincott.

Davidson, J. and Templin, T. (1986) Determinants of success among professional golfers. *Research Quarterly for Exercise and Sport*, **57**.

Ehrenberg, R. G. and Bognanno, M. L. (1990) The incentive effects of tournaments revisited: evidence from the European PGA Tour. *Industrial and Labor Relations Review*, **43**.

Ehrenberg, R. G. and Bognanno, M. L. (1991) Do tournaments have incentive effects? *Journal of Political Economy*, **98**, 1307–1324.

Ewen, G. H. (1978) What the new multi-ball allowances mean to you. *USGA Golf Journal*, January/February.

Hale, T. and Hale, G. (1990) Lies, damned lies and statistics in golf. In Cochran, A. J. (ed.) *Science and Golf: Proceedings of the First World Scientific Congress of Golf*. London: E. and F. N. Spon.

Hale, T., Harper, V. and Herb, J. (1994) The Ryder Cup: an analysis of relative performance 1980–1993. In Cochran, A. J. and Farrally, M. R. (eds.) *Science and Golf II*. London: E. and F. N. Spon, pp. 205–209.

Heiny, R. L. and Crosswhite, C. E. (1986) Best-ball events in golf: an application of the multinomial distribution. *The American Statistician*, **40**, 316–317.

Jones, R. (1990) A correlation analysis of the Professional Golf Association (USA) statistical rankings for 1988. In Cochran, A. J. (ed.) *Science and Golf: Proceedings of the First World Scientific Congress of Golf*. London: E. and F. N. Spon.

Knuth, D. (1995) Tournament point system: a system for making tournaments equitable. Published at www.usga.org/handicap/.

Knuth, D. (1996) How well should you play? Published at www.usga.org/handicap.

Knuth, D., Scheid, F. and Engel, F. P. (1994) Outlier identification procedure for reduction of handicaps. In Cochran, A. J. and Farrally, M. R. (eds.) *Science and Golf II*. London: E. and F. N. Spon, pp. 228–233.

Larkey, P. D. (1990a) A batting average for tour golfers. *Golf World*, 14 December. (Reprinted in *Australian Golf Digest*, 1991, pp. 108–111 and modified version, Who's the Real No. 1?, published in *Golf Weekly Illustrated*, **3**(4), 31 Jan–6 Feb 1991, UK, pp. 20–23.)

Larkey, P. D. (1990b) Fair bets on winners in professional golf. *Chance*, **3**(4), 24–26.

Larkey, P. D. (1991a) A better way to find the top scorer. *Golf World*, 11 January, pp. 73–74.

Larkey, P. D. (1991b) A distinction that money can't buy. *Golf World*, 25 January, pp. 112–113.

Larkey, P. D. (1991c) Taking the measure of top ten finishes. *Golf World*, February 1.

Larkey, P. D. (1991d) How to measure strength of field. *Golf World*, 29 March, pp. 46–47.

Larkey, P. D. (1992) All the numbers back up the bear. *Golf World*, 3 April, pp. 83–86.

Larkey, P. D. (1994) Comparing players in professional golf. In Cochran, A. J. and Farrally, M. R. (eds.) *Science and Golf II*. London: E. and F. N. Spon, pp. 193–198.

Larkey, P. D. and Smith, A. A. Jr. (1998) All around improvements. In Cochran, A. J. and Farrally, M. R. (eds.) *Science and Golf III*. London: E. and F. N. Spon.

Mosteller, F. and Youtz, C. (1992) Professional golf scores are Poisson on the final days. *1992 Proceedings of the Section on Statistics in Sports*, American Statistical Association, pp. 39–51.

Mosteller, F. and Youtz, C. (1993) Where eagles fly. *Chance*, **6**(2), 37–42.

O'Muircheartaigh, I. G. and Sheil, J. (1983) Fore or five? The indexing of a golf course. *Applied Statistics*, **32**.

Riccio, L. J. (1990) A statistical analysis of the average golfer. In Cochran, A. J. (ed.) *Science and Golf: Proceedings of the First World Scientific Congress of Golf*. London: E. and F. N. Spon.

Riccio, L. J. (1994) The ageing of a great player; Tom Watson's play. In Cochran, A. J. and Farrally, M. R. (eds.) (1994) *Science and Golf II*. London: E. and F. N. Spon, pp. 210–215.

Scheid, F. (1971) You're not getting enough strokes. *Golf Digest*, June (reprinted in *The Best of Golf Digest*, 1975, pp. 32–33).

Scheid, F. (1972a) A least-squares family of cubic curves with an application to golf handicapping. *SIAM Journal on Applied Mathematics*, **22**(1), 77–83.

Scheid, F. (1972b) A basis for golf handicapping. *SIAM Conference*, Austin, Texas, 15 pp.

Scheid, F. (1973) Does your handicap hold up on tougher courses? *Golf Digest*, October, 31–33.

Scheid, F. (1975) A nonlinear feature of golf course rating. TIMS-ORSA Meeting, Kyoto, Japan, 10 pp.

Scheid, F. (1977) An evaluation of the handicap system of the USGA. In Ladany, S. P. and Machol, R. E. (eds.) *Optimal Strategies in Sports*, pp. 151–155.

Scheid, F. (1978) The search for the perfect handicap. *Proceedings of the Winter Simulation Conference*.

Scheid, F. (1979) Golf competition between individuals. *Proceedings of the Winter Simulation Conference*, 15 pp.

Scheid, F. (1980a) Simulating golf to find an accurate ability measure. *Simulation '80*, Interlaken, Switzerland.

Scheid, F. (1980b) How to measure golfing ability, current status. *EURO IV, the Fourth European Congress on Operations Research*, Cambridge University, UK, 14 pp.

Scheid, F. (1990) On the normality and independence of golf scores, with applications. In Cochran, A. J. (ed.) *Science and Golf: Proceedings of the First World Scientific Congress of Golf*. London: E. and F. N. Spon.

Scheid, F. (1994) The search for the perfect handicap. In Cochran, A. J. and Farrally, M. R. (eds.) *Science and Golf II*. London: E. and F. N. Spon, pp. 222–227.

Scheid, F. and Calvin, L. (1995) Adjusting golf handicaps for the difficulty of the course. *Proceedings of the Joint Statistical Meetings*, Orlando.

United States Golf Association (all documents listed are available at www. usga.org).

USGA Handicap Manual (www.usga.org/handicap/manual/index.html).

USGA Position paper on the equitable stroke control procedure (www.usga.org/handicap/esctest.html).

Wiseman, F., Chatterjee, S., Wiseman, D. and Chatterjee, N. S. (1994) An analysis of 1992 performance statistics for players on the US PGA, Senior PGA and LPGA Tours. In Cochran, A. J. and Farrally, M. R. (eds.) *Science and Golf II*. London: E. and F. N. Spon, pp. 199–204.

World Golf Ranking System (1997) www.golfweb.com/ga97/pga/ranking/.

7

Performance Indices for Multivariate Ice Hockey Statistics

Bill Williams and David Williams

Hunter College and The St Louis Blues, USA

7.1 Hockey on ice

It cannot be surprising that the origins of today's ice hockey game can be traced back to Canadian winters in the 19th century. Not only is much of Canada very cold during winter but, over a century ago, the country was much more agrarian; so the Canadian people were ready for a game to play on their abundant, frozen ponds.

In the mid-1800s, chaos was the best description of the game. The 'rules' were local and ad hoc. Games were sometimes played with dozens of players on each side! There is still disagreement about many aspects of the beginnings of early ice hockey. Nevertheless, in the 1880s, students at McGill University in Montreal and Queens University in Kingston began to play an organized game of hockey with consistent rules.

Shortly afterwards, leagues were formed in both Ontario and Quebec. In 1883, Baron Stanley, the governor-general of Canada, purchased a 'cup' for less than $50 to be awarded to the amateur champions of Canada. A team from the Montreal Amateur Athletic Association was the first winner. From that day to this, winning the Stanley Cup, currently awarded to the champions of the National Hockey League (NHL), has represented the pinnacle of hockey achievement. Nearly all players, at some time, envision their names engraved on the Cup as champions of all of hockey. Over the years, it has been to Russia, endured a trip to the bottom of the St Lawrence River, and had a vast mix of liquids consumed from it.

The early players were hardy men. Tough physical play was then, as it is now, part of the game of ice hockey. This has drawn criticism, on occasion, but the reluctance to remove this aspect of the game has a very long history. An Ottawa newspaper, describing the first Stanley Cup championship game in 1884, commented that, 'general rabble predominated'. A later remark, attributed to Conn Smythe, an early NHL mogul with the Toronto Maple Leafs, 'If you can't beat 'em in the alley, you can't beat 'em on the ice', has not been forgotten. The game is still not for the faint of heart. In all fairness, the sport does not have a more physical history than some other sports, especially if fan contributions are considered!

The National Hockey League, the world's premier league, formed on 22 November 1917 with four teams. Toronto was the only team that played on artificial ice! Until the 1960s, for the most part, the NHL consisted of six teams: Boston, Chicago, Detroit, Montreal, New York (Rangers) and Toronto. Most of the early players were Canadian. In 1966, the league doubled in size and since that time has been the fastest growing professional sport. This season, 1997/98, the NHL has 26 teams and has announced future expansion to 30 teams.

While ice hockey has grown quickly in North America, it has grown even more rapidly elsewhere, particularly in Europe. There are now excellent European leagues; especially in Finland, Sweden and Russia. Many countries now enter ice hockey teams in the winter Olympics and produce outstanding hockey players able to play in the NHL. The availability of this large pool of players (in combination with the infusion of television money to lure them from their homelands) has been a major factor in the expansion of the NHL.

Additional details of the history of the NHL can be found in the *National Hockey League Official Guide and Record Book, 1996–97* (The National Hockey League, 1996).

7.2   A review of the statistical research

Applications of the methods of mathematical statistics to hockey data have been scarce. This appears to be changing, perhaps paralleling the increasing popularity of hockey. Twenty years ago, Morris (1973) and Mullet (1977) provided two of the earliest applications. Mullet showed that the goals scored by and against teams in the NHL are surprisingly well described by Poisson distributions and, even more surprisingly, that the goals for and against a team seem to be independent. He used these results to predict game outcomes quite accurately. Recently, Lock (1997) applied a variation of this Poisson model technique to college hockey and was also able to predict game outcomes accurately. Danehy and Lock (1993) used regression models to develop ratings of the National Collegiate Athletic Association (NCAA) Division I Men's Ice Hockey teams. A rating system for individual players is developed in the later sections of this chapter.

As in other sports (soccer and American football, in particular), the method of resolving games tied at the end of regulation time has been a source of controversy. Even today, various methods co-exist. During the regular season, the NHL plays a single 'sudden-death' overtime period that ends as soon as one goal is scored; if no goal is scored within five minutes, the game ends in a tie. However, during the NHL playoffs, there are no ties; a game continues until one team scores the winning goal. The International Hockey League (a North American league) and most international competitions, including the Olympics, use a different method called a 'shoot-out'. In a shoot-out, each team, in turn, sends a player to attack the other team's goal which is guarded only by the goalkeeper. Goals scored in a shoot-out determine the winning team. Morris (1973) began the statistical analysis of the different methods for settling ties that was only recently resumed by Liu and Schutz (1994) and Hurley (1995). These studies find that the stronger team does tend to win the game, but Hurley also found that the shoot-out gives the weaker team a better chance to win. Liu and Schutz found that doubling the current five minute overtime would settle 60% to 65% of the games tied at the end of regulation time. Twenty minutes of overtime would settle 80% to 85%.

Fans do not like ties and seem to find the shoot-out exciting; players do not like ties either, but they also feel that shoot-outs often lead to random outcomes. In an attempt to satisfy both groups, Hurley analyzed some interesting combinations of sudden death and the shoot-out to find the means and variances of expected game durations. Arguments for and against the shoot-out are still dominated by purist arguments such as 'real hockey games are settled by team play!' In this issue and others, hockey officials so far seem uninfluenced by analytic arguments about the game; the last NHL rule change in settling ties was in 1983, when the five-minute overtime was introduced. Hopefully, this will change soon in light of promising new research. For example, an innovative new method for measuring power play and penalty-killing efficiency has considerable intuitive appeal (Anderson-Cook and Robles, 1997).

One of the authors (David Williams) played in the NHL for four years. Applying his experience, we addressed two statistical issues uppermost in player interest and concern. A salary study (Williams and Williams, 1997) showed that the salaries of NHL players are strongly related to the countries of their origin. We argued that the league options that players have within their own countries are better in some countries than others. The result is that more money is required to lure the players with the better options into the NHL. A curious, tangential finding of this salary study indicated that NHL talent scouts, as they roam the world searching for ice hockey talent, are very consistent in their draft selections.

The plus/minus statistic concerns players; in fact, it frequently annoys them. A player's plus/minus is the number of goals scored by his team minus the number scored against his team while he is on the ice. Consequently, the reason for player concern is that this statistic depends on much more than a player's own performance; it also depends heavily on the way that each player is used by the coach and the general performance of the player's team. In another study, we demonstrated how a player's plus/minus should be compared from team to team (Williams and Williams, 1996). Unfortunately, current data do not allow adjustment of this statistic for the different ways in which team coaches use their players.

7.3 Statistical performance evaluation in the National Hockey League

Hockey is very much a team sport and, as such, does not lend itself easily to numerical evaluation of individual players. In varying degrees, every player statistic is correlated with the quality of the player's team. On a weak team, even the most talented player will have difficulty scoring goals, and every regular player on the team is likely to have a negative plus/minus. This is discussed more completely later. Nevertheless, many statistics on individual player performance are gathered, published and analyzed.

The statistics gathered on the teams and players in the North American hockey leagues are quite similar. However, the interpretation of these statistics is not always the same. The Quebec League, for aspiring junior players 18 to 21 years old, is considered to be a wide-open high-scoring league. In this league, a young player would be well advised to score a large number of goals in order to draw attention. In contrast, the Western League, another junior league, is considered to be physically tough and so a hopeful player needs to make sure that his statistics indicate an 'appropriate' response to this particular aspect of the game (e.g. a high number of penalty minutes). This chapter focuses on the NHL in which the world's best players currently perform.

While hockey writers focus on their favorite statistics and publishers print books filled with them, an interesting question remains outstanding: 'How do NHL players and teams use performance statistics internally?' Perhaps more than the public realizes, individual players often have statistical targets written into their contracts. These targets may be specific to the player or may be based on team performance. A defenseman may receive a bonus for games in which the team's opposition is held to one goal or less, or if he is elected to the league All-Star team. A forward may receive a bonus if he scores a stated number of goals during the season. There are many different bonus clauses in player contracts.

Team 'segment' bonuses, which are not part of the players' contracts, are common. All NHL teams use them and some minor league teams have them too. These bonuses are given if a team achieves specified statistical goals during the segment, which is usually five games but is sometimes ten. To illustrate, all players may receive a cash bonus if the team wins a stated number of games, or perhaps if the team holds the opposition below a stated number of goals during the segment. These segment bonuses are used to influence team play, particularly to keep the team playing consistently during the long season.

Salary arbitration is another important internal use of statistics. Arbitration is a defined procedure for settling salary disputes in the case of disagreement between a player and his team. During these negotiations, the use of statistics is adversarial and is not particularly 'academically' oriented. While a player's statistics will be part of the agenda, the actual meeting is a legal one dominated by lawyers and legal protocols. How statistical analyses affect an arbitrator in the privacy of his own deliberations is speculative, but the analyses may well miss the mark if they are overly complex.

The players have an additional problem regarding the use of statistics, sometimes they cannot find out what they are! Generally, the teams gather the game performance data and are not necessarily quick to share them with the players. Some players in the NHL have performance bonuses for total on-ice time during the season. This is not an easy statistic to keep and sometimes players with such bonus clauses have found that their teams are not eager to keep them informed of their progress. The NHL teams keep many statistics that are not systematically shared with the public or players. In one arbitration meeting, a player was surprised to find that he was not scoring enough goals during the last two minutes of games!

How do the coaches use statistics? This varies considerably, even on the same team during the same year. This is particularly true of the plus/minus. Sometimes, players are told not to be concerned about lower plus/minuses and at other times they are criticized for them (Williams and Williams, 1996).

The remainder of this chapter introduces a methodology to evaluate individual players based on multivariate statistical methods. Using this evaluation, it is interesting to observe and speculate about the fact that player salaries are highly correlated with their offensive points, but are not correlated at all with any measure of defensive play (Williams and Williams, 1997).

7.4 Did Wayne Gretzky have a poor season in 1992/93?
A question of multiple player statistics

In sports, writers, fans and even team managements all evaluate players with their favorite statistics. Disagreements about the importance of the various statistics are common. After more than 100 years, baseball fans still argue about the significance of slugging percentages and batting averages. This kind of dispute is common to most sports and hockey is certainly not an exception. In hockey, goals scored by a player may be the most important statistic because, in the last analysis, scoring goals wins games. On the other hand, many excellent goal scorers rely heavily on their play-making linemates to get the puck to them in advantageous positions – so the argument begins that assists are equally, or even more, important.

On any team, not every player has the same role and this may affect the interpretation and importance of the individual statistics. For NHL players who act as enforcers, the number of major fighting penalties is more important than goals. Any goals scored by them are simply bonuses to their teams. In NHL salary arbitration, players and management typically focus on very different statistics since their financial perspectives are quite different. For most players, the relative importance of an individual statistic is subjective.

Statistical analysis is eased by consistent interpretation of data (e.g. agreement that 'bigger' numbers are better, or worse, than 'smaller' ones). Unfortunately, in hockey, this is not always the case; some statistics are alternately viewed from different ends of the telescope. Further ambiguity is created by the fact that hockey is a team game and a player's statistics are affected by both the individual player's performance *and* that of his team-mates. As a result of this ambiguity, evaluation of hockey players has been dominated by subjective professional judgment, perhaps influenced by a small number of selected statistics.

The ambiguity that exists in the selection and interpretation of sports statistics, and hockey in particular, raises the question of whether the methods of mathematical statistics might be used to compress and summarize the multiple observations made on each NHL player into *meaningful* performance indices. This section describes an example of the development of indices that not only have useful interpretations, but also have the desirable property that they are totally data driven: they are not subjective. While there has been little use of the advanced methods of statistics in the analysis of hockey, these methods have been used in actual NHL salary arbitration cases. We shall not use such a case for illustration, but rather, will evaluate a major controversy of the 1992/93 National Hockey League season.

In 1992/93, Wayne Gretzky, one of the National Hockey League's all-time great players, scored 'only' 16 goals and 49 assists, well below his best season and even below his average season performance. Not surprisingly, Gretzky expressed dissatisfaction and speculated about retirement. Could it be true that this great player, then a member of the Los Angeles Kings, had suddenly slipped that much?

Although Gretzky is certainly the focus here, the methods described are general and can be applied to any hockey player in any league. The analysis demonstrates the strengths of exploratory data analysis (Tukey, 1977) and methods of multivariate statistical methodology in developing useful models with intuitive appeal.

7.5   Player evaluation with respect to individual variables

How should we evaluate hockey players? Since teams win games by scoring goals, it seems clear that scoring more goals is better for both a player and the team. Indeed, the number of goals scored by a player may be his most important, single statistic. So this evaluation of Wayne Gretzky's season starts by looking at goals scored by the individual 1992/93 Los Angeles Kings players. Because of the considerable variation in the data on players who appeared in fewer than 15 games, only players who appeared in more than 15 games during the season are included.

7.5.1   Goals

A ranking of Kings players in 1992/93 according to the number of goals scored shows Luc Robitaille at the top with 63 goals, followed by Granato and Carson with 37 goals each. Gretzky was ninth, well behind, with 16 goals. Even Rob Blake, a defenseman, scored as many as Gretzky; so considering goals alone, perhaps Gretzky's fears were well-founded.

However, while Gretzky does hold most of the NHL's goal-scoring records, his play-making ability may well be the outstanding feature of his game. In hockey, many goals are scored easily by a player after most of the effort has been supplied by a team-mate. In fact, some prolific goal-scoring players are considered to be one-dimensional, with perhaps a very accurate shot, but with little ability to move the puck into a scoring position. This is recognized in hockey and so up to two players handling the puck immediately prior to a goal may be credited with 'assists'. Some players excel at this aspect of the game and are highly paid for it; so to evaluate Gretzky properly, assists must be studied as well as goals.

7.5.2   Assists

A ranking of Kings players in 1992/93 according to the number of assists shows that, as with goals, Robitaille was at the top, Kurri was second, but now Gretzky has risen to third. So how should Gretzky's third place in assists be balanced with his ninth place in goals? In an overall evaluation, should he be ranked third on the team, or closer to ninth, as his goals would suggest? Since it is not clear how to give relative weights to goals and assists, it is not clear where Gretzky stands. Equal weighting would suggest that he be ranked sixth, but why equal weighting? And, even though we may agree that goals and passes that set-up goals are the most important plays, hockey fans certainly know that there is more to the game than just goals and assists.

7.5.3 Penalties

The number of penalties a player incurs is also a relevant statistic. Fans and coaches get upset with 'dumb' penalties, but not every penalty is considered 'dumb'. In a game of intimidation like hockey, some players are expected to get penalties, especially penalties associated with certain types of physical infractions. Popular writers (e.g. Dryden, 1994) have even developed performance indices which evaluate a player's penalties positively, the more penalties the better. General managers of NHL teams also often interpret penalty minutes positively, but it is difficult to make this argument for all players. Most penalties result in power plays for the opposition during which the penalized team must play short-handed. Opponent's goals are scored at a higher rate during power plays than when the teams play at even strength. On average, two-minute power plays result in goals scored by the opposition about 20% of the time. So 300 minutes in penalties against a player could result in as many as 30 goals for opponents. A team that receives many penalties is likely to lose games by doing so.

   This negative interpretation of penalty minutes may be unfair to the enforcers in the NHL who are paid to intimidate other players and break certain rules. In a study covering more teams, enforcers perhaps should be identified and studied separately, but this was not done here because only two players, Marty McSorley and Warren Rychel – who are not the focus of this analysis – would possibly be affected by it.

   A ranking of Kings players in 1992/93 shows that Gretzky had the smallest number of penalty minutes on the entire team. This is a strong positive for Gretzky because he obtained his goals and assists without putting his team in many short-handed situations. Still, we can hardly rank him as the number one Kings player on this basis alone. Not only that, but there are many more statistics available for exploration.

7.5.4 Plus/minus

Goals and assists are very important in hockey and high scoring players are paid well for it. But sometimes, teams score easy goals *against* another team's best offensive players because these players do not play well defensively. In fact, some offensive players appear to have very little interest in defending their own end of the rink. Since it takes good defense as well as good offense to achieve victory, hockey uses a statistic called 'plus/minus' which purports to measure a player's offensive versus defensive abilities.

   When a team scores an even strength goal, every player on the ice for the scoring team gets a 'plus' and every player on the ice for the other team gets a 'minus'. The accumulated net difference between a player's pluses and minuses is referred to as his 'plus/minus'. A negative plus/minus suggests that a player may not be paying attention to his defensive responsibilities and may not be valuable even if he is a high scorer. Unfortunately, interpretations of this statistic can also be ambiguous, because some players who are very good at defensive play are consistently used by coaches against the other teams' most offensively-skilled players. The result is that these players get few goals, but do collect minuses. Also, some defensemen are known to be very slow leaving the ice for player changes when the forwards on their team are in the middle of promising rushes into the other team's territory; this tactic tends to inflate their plus/minus ratings. Nevertheless, all things considered, a higher plus/minus is better than a lower one.

So how did the Kings do while Robitaille, Kurri and Gretzky were on the ice? A ranking of Kings players in 1992/93 with respect to plus/minus shows that Gretzky is eighth; Robitaille and Kurri are near the top, but behind whom? So the evaluation of Gretzky continues to be ambiguous – he was ninth in goals, third in assists, first in penalties and now we see that he was eighth in 'plus/minus'. So where we began with one statistic, goals, we now have four, all of which are potentially important candidates for integration into an overall evaluation of Gretzky's 1992/93 season.

7.5.5   Per season or per game?

Gretzky was injured in 1992/93 and played in only 45 games. In contrast, Robitaille played in all 84 games and, as a result, had much more playing time to assemble his point totals. How should their statistics be compared? How should we adjust the data for this playing time discrepancy? Or should we adjust at all? While injuries are common in hockey, a smaller number of games played during the season often reflects a player who is not really part of the regular line-up. Certainly, what any player's season would have been like had he played more games is completely a matter of speculation, and not necessarily a matter for simple extrapolation. In professional hockey circles, it is virtually standard to look at a player's performance statistics over a season, *without* regard for the reasons for playing in a limited fraction of the schedule. And, it must be admitted, there is a certain cold logic and fairness in this view. So initially, our analysis is based on season totals.

7.6   Player evaluation with respect to multiple variables

To overcome the ambiguity that results from examining a number of statistics one at a time, we need a methodology to assess the variables simultaneously in a single, comprehensive overview. To this end, it is useful to study first the variables two at a time rather than starting with all available statistics.

Figure 7.1 displays a log–log scatterplot of the number of goals scored by each of the Kings players against their assists. The top three Kings players, Gretzky, Robitaille and Kurri, are labeled in this and each of the subsequent plots. Robitaille is at the top right corner of Fig. 7.1 because he led the team in both goals and assists. In fact, Robitaille also led the team in shots and would also be in the top right of plots of shots against both goals and assists.

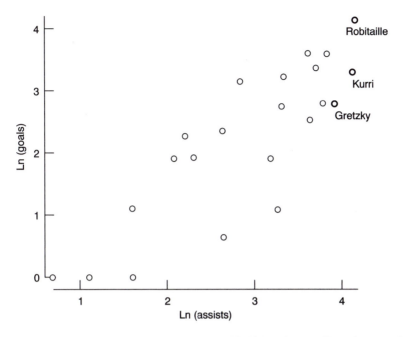

Figure 7.1 Goals versus assists: players on the 1992/93 Los Angeles Kings (redrawn from Williams, 1994).

So clearly on the basis of these offensive statistics, goals, assists and shots, Robitaille had, unambiguously, the best season.

The correlations between pairs of the three offense statistics above are all over 0.9. The high correlations between goals and shots (0.91) and assists and shots (0.95) are particularly interesting because the latter is persistently higher, suggesting that if a player shoots at the goal he is more likely to get an assist than a goal. While this fact does not seem to be widely known, it will not surprise Ray Bourque fans because the stellar defenseman often remarks that he shoots at the net expecting *rebounds*, not goals!

That goals, assists and shots are all very closely related is not surprising, because they all measure related aspects of offensive play. But it is exactly this type of strong relationship that can be exploited to reduce the many available statistics into useful performance indices. The following simple example illustrates this.

In Fig. 7.1, the high correlation between goals and assists suggests that the orthogonal axes be rotated so that one of the axes goes through the data points in the direction of the maximum spread of the points. The direction of this line then has maximum variability among the players and so has maximum discriminating ability among them. Distance, call it P, along this new axis (not shown) gives almost complete information on both goals and assists and so, in this sense, this direction (called the first principal component) may reasonably be described as the *single* piece of information contained in Fig. 7.1.

The equation of the new axis is a linear function of both goals and assists. Since P increases with an increase in goals and assists, which are both positive measures of performance, we can interpret larger P values as better on-ice performances. In this sense, P is an index of player performance which reflects a player's offensive production in both goals and assists.

The percentage variability associated with each of the axes can be measured; in this illustrative example, the first component explains 95% of the total variability. So in the event that we were really restricted to just goals and assists, the single dimension P would capture nearly all of the potential to distinguish among the players.

Any believable index should be strongly related to the input statistics which it purports to replace. In this example, the correlations of P with goals and assists are high, 0.821 and 0.896, respectively. So relatively little is lost by replacing the two statistics, goals and assists, by the single statistic, P. Furthermore, statistical theory shows that the weighting of goals and assists in P is proportional to these correlation coefficients, which means that assists are weighted slightly more heavily in P than goals.

Finally, if P is consistent with the fact that Robitaille was first in both goals and assists, Robitaille should be highest on the P index. He is. Kurri was second in assists and fifth in goals and Gretzky was ninth and third. How does P rank these players? The first six players, ranked by P, are in order: Robitaille, Kurri, Granato, Carson, Donnelly and Gretzky. Since P weights goals and assists almost equally, it is not surprising that the P ranking is the same as the ranking of players by total season points, goals *plus* assists.

The principal component procedure can be applied no matter how many different, original input variables exist at the outset. This is very important because it aids us in determining which variables are most important and enables us not to be seriously impeded by the complexity caused by including many on-ice statistics. With the use of many input variables, additional principal components are determined. Each component adds more information, but less than the preceding components. Formally complete descriptions of principal component analysis can be found in many statistics books, e.g. Johnson and Wichern (1992) and Searle (1982).

7.7   The complete statistical ice hockey player

7.7.1   Total statistics

The basic hockey statistics used to assess players' performances are: penalty minutes, plus/minus, and shots-on-goal along with goals and assists (which were both classified by whether they were obtained while the team was playing at even strength, with a power play advantage, or short-handed). Using these data, the principal components (as illustrated in the previous section) were calculated. The first principal component accounts for 67.7% of the total variability in the base statistics and the second component accounts for 17.6%, for a comfortable total of 85.4%. The remaining 14.6% of variability was spread over the remaining principal components. This effectively reduces the dimensionality of the data to two and, as we shall show, both of these components have very appealing hockey interpretations.

The highest correlations of the base statistics with the first principal component are goals (0.946), assists (0.954), shots (0.945) and power play goals (0.879).

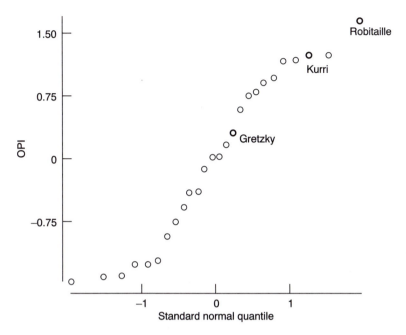

Figure 7.2   Normal probability plot of Offensive Performance Index (OPI): players on the 1992/93 Los Angeles Kings (redrawn from Williams, 1994).

Since the weight of each variable is proportional to these correlations, we know that the first principal component involves these variables with the heaviest weights. Each of these heavily weighted variables is a measure of offense. Consequently, we interpret the first principal component as an Offensive Performance Index (OPI).

As designed into the study, the better performing players will always appear towards the upper right-hand corner of the graphs. Not surprisingly then, in a normal probability plot of OPI (Fig. 7.2), Robitaille has the highest rating on offense, Kurri is third, but Gretzky's OPI rating places him 11th! Does this 'low' ranking of Gretzky counter reality, or did the Great One really have an off-year? For Gretzky fans, this result demands further study.[1]

Figure 7.3 shows a normal probability plot of the second principal component. In this dimension, Gretzky has the top score which raises the question of what aspect of the game is reflected in it. Why does Gretzky rank so high? The second principal component is most highly correlated with (negated) penalty minutes (0.894) and 'plus/minus' (0.488), which means that these are the variables most heavily weighted. Other variables receive much smaller weights. This component is an efficiency index which rates players highly whose playing style does not yield many points or scoring opportunities to

[1] A Normal probability plot for a set of values plots each value against the standard Normal quantile for its rank in the set. Points in a straight line indicate that the set of values follows a Normal distribution. Normal probability plots are not necessary simply to rank the players, but the resultant distributional observations are often very informative for making year-to-year and team-to-team comparisons. For that reason, the authors use them routinely.

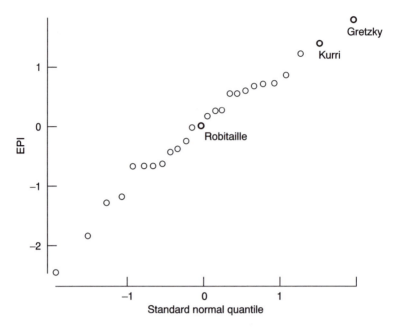

Figure 7.3 Normal probability plot of Efficiency Performance Index (EPI): players on the 1992/93 Los Angeles Kings (redrawn from Williams, 1994).

the opposing team. As a result, we have labeled this dimension, EPI, for Efficiency Performance Index. Gretzky was the clear team leader.

Figure 7.4 displays the offense index, OPI plotted against the efficiency index, EPI. (Keep in mind that OPI and EPI are uncorrelated.) Since the better performing players will have higher OPI indices and/or higher EPI indices, their plot points are towards the upper right corner of the figure. Gretzky is certainly in the upper right corner, although he is not rated highest in *both* dimensions. So, judged on the season as a whole, Gretzky was not the top Kings' offensive player, although he did have value as their player with the highest efficiency.

7.7.2   Per game statistics

The analysis so far compares the Kings players on the basis of *total* season statistics. So considering that Gretzky played in only 45 games, barely over half the season, the fact that Gretzky is near the top at all should be considered a big plus for him. But this does re-raise the question, how would the analysis turn out if repeated for the same variables, but rather on a per game basis? It seems reasonable to expect that the strength of Gretzky's known offensive ability will be apparent in such an analysis and will be reflected in a higher OPI rating. Simultaneously, there is no obvious reason to suspect that a per game analysis would change his top EPI rating.

Each of the input variables was divided by the number of games played by each player and the principal components recalculated. The results show that Gretzky still has the top efficiency rating, but counter to our earlier intuition,

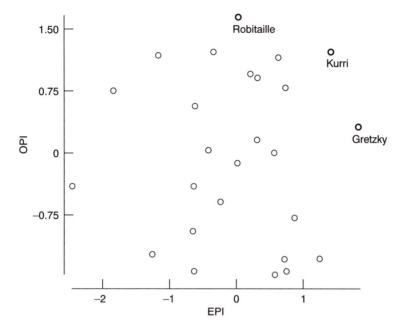

Figure 7.4 Offensive Performance Index (OPI) versus Efficiency Performance Index (EPI): players on the 1992/93 Los Angeles Kings (redrawn from Williams, 1994).

he moved up only slightly on the offense rating! Must we now admit that Gretzky had an off-year, and if we do not admit it, what reasons can be found to deny it? A review of the data reveals the extremely surprising fact that Gretzky scored *no* power play goals during the 1992/93 season. Since the number of power play goals is heavily weighted in OPI, how far did Gretzky drop in the offense ratings solely for this reason?

To determine more precisely the effect of Gretzky scoring no power play goals, this statistic was simply removed as an input variable and all the indices recalculated using the same definitions of OPI and EPI. Gretzky is rated highest on both OPI and EPI. In the complete year analysis, Gretzky has been clearly downrated by his failure to score any power play goals. Is this reasonable and fair? Regrettably (for Gretzky fans), this is a data driven study and to omit variables selectively is counter to the study goal of compressing *all* the relevant data into summarizing indices. This sensitivity analysis does identify a weakness in Gretzky's offensive game in 1992/93.

7.8   Did Gretzky have a bad year?

If a 'bad' year for Gretzky is not leading the team in offense, then Gretzky's offense did not lead the Kings' team in 1992/93, Robitaille's did. But in spite of missing 39 games, Gretzky's efficiency was the highest on the team; and he produced points without the damaging side effects of penalties and allowing goals by the other team – not a trivial accomplishment.

Even though Gretzky's OPI rating was hurt badly by the anomaly that he scored no power play goals in 1992/93, the analysis strongly suggests that his ability was still there. Historically, we now know that Gretzky eventually also came to this conclusion and returned the following season to his accustomed position, leading the entire National Hockey League in total points.

Gretzky's play during the 1992/93 season has been compared to the other players on the Kings team. We have not analyzed whether Gretzky's ability is leaving him by comparing his 1992/93 season with his earlier years. Certainly, he did not score the phenomenal number of points that he did during his earlier years in Edmonton, but then he was no longer surrounded by the great Oiler supporting cast. Furthermore, comparative team analyses are much more complex (e.g. Williams and Williams, 1996).

7.9   Conclusions

This chapter has described several research areas in statistics for ice hockey. However, its main goal has been the description of standard statistical techniques that can be applied to develop performance indices. These techniques have three important advantages.

(1)   *All* of the available input statistics may be included. This is an important property that can be used to eliminate any arguments that particular statistics were, or were not, included in an analysis.
(2)   The indices do not require subjective, arbitrary weightings of the base statistics. The methodology determines the importance of every variable, and does it in such a way to maximize the numerical discrimination among players.
(3)   The derived indices are highly related to the base variables and so are intuitively very satisfying.

Finally, it is important to remark that this is a numerical analysis. Hockey is a physically intense team game that depends on many factors that are not measurable at all, let alone without ambiguity. Nonetheless, the application of these techniques in NHL salary arbitration cases demonstrates their usefulness and the need for further research in this area.

References

Anderson-Cook, C. M. and Robles, R. (1997) Measuring hockey powerplay and penalty killing efficiency: a new approach. Presentation at *1997 Joint Statistical Meetings*. Anaheim, CA.

Danehy, T. J. and Lock, R. H. (1993) Statistical methods for rating college hockey teams. *1993 Proceedings of the Section on Statistics in Sports*, American Statistical Association, 4–7.

Dryden, S. (1994) *The Hockey News 1993–1994 Yearbook*. Toronto: Transcontinental Sports Publications, **46**(2), 11–12.

Hurley, W. (1995) Overtime or shootout: deciding ties in hockey. *Chance*, **8**(1), 19–22.

Johnson, R. A. and Wichern, D. W. (1992) *Applied Multivariate Statistical Analysis*, 3rd edn, Englewood Cliffs, NJ: Prentice Hall.

Liu, Y. and Schutz, R. W. (1994). Overtime in the National Hockey League: is it a valid tie-breaking procedure? *1994 Proceedings of the Section on Statistics in Sports*, American Statistical Association, 55–60.

Lock, R. H. (1997) Using a Poisson model to rate teams and predict scores in ice hockey. Presentation at *1997 Joint Statistical Meetings*. Anaheim, CA.

Morris, C. (1973) Breaking deadlocks in hockey: prediction and correlation. In F. Mosteller (ed.), *Statistics by Example, Detecting Patterns*. Reading, MA: Addison Wesley, 119–140.

Mullet, G. M. (1977) Simeon Poisson and the National Hockey League. *The American Statistician*, **31**(1), 8–12.

The National Hockey League (1996) *National Hockey League Official Guide and Record Book 1996–97*. Chicago: Triumph Books.

Searle, S. R. (1982) *Matrix Algebra Useful for Statistics*. New York: John Wiley and Sons.

Tukey, J. W. (1977) *Exploratory Data Analysis*. Reading, MA: Addison Wesley.

Williams, D. and Williams, W. H. (1996) Who won the 1995/96 NHL plus/minus award? *1996 Proceedings of the Section on Statistics in Sports*, American Statistical Association, 49–54.

Williams, W. H. (1994) Performance indices for on-ice hockey statistics. *1994 Proceedings of the Section on Statistics in Sports*, American Statistical Association, 50–54.

Williams, W. H. and Williams, D. (1997) Are salaries in the National Hockey League related to nationality? *Chance*, **10**(3), 20–24.

8

Developing Strategies in Tennis

John S. Croucher
Macquarie University, Australia

8.1 Introduction

Tennis has filled the research literature over the years with an abundance of mathematical papers attempting to analyse the game from a variety of perspectives. Many papers are theoretical in nature and based on arguable assumptions, while others draw conclusions using data recorded from actual matches. In recent years, the advent of sophisticated recording techniques using computers has enhanced the way in which tennis can be analysed, with every facet of the game now being examined in minute detail.

This chapter traces the development of tennis research over the past 25 years, including some of the classical findings that are referenced time and again in modern studies. Only the essential mathematical equations will be shown, with the emphasis being on the conceptual results. The interested reader is encouraged to explore the appropriate reference for a fuller explanation.

8.2 Service strategies

An important area of study in tennis is the effectiveness of a player's serve. In professional tennis in particular, players are expected to win their service games and failure to do so results in a 'break' of their serve that can lead to the loss of the set. It is not surprising that much research has been directed at analysing the tennis serve, not only service strategies but also ways in which the return of serve can neutralize a server's advantage.

8.2.1 Strong and weak service strategies

In one of the earliest papers of its type, Kemeny and Snell (1960) used a Markov chain approach to model a single game of tennis. Some time later Hsi and Burych (1971) presented a probability model for games involving two players, in which they used the game of tennis as one of their illustrations and calculated the probability that one player wins a single set of 'classical' (i.e. non tie-breaker) tennis.

As is the case with much of the theory-based research, it was assumed that the probability that a player wins a point on serve is constant throughout the match and that the points played are independent. While these assertions may not be entirely accurate in all cases, they certainly permitted the development of a multitude of mathematical results that provided some interesting, if debatable, conclusions. To provide some ammunition in support of these assumptions, Pollard (1983) examined 5503 points played in 35 championship matches and showed that neither of these assumptions could be rejected on statistical grounds.

Two opponents on the professional tennis circuit are nearly certain to meet each other several times during the year under relatively similar conditions (although, presumably, different playing surfaces were also taken into account). So, it was claimed by George (1973) that the required probabilities could be estimated with 'reasonable precision'. George also used these probabilities to uncover a service strategy that maximized the probability of a player winning a point on serve. Although a simple attempt to do this was made two years earlier by Gale (1971), it is the paper by George that is considered by many to be the pioneer work in the field and is worthy of some discussion here.

The rules of tennis give a player two chances to make a proper service on each point. George correctly claimed that most experienced players approach the first serve differently from the second, which is only used if the first serve is faulted. As a result, tennis players usually possess two types of serve. One, labelled the 'strong' serve, is traditionally used on the first service. This serve generally gives a high probability of winning the point to the server, given that the serve is good. The second serve, labelled the 'weak' serve, is usually reserved for the service following an initial fault. Although this serve gives a reduced probability of winning the point given that it is good, it has the advantage of a higher probability of actually being good. The relative efficacy of the two serves can vary markedly from player to player.

A probability model for winning a service point was developed for this situation using the following definitions:

$P(A)$ = the probability of the server winning the point (Event A)

$P(S)$ = the probability of a non-faulted strong serve (Event S)

$P(W)$ = the probability of a non-faulted weak serve (Event W)

$P(A|S)$ = the conditional probability that the server wins the point if the serve is strong and not faulted

$P(A|W)$ = the conditional probability that the server wins the point if the serve is weak and not faulted

$P(AS)$ = the probability of the player serving a non-faulted strong serve *and* winning the point

= $P(A|S)P(S)$

$P(AW)$ = the probability of the player serving a non-faulted weak serve *and* winning the point

= $P(A|W)P(W)$

If a player follows the usual strategy of using an initial strong serve followed by a weak serve if the initial serve is faulted, it follows that the probability of the server winning the point is:

$$P(A) = P(A|S)P(S) + P(A|W)P(W)[1 - P(S)] \tag{8.1}$$

Table 8.1 Probability model of winning a point according to service strategy

Strategy	First serve	Second serve	P(A)
SW	Strong	Weak	$P(AS) + P(AW)[1 - P(S)]$
SS	Strong	Strong	$P(AS)[2 - P(S)]$
WW	Weak	Weak	$P(AW)[2 - P(W)]$
WS	Weak	Strong	$P(AW) + P(AS)[1 - P(W)]$

However, the above strategy is not the only one available, since there are four possible combinations of sequences of strong and weak serves. The probabilities of winning the point for each of these combinations are shown in Table 8.1.

Since it is reasonable to assume that $P(A|S) \geqslant P(A|W)$, it follows that Strategy WS will always be inferior to Strategy SW and should never be used. If we define

$$R = [1 + P(S) - P(W)]^{-1} \quad \text{and} \quad Z = P(AW)/P(AS)$$

then, from Table 8.1, it can be shown that

$$\text{Strategy SW is optimal if } \quad 1 \leqslant Z < R$$

$$\text{Strategy SS is optimal if } \quad Z \leqslant 1 < R$$

$$\text{Strategy WW is optimal if } \quad 1 \leqslant R < Z$$

These optimality regions are shown in Fig. 8.1. Unfortunately for George, in the early 1970s no detailed data were routinely kept on professional matches, and so he was obliged to record information personally from two matches to illustrate his point. (This was one of the first attempts at data recording techniques, which are considerably more sophisticated today.) The first of these CBS Tennis Championship matches played on clay was held on 27 August 1971 between John Newcombe (Australia) and Arthur Ashe (USA). The second match was the final played next day between Newcombe and Rod Laver (Australia). New-combe won the first match 6-4, 7-5 while Laver won the final 6-2, 6-4. Hence, serving data were collected for two of Newcombe's matches (204 service points) and one each for Laver (85 service points) and Ashe (86 service points).

Based on the estimates from these games, it was found that $1 < Z < R$ in all cases, and so Strategy SW was optimal for all three players. However, only in Laver's match with Newcombe was the evidence reasonably strong, with his $P(A) = 0.66$ for Strategy SW, while only being 0.54 for Strategy SS. For Ashe v. Newcombe, the $P(A)$ values for Strategy SW were 0.56 and 0.63 for the two players, respectively, while for Strategy SS they were 0.53 and 0.60, respectively. In the final, Newcombe's $P(A)$ was 0.52 using Strategy SW and 0.50 using Strategy SS.

A full account of the probabilities in all matches can be found in George's paper, but what made this research of particular interest was its combination of theoretical work with at least a crude attempt to record actual data to produce an interesting result. Indeed, if enough data were collected on modern players, Strategy SS using a second strong serve could well be an option that should be used more often depending on the opponent.

Several years later, King and Baker (1979) employed George's results to rank the strategies using more extensive data, this time from ten matches among

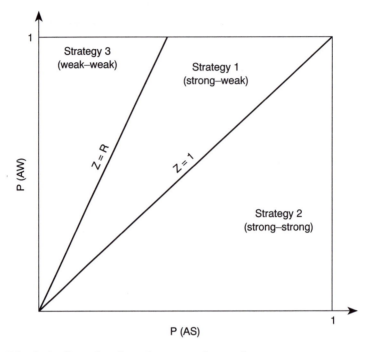

Figure 8.1   Optimality regions for various types of strategies.

world-class women players during the 1976 Virginia Slims tournament played on indoor courts. Once again the authors recorded the data by hand. Interestingly, they reached a similar conclusion to George in that they found the usual strong serve–weak serve strategy was by no means *clearly* superior to the other strategies. Indeed, they concluded that by adopting the conventional SW Strategy, players actually reduce their match-winning chances if they do not consider maximizing their own strengths relative to those of their opponents.

As an example, players strong in volley execution may be wise to use an SS strategy throughout a match since they have an increased chance of success by coming into the net following a strong serve. For players using the conventional SW strategy, improvement of rallying strength following a successful strong service ($P(A|S)$) was clearly the most potent factor in improving point-winning probability. In all cases, such improvements would have been much more effective than improving proportions of good first services $P(S)$.

8.2.2   Service strategies within the context of a match

The importance of the effect of improving service point-winning probability in relation to the probability of winning the match was already well-known at the time of these papers (see Carter and Crews, 1974; Hsi and Burych, 1971; Weinberg *et al.*, 1980). In a more recent analysis of serving strategies, Croucher (1995) used video replays to investigate the serving performance of the two Men's Wimbledon finalists in 1994, namely Pete Sampras (a right-handed

Table 8.2 Points won (Sampras v. Ivanisevic, 1994 Wimbledon Men's Singles Final).

Player	Set 1	Set 2	Set 3	Total
Sampras	50	44	24	118
Ivanisevic	44	37	7	88
Total	94	81	31	206

player from the USA) and Goran Ivanisevic (a left-handed player from Croatia). In this match, Sampras was heavily favoured (as the No. 1 seed) while Ivanisevic (the No. 4 seed) was only given an outside chance. The match was notable for its unusual scoreline of 7-6, 7-6, 6-0 to Sampras. (These two players met again in the 1995 semi-final where Sampras won a much closer contest 7-6, 4-6, 6-3, 4-6, 6-3.) This paper illustrates how match analysis can be used to develop strategies for play, and some of the relevant details of the contest are discussed below.

The total points won in each set by each player are shown in Table 8.2. Overall, 206 points were played, of which 118 (57%) were won by Sampras and 88 (43%) by Ivanisevic. However, Sampras won 24/31 (77%) of the points in the final set to emphasize his dominance at the finish.

Overall, Sampras won 76 (75.2%) of his 101 service points, while Ivanisevic won 63 (60%) of his 105 serves. Interestingly, 25 (40%) of Ivanisevic's winning service points were due to aces, emphasizing his reliance on a strong first serve. Croucher then examined the match videotape in more detail to uncover any strengths or weaknesses in serving patterns. In particular, service points were classified according to whether they were an ace, double fault, and made from the backhand or forehand court. The results are shown in Table 8.3, including from which court the service was made, which was a significant factor since the proportions of points won in each case varied markedly.

From a tactical point of view, with Sampras being right-handed and Ivanisevic being left-handed, serving from the forehand court or backhand court may not bear the same significance for each player. In particular, Sampras won 80% of points he served from the forehand court but only 70% from the backhand court, while Ivanisevic won 68% from the forehand court but only 52% from the backhand court. Hence, for both players, serving from the forehand court brought the largest success, although Ivanisevic served nearly two-thirds of his aces from the backhand court. However, when he could not produce an

Table 8.3 Percentage of service point outcomes for Sampras v. Ivanisevic (1994 Wimbledon Men's Singles Final)

Server	Court served from	Points won		Points lost	
		Ace	Other	Double fault	Other
Sampras	Forehand	14.8%	66.7%	5.5%	13.0%
	Backhand	19.2%	51.1%	4.2%	25.5%
Ivanisevic	Forehand	17.0%	50.9%	0.0%	32.1%
	Backhand	30.8%	19.2%	5.8%	44.2%

Table 8.4   Service points won–lost by Sampras (1994 Wimbledon Men's Singles Final)

From	To		
	Ivanisevic's forehand	*Ivanisevic's backhand*	*Total*
Forehand court	16-0	27-11	43-11
Backhand court	7-1	26-13	33-14
Total	23-1	53-24	76-25

ace from there, he won only 11 of the remaining 27 points (41%), thus placing an extremely heavy reliance on his producing a clean winning first serve.

One of the major deficiencies in Ivanisevic's service game was his poor record in winning points on his second serve from the backhand court. In fact, of his 19 serves from that court he managed to win only six points (32%) compared to winning 13/21 (62%) of second serves from the forehand court. Interestingly, in his first five service games of the match, Ivanisevic lost the second point played (all served from the backhand court) in each game (every time trying to serve to Sampras' backhand), until finally winning the second point of his sixth service game with an ace out wide to the backhand. Ivanisevic should not have persisted with such a poor tactic throughout the match. A superior strategy would have been to try a stronger second serve from the backhand court and randomly mixing the direction by occasionally serving to Sampras' forehand.

The combination of service court and direction of serve for each player is summarized in Tables 8.4 and 8.5. Table 8.4 shows Sampras' outstanding success rate for serving to Ivanisevic's forehand, by winning all 16 points from the forehand court and 7 out of 8 from the backhand court for an overall success rate of 96%. This was in marked contrast to his efforts at serving to Ivanisevic's backhand in winning 27/38 (71%) serves he made from the forehand court and 26/39 (67%) made from the backhand court. Given his relatively lower success rate serving to Ivanisevic's backhand it would have been a much better strategy for Sampras to serve more often to Ivanisevic's forehand, clearly his weak point. In fact, all of Sampras' five double faults were attempts to serve to Ivanisevic's backhand.

From Table 8.5 it can be seen that Ivanisevic had equal success serving to Sampras' forehand (59% success) or backhand (60% success). However, there were significant differences in combinations of serves, where he did poorly when serving to Sampras' forehand from the forehand court and Sampras' backhand from the backhand court in winning only 25/53 (47%) of points. This contrasted sharply to when he reversed the direction and served

Table 8.5   Service points won–lost by Ivanisevic (1994 Wimbledon Men's Singles Final)

From	To		
	Sampras' forehand	*Sampras' backhand*	*Total*
Forehand court	8-8	27-9	35-17
Backhand court	11-5	17-20	28-25
Total	19-13	44-29	63-42

to Sampras' backhand from the forehand court, where he won 38/52 (73%) of points.

Second serves were also a major factor in the match, with Ivanisevic winning only 19/40 (48%), while Sampras won 31/51 (61%). Curiously, Ivanisevic made only two attempts to serve to Sampras' forehand when making a second serve from the backhand court (both times early in the second set), and in doing so won one and lost one. The remaining 17 such serves he directed at Sampras' backhand for a poor success rate of only 29%.

8.3   Probability of success

As well as considering the merits of various service strategies, there has been much research over the years on the theoretical probability of winning tennis matches, although much of this has not related to actual match data. Carter and Crews (1974) and Hsi and Burych (1971) both formulated games of tennis as mathematical models where the probability of winning a point is constant for each player. Fischer (1980) presented a sophisticated analysis in which he gave probabilities of winning both a set (both tie-break and advantage) and a match in terms of the probability of winning an individual game. Once again the independence of points is assumed, and the author uses an 'average' probability of a player winning an individual point, irrespective of serving or receiving. Since these probabilities are almost certainly different in practice, the results are of limited value.

While most papers developed probabilities for success from the commencement of a match, Croucher (1986) considered these probabilities from various starting points. That is, he calculated the probabilities that a server would win a game from each of the 16 possible scorelines. Table 8.6 presents these values given the probability p of winning an individual point on serve ranging between 0.30 and 0.80. (These probabilities should be sufficient for most players – Pollard (1983) found in his study of professional tennis players that the winners of matches had an average probability of 0.71 while the losers averaged 0.62.)

According to Table 8.6, the probability of the server winning from:

- 0-30 is *always* greater than from 15-40;
- 40-15 is *always* greater than from 30-0;
- 15-30 is *always* greater than from 30-40;
- 40-30 is *always* greater than from 30-15;
- 30-30 (or deuce) is greater than from 15-15 if $p < 0.50$, or less than if $p > 0.50$;
- 30-15 is greater than from 15-0 if $p > 0.63$ and less than if $p < 0.63$.

Proper analysis of the mathematical theory of tennis would not be complete without a discussion of the importance of each point in a tennis match. In what has become a popular definition of importance I, Morris (1977) based the following definition on the difference between two conditional probabilities

$$I(\text{point}) = P(\text{server wins game}|\text{server wins point})$$

$$- P(\text{server wins game}|\text{server loses point}) \qquad (8.2)$$

Table 8.6 The probability of a server winning a game from various scorelines

					Probability p of the server winning a point						
Current score	0.30	0.35	0.40	0.45	0.50	0.55	0.60	0.65	0.70	0.75	0.80
0-0	0.099	0.170	0.264	0.377	0.500	0.623	0.736	0.830	0.901	0.949	0.978
15-0	0.211	0.311	0.424	0.542	0.656	0.758	0.842	0.905	0.949	0.976	0.990
30-0	0.412	0.523	0.631	0.742	0.813	0.879	0.927	0.960	0.980	0.991	0.997
40-0	0.710	0.787	0.850	0.900	0.938	0.963	0.980	0.990	0.996	0.998	0.999
0-15	0.051	0.095	0.158	0.242	0.344	0.458	0.576	0.689	0.789	0.870	0.930
15-15	0.125	0.296	0.286	0.389	0.500	0.611	0.714	0.804	0.875	0.928	0.964
30-15	0.284	0.381	0.485	0.589	0.688	0.775	0.847	0.903	0.944	0.970	0.986
40-15	0.586	0.672	0.751	0.819	0.875	0.919	0.951	0.972	0.986	0.994	0.998
0-30	0.020	0.040	0.073	0.121	0.188	0.271	0.369	0.477	0.588	0.696	0.795
15-30	0.056	0.097	0.153	0.225	0.313	0.411	0.515	0.619	0.716	0.802	0.873
30-30	0.155	0.225	0.308	0.401	0.500	0.599	0.692	0.775	0.844	0.900	0.941
40-30	0.409	0.496	0.585	0.671	0.750	0.820	0.877	0.921	0.953	0.975	0.988
0-40	0.004	0.009	0.020	0.037	0.063	0.100	0.150	0.213	0.290	0.380	0.481
15-40	0.013	0.028	0.049	0.081	0.125	0.181	0.249	0.328	0.414	0.506	0.602
30-40	0.047	0.079	0.123	0.180	0.250	0.329	0.415	0.504	0.592	0.675	0.753
40-40	0.155	0.225	0.308	0.401	0.500	0.599	0.692	0.775	0.844	0.900	0.941

Croucher (1986) gives a complete list of the importance of each point for values of *p*, ranging between 0.30 and 0.80. Since the receiver's probabilities are the complement of those of the server, every point is equally important to both players. A summary of the relative importance of points is shown below.

- The first point (0-0) is always of only average importance. That is, it never ranks highly or near the bottom for any value of *p*.
- The point 30-40 has top ranking for $p \geqslant 0.50$ but has decreasing importance as *p* falls below 0.50.
- The point 40-30 has top ranking for $p \leqslant 0.50$ but has decreasing importance as *p* rises above 0.50.
- No point has a consistently high (or low) ranking for *all* values of *p*. However, the points 30-30 and deuce never rank below 6th out of 16.
- For $p \geqslant 0.50$, those points where the server is trailing rank highly, while for $p < 0.50$ those points where the server is ahead rank highly.
- The point 30-40 always ranks higher than the point 15-30.

This last conclusion can be proven as follows. Define:

I_{sr} = the importance of a point when the server has score *s* and the receiver score *r*

P_{sr} = the probability that the server will win the game when the score is *s* to *r*

In this context, we use the values of *s* and *r* to be 0, 1, 2, 3, 4 where they represent the scores of 0, 15, 30, 40 and game, respectively. For the score of 15-30 we have:

$$I_{12} = P_{22} - P_{13} \qquad (8.3)$$

and for the score 30-40 we have:

$$I_{23} = P_{33} - P_{24} \qquad (8.4)$$

Since $P_{24} = 0$ as the server has lost, it follows from (8.4) that $I_{23} = P_{33}$. Also, since a game must be won by a margin of two points or more, $P_{22} = P_{33} = I_{23}$. Substituting into (8.3) yields:

$$I_{12} = I_{23} - P_{13} \qquad (8.5)$$

Since $P_{13} > 0$, it follows from (8.5) that $I_{23} > I_{12}$, or that the point 30-40 is always more important than the point 15-30.

Using the more complex notion of 'time-importance', in which the importance of a point is weighted by the expected number of times the point is played, Morris develops several interesting conclusions. First, he claims that, even though more points are served into the forehand court than into the backhand court, the same total time-importance is associated with both sides of the court. It follows that the higher *average* time-importance is experienced in the backhand court and suggests the reason why doubles teams are advised to have the more experienced or stronger player on the left side. (This ignores whether the players might be a left-hand and right-hand combination, which would presumably be another factor.)

Secondly, Morris claims that the total time-importances associated with even-numbered and odd-numbered service games are equal. Since there are more odd-numbered service games, the player who serves first generally will serve under less pressure. There is some logic in this statement, since professionals are always expected to win their own service. Winning the toss, a nervous starter may elect to receive, since there may be a high probability that service will be broken in the first game but, having settled down, a lower probability by the second game when it will be his/her turn to serve first.

The winner of a 'classical' or 'advantage' set must get two games ahead once the game score reaches 6-6. This method often produces very long matches. The tennis tie-breaker, introduced to shorten the length of matches, has had its desired impact, and its effect has been examined in research by Croucher (1982) and Pollard (1983). The tie-breaker is played when the game score reaches 6-6 in a set. Some of the more interesting results from these papers surrounding tie-break sets are as follows.

(1)   For two unequal players, the probability that the better player wins the set is greater for the classical version.
(2)   The expected value and variance of the duration of a match (in points played) are always smaller for the tie-break version.
(3)   If both players have a probability p of winning service points between 0.50 and 0.60, then the classical and tie-break sets have similar characteristics and so the tie-breaker rule is ineffective. While this case applies for most women's matches where p does lie in this range, for most men, especially on grass surfaces, the value of p is almost always greater than 0.60.

8.4   Data recording

The field of notational analysis in all sports is one of growing importance as players and coaches try to gain a competitive edge. The idea is to find an effective way in which to record accurately appropriate information from a match in a way that is going to be of real benefit. In some circumstances immediate feedback may be required (e.g. recording data on your next opponent), while on other occasions it may not be necessary to analyse the data on the spot (e.g. collecting information for a database).

A number of aspects of tennis require examination and proper analysis that can only be achieved through the collection of appropriate data. Details can be recorded that clearly show patterns of play, along with the movement of players throughout the points. Of particular interest is the identification of the strengths and weaknesses of the players and the types of situations that bring these to the fore. Once these are known, a coach can recommend appropriate actions and tactics that will maximize the expected advantage to the player.

Recording information from tennis matches (or, indeed, any sport) can range from the simple use of pencil and paper to the employment of video cameras linked to computers. For many recreational tennis players, the former method may be quite sufficient as a friend could record basic information including, for example, the types of unforced errors made, serving statistics and data on particular strokes such as lobs and passing shots.

The professional level requires more elaborate analysis, not only on the player's own game but on that of the opponent. A player would be most unwise to enter a match with little or no information about the strengths or weaknesses of the opponent, who almost certainly has those facts on the player's game. In some of the more sophisticated techniques, data are recorded either directly into a computer or on data sheets that are placed into a specially designed database written for this purpose. These video/computer systems provide comprehensive sports analysis by transferring a videotaped sequence to computer memory. Once transferred, the sequence can be immediately retrieved and replayed for individualized coaching and instruction.

The data entry person makes one complete pass through the match in which relevant information or specific incidents are noted along with the time they occurred. Once this has been done, a coach can then later type in, for example, 'passing shots', and the video recorder will intelligently go to those incidents in the match where passing shots were in evidence.

One of the earliest and long-lasting techniques in the important area of data analysis in tennis illustrates the idea behind the modern data analytic techniques of developing strategies. In 1982, a former South African amateur, Bill Jacobsen, began to record his son's tennis matches using hand-collected data that he fed into a microcomputer. He soon designed a 4 lb portable computer with which a single observer could easily record ten times as much information. The system, now known as CompuTennis, was subsequently awarded a contract by the US Tennis Association to chart matches for all four national teams – Davis Cup, Wightman Cup, Federation Cup and Davis Cup under 18.

CompuTennis does not attempt to cover every shot in a match since players can rally for minutes at a time. Instead, the observer records only key strokes such as the serve, return of serve and the sequence of shots (up to 5 or 6) that lead to the end of the point.

The data can be fed into the computer by one of three ways:

(1)   at courtside while play is in progress;
(2)   from a videotape replay of the match;
(3)   by hand using CompuTennis hand charts.

Each key on the keyboard is programmed for a particular function to collect five types of data:

(1)   the shot type;
(2)   the stroke description;
(3)   the direction of the shot;
(4)   the result of the shot;
(5)   the location of each player at the end of each point.

The system has the capacity to record shot descriptions (e.g. half volley, passing shot), whether the shot is forehand or backhand, the zones of the court travelled, and the result (i.e. forcing shot, error). There are console keys on which to record ten other statistics such as the number of times a player runs around his backhand.

Some of the fascinating statistics to be uncovered by CompuTennis are listed below. (See CompuTennis, 1996.)

•   In 99% of matches, the player who wins the most points in the match wins the match.

Table 8.7 Time of points for three Grand Slam championships in 1991

	French Open (clay)	US Open (hard)	Wimbledon (grass)
Men			
No. of points	279	155	214
Avg. time per point	10.0 s	7.6 s	2.6 s
Actual play time	50 min	19 min	9 min
Avg. rest time between points	21.48 s	26.17 s	27.36 s
Total rest time	2 hr 30 min	1 hr 39 min	2 hr 22 min
Play time per hour	14 min 56 s	9 min 58 s	3 min 42 s
Women			
No. of points	122	119	200
Avg. time per point	11.0 s	5.3 s	5.9 s
Actual play time	24 min	11 min	20 min
Avg. rest time between points	21.9 s	16.7 s	21.5 s
Total rest time	1 hr 7 min	56 min	1 hr 48 min
Play time per hour	15 min 43 s	9 min 41 s	9 min 18 s

Source: CompuTennis

- On 95.5% of occasions, the player who has the first match point will win the match.
- In matches played on grass, more than 80% of the points are over within three seconds.
- In the Wimbledon 1991 Men's Final between the two big servers Boris Becker and Michael Stich, the average length of a point was 2.6 seconds. This translated into actual playing time per hour of only three minutes 42 seconds. Only 9 minutes of the 2 hour 31 minute match was actually spent playing tennis.
- By contrast, the Monica Seles v. Martina Navratilova US Open final in 1991 saw an average point length of 5.3 seconds, with an average playing time of 9 minutes 41 seconds per hour. This long playing time did not suit the older Navratilova who did not give herself enough time to recover between points.

Comparisons of several key statistics for three Grand Slam championship matches in 1991 are shown in Table 8.7. The reader will be able to make other interesting comparisons between not only male and female matches but between the type of surface on which the match was played. In particular, note the significantly longer playing time for each point on clay, along with the longer playing time per hour.

A summary of the CompuTennis data (Jacobsen, 1993) on the number of unforced errors made by men and women on different surfaces is shown in Fig. 8.2. The data were taken from a selection of professional tennis players in various tournaments up to 1993. Percentages are higher on all surfaces for women than men. Clay courts produce the highest values since the points are much longer, producing more opportunity for a greater variety of strokes and tactics. As a result, tactical and fatigue errors tend to be higher on this surface. On the other hand, the tactical demands are fewer on grass; players are limited to fewer possible successful strategies or stroke selections, because the speed of the court gives players little time to think during the points.

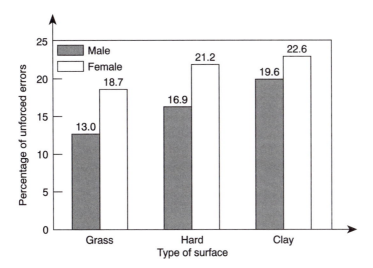

Figure 8.2   The percentage of unforced errors made by men and women professional tennis players on different surfaces (redrawn from Jacobsen, 1993).

Many other aspects can be examined by match analysis, including whether players should serve differently to the forehand or backhand court, the importance of the ability to serve an ace or avoid double faults, the ability to return a clean winner off a serve and a measure of how aggressive a player is. These properties of a player's game are among those considered by CompuTennis and illustrate the power of proper data collection to develop winning strategies effectively. In fact, CompuTennis has developed the concept of an 'Aggressive Margin' that attempts to explain why a particular player has an edge in a crucial match, and measures a player's improvement from match to match or against other players (see Jacobsen, 1994).

8.5   Remarks

There is no doubt that the game of tennis, like many other sports, is undergoing revolutionary change with regard to the amount of information that can be gathered and analysed. While much research has centred on singles play, there is ample scope for proper investigation of doubles play in which extra variables involving the interaction between playing partners are clearly significant factors. One area for further investigation would be to characterize optimal combinations of players to maximize success in doubles and an examination of patterns of play. It would be particularly instructive, for example, to analyse video coverage of the Australian doubles combination of Mark Woodforde and Todd Woodbridge to discover what factors are relevant in making them the most successful doubles combination of all time.

Other areas for tennis research include the psychological profile of players, comparison of different types of match preparation and training techniques and the resultant effect of experiencing a perceived 'bad' line call. Different styles of play can also be compared in order to find optimal strategies for a particular situation, depending upon the opponent. For example, CompuTennis research shows that, since 1945, less than 30% of the winners of the Men's French Open (played on clay) were 'net rushers', while the last female net rusher to win was Martina Navratilova in 1984. This leads to the question of what tactics are most effective for different surfaces. There is even room for statistical analysis of the most effective way to rank professional players and whether the current ATP (Association of Tennis Professionals) method is the most reasonable. And if one wants to examine the underlying principles of the tennis scoring system itself, the groundwork has already been laid by Schutz (1970) and Miles (1984). In particular, the latter claims that, largely because the proportions of service points won by both players is so high in top men's tennis, the efficiency of the traditional scoring system (i.e. its ability to identify the superior player) is unduly low. Indeed, he argues that the introduction of the tennis tie-breaker has further reduced efficiency and he proposes a simple model to make the scoring more efficient.

References

Carter, Jr. W. H. and Crews, S. L. (1974) An analysis of the game of tennis. *The American Statistician*, **28**, 130–134.

CompuTennis (1996) *Current Topics and Technical Briefs*. Los Altos, CA: CompuTennis.

Croucher, J. S. (1982) The effect of the tennis tie-breaker. *Research Quarterly for Exercise and Sport*, **53**, 336–339.

Croucher, J. S. (1986) The conditional probability of winning games of tennis. *Research Quarterly for Exercise and Sport*, **57**(1), 23–26.

Croucher, J. S. (1995) Replaying the 1994 Wimbledon men's singles final. *New Zealand Statistician*, **30**(1), 2–8.

Fischer, G. (1980) Exercise in probability and statistics, or the probability of winning at tennis. *American Journal of Physics*, **48**(1), 14–19.

Gale, D. (1971) Optimal strategy for serving in tennis. *Mathematics Magazine*, **5**, 197–199.

George, S. L. (1973) Optimal strategy in tennis. *Journal of the Royal Statistical Society Series C*, **22**, 97–104.

Hsi, B. P. and Burych, D. M. (1971) Games of two players. *Journal of the Royal Statistical Society Series C*, **20**, 86–92.

Jacobsen, W. (1993) All unforced errors are not the same. *Sports Science for Tennis*, Summer Edition, 1–3.

Jacobsen, W. (1994) Getting aggressive. *Tennis Match*, Sept/Oct, 17.

Kemeny, J. G. and Snell, J. L. (1960) *Finite Markov Chains*. Princeton: Van Nostrand.

King, H. A. and Baker, J. A. (1979) Statistical analysis of service and match-play strategies in tennis. *Canadian Journal of Applied Sports Science*, **4**, 298–301.

Miles, R. (1984) Symmetric sequential analysis: the efficiencies of sports scoring systems (with particular reference to those of tennis). *Journal of the Royal Statistical Society Series B*, **46**(1), 93–108.

Morris, C. (1977) The most important point in tennis. In S. P. Ladany and R. E. Machol (eds), *Optimal Strategies in Sports*. pp. 131–140. New York: North-Holland.

Pollard, G. H. (1983) An analysis of classical and tie-breaker tennis. *Australian Journal of Statistics*, **25**, 496–505.

Schutz, R. W. (1970) A mathematical model for evaluating scoring systems with specific reference to tennis. *Research Quarterly for Exercise and Sport*, **41**, 552–561.

Weinberg, R. *et al.* (1980) Influence of cognitive strategies on tennis serves. *Perception and Motor Skills*, **50**, 663–666.

9

Statistical Modelling in Track and Field

Robert W. Schutz and Yuanlong Liu
University of British Columbia, Canada

9.1 Introduction

The sport of track and field, or athletics as it is known in most countries outside of North America, is undoubtedly the oldest of the competitive sports covered in this book. Its recorded history goes back to at least 776 BC when the original Olympic Games were held in Greece, and although a sprint (approximately 192 m) was the only event in the inaugural games, other events soon followed (javelin, discus, long jump). However, organized competitive running, jumping and throwing were virtually non-existent from the end of the Greek Games (circa 300 AD) until approximately 1830, and were not widely recognized until the modern Olympic Games began in 1896 (Queracentani, 1964). Despite this long history of competition, the statistical analyses of track and field performances are restricted to data from the 20th century. Written histories of early races report only the winners, but no records were kept of times (probably because of the unavailability of suitable timing devices).

The formation of the International Amateur Athletic Foundation (IAAF) in 1913 signalled the beginning of a controlled and rigorous system for measuring, verifying and recording world record performances in track and field events. The IAAF accepted some 'records' set at well-organized meets prior to 1913, and has updated these official world records annually since 1914. The earliest officially recognized record was set in 1896 by B. Wafers, with the time of 21.2 s in the rarely run 200 m straight race. For most men's events, formal records are available from approximately 1905. Data for women are not nearly so complete, primarily because the IAAF would not recognize women's participation at all for many years, and then later in only a few events. Some women's track and field events were incorporated into the Olympics in 1928, but endurance events such as the 1500 m and 5000 m are relatively recent additions to the Games. The earliest IAAF record for women is in the 880 yards when Mary Lines set a record of 2 min, 26.6 s (2:26.6) in 1922, and continuous records of women's performances are generally available from 1950 onwards (except for the longer distances like the 3000 m where we have data only from 1960).

The nature and general availability of these data have resulted in their extensive use by researchers, teachers and sports enthusiasts. The data are unique in that they: (1) possess a meaning that is apparent to most people; (2) are collected under very constant and controlled conditions (carefully monitored by the IAAF), and thus are very accurate and reliable; (3) are recorded with great precision (e.g. to the hundredth of a second in races), and thus permit very fine differentiation of change and/or differences; (4) are both longitudinal (90 years for men's records) and cross-sectional (over different distances and across gender); and (5) are publicly available at no cost (e.g. a full listing of the progression of men's and women's world records is available at www.uta.fi/~csmipe/sport). Thus, they provide wonderful data sets to test statistical models of change as well as being a rich source of information for studying the limits of human locomotion. Lloyd, a physiologist writing in 1966, said this about analyzing world records in running:

> One beauty of this field is that the athletic authorities of the world provide one with scientifically-impeccable data, usually accurate to a fraction of one per cent, and the guinea pigs select, house and feed themselves. The scientist merely has to sit back and collect the results, ... (Lloyd, 1966, p. 515)

We doubt if the runners would appreciate being referred to as guinea pigs, but apart from that, Lloyd's statement is undoubtedly very true. It is a very rich and accurate data set.

These data, the world and Olympic record times and distances, have been the subject of much of the statistical analyses of track and field performance. Although other sports have been analyzed statistically with respect to: (1) team decision strategies (e.g. batting order in baseball, time to pull the goalie in ice hockey); (2) individual decision strategies (e.g. shot sequences in squash, shot selection in basketball); (3) management strategies (e.g. player drafting, ticket pricing); (4) scheduling and scoring procedures (e.g. tournament structures, handicapping golf); and (5) prediction (e.g. predicting game outcomes, forecasting records), the vast majority of the track and field publications have dealt only with prediction. Surprisingly, given this narrow focus, there has been a considerable number of statistical articles written on the topic of track and field. An analysis of a bibliography of 577 'statistics in sports' articles (Schutz, 1994) revealed that 242 of these papers dealt almost exclusively with six sports: baseball, track and field, American football, basketball, tennis and ice hockey, with baseball having the most articles (34%) and track and field the second most (22%). Of the track and field articles, 62% focused on the prediction of records – a much larger proportion than in other sports. The relatively few papers on strategy in track and field is not surprising, given that the strategies used by runners (and to a lesser extent by the field event athletes) are not quantified or recorded, and indeed are often not apparent to the spectator. Thus, the only data we have to analyze are the time or distances of the final outcomes.

Given the above, the primary focus of this chapter is on predicting future performances in track and field, including an examination of the work done on comparing male and female performances. There has been some work done in the area of optimal strategies and on scoring systems, and those works will be examined first. Additionally, there has been a considerable amount of work done on attempting to develop mathematical models of running, jumping and throwing performances. These models are more of a 'mathematics of sport' nature than a 'statistics of sport' nature, in that they often involve no probabilistic components. Because they have relevance for predicting records, these models will be reviewed here also. Finally, it should be noted that the literature contains a number of articles which use track and field data to explain or test a statistical methodology (e.g. Dawkins *et al.*, 1994; Naik and Khattree, 1996), but as these papers do not add substantially to our understanding of track and field performance *per se*, they are not included in the reviews in this chapter.

9.2 Optimal strategies and scoring systems

9.2.1 Optimal strategies

As noted in Section 9.1, although strategy plays an important goal in some track events (most notably the distance events), we have no measures of this strategy and thus it is impossible to analyze it quantitatively. Keller (1974), however, proposed an 'optimal running strategy' on purely theoretical grounds. Using Newton's second law, Keller derived equations for the optimal velocity curve to run a sprint (defined as less than 291 m) and other distances. Simply put, his optimal strategy for completing a non-sprint race in the shortest possible time is for the runner to accelerate as fast as possible to an optimal velocity and then maintain this velocity until his/her oxygen supply runs out – which should occur just before the finish line, thus resulting in a slight reduction in velocity in the last second or two.

Keller's model has received considerable attention and appears to be accepted by most scientists. In practice, however, it does not seem to have much empirical support, as athletes rarely employ a strategy of constant velocity throughout an entire race of 800 m or longer. Statistically, the model suggests that runners can accelerate to their optimal velocity in a very short period of time (less than two seconds for a 400 m runner), but 400 m runners take considerably longer than that to reach constant velocity. Also, Keller's model would predict that Donovan Bailey, the world record holder in the 100 m, could run the 200 m in a time of 17.72 s (Tibshirani, 1997). Given that this would be 1.60 s below the current world record, and over 2.0 s faster than Bailey's best time for this race, it is clearly a highly improbable event! Thus, while Keller's model is an interesting model of running and has validity in some applications (e.g. Pritchard and Pritchard, 1994), it may not be physiologically and biomechanically sound, or it is one which athletes do not choose to use as a race strategy. Perhaps this is because winning races and setting records involves more than physiology and mechanics – the psychological components dictate different strategies.

An example of optimal strategy in a field event is given by Sphicas and Ladany (1976). They used dynamic programming to determine the 'optimal aiming line' for the long jump – that is, how close to the edge of the take-off board should a jumper aim for in order to maximize the probability of hitting the exact spot while minimizing the probability of fouling (stepping over the board). They showed that the optimal solution is a function of a number of factors; namely, the number of jump attempts remaining, the longest jump recorded so far in the competition, the maximum expected distance, the distance from the aiming line to the official take-off line, the distance from the actual take-off point to landing, and the distance between the actual take-off and the aiming line. Their results indicated that a stationary model was virtually just as good as a dynamic model, that is, a jumper could establish an optimal constant aiming line for all jumps regardless of the state of the factors listed above. The important factor for an athlete is knowledge of the expected variability in the distance between the actual take-off point and the aiming line. What they clearly show is that a competitor should never aim for the official take-off line, but rather aim for a point two to four inches behind it.

Other examples of optimal strategies are the assignment of runners to a relay team (Heffley, 1977) and the optimal angle of release for a throwing event such as the shot or hammer (Townend, 1984).

9.2.2   Scoring systems

Since the measured time or distance alone usually determines the outcome, very few scoring systems are used in track and field events. However, the men's ten-event decathlon (100 m, long jump, shot, high jump, 400 m, 110 m hurdles, discus, pole vault, javelin and 1500 m) and the women's seven-event heptathlon (100 m hurdles, shot, high jump, 200 m, long jump, javelin and 800 m) involve scoring procedures, as do non-Olympic running events such as team cross-country races (e.g. Sidney and Sidney, 1976) and triathlons (e.g. Wainer and De Veaux, 1994).

In the decathlon, approximately 1000 points are awarded for an outstanding performance, with the current world record holder averaging approximately 900 points per event. The scoring method is a criterion-referenced system; each contestant is awarded points according to his time or distance, regardless of where he places relative to the rest of the competitors. The system was originally set up so that 1000 points for a single event was an almost unattainable score (the average was under 600 points per event in 1911), and every event counted equally towards the total score. However, as the decathletes became more skilled at some events, the relative point values became distorted and thus the scoring system was changed (in 1934, 1952, 1963 and 1984). We have done some work on examining the validity of the present scoring system and found it to be a valid system – that is, relative to the world records for the ten individual events, the decathlon points for each event represent a fair system. Tidow (1989) has also done some work in this area.

The women's heptathlon is scored in a similar manner to the decathlon, although the average points per event is 1200 (in comparison to 900 for the men). It too has undergone many changes over the years, but in the order and make-up of the events and not in the scoring system itself.

9.3 Modelling present and future performances

9.3.1 Historical overview

Interest in men's and women's abilities and accomplishments in running, jumping and throwing has a long and diversified history. Various studies have dealt with: (1) the mathematical relationship between time and distance (or velocity and distance); (2) the mathematical relationship between physiological function and performance (time, velocity) in distance running; and (3) the prediction of future performances, perhaps even 'ultimate' performance.

Even before the IAAF had established its formal world records, Kennelly (1906), an engineer, examined the relationship between velocity and distance for walking, running, rowing, skating and swimming. He showed that velocity versus distance plotted on a log-log scale yielded a linear relationship that was reasonably stable over all types of events. In general, doubling the distance resulted in the time increasing by a factor of 2.18 (which Kennelly, 1926, later adjusted slightly to 2.11). The noted physiologist and Nobel Laureate, A. V. Hill (1925), applied the concepts of oxygen intake and oxygen debt to provide a physiological explanation and mathematical expression of the time–distance relationship for running. Brutus Hamilton (1934), a California track and field coach, issued a much publicized list of 'ultimate performances' or 'perfect records' in the men's Olympic track and field events – all of which were surpassed by 1970. Henry (1954), a physical educator, modelled the velocity–distance relationship across all running events with a formula based on theoretical and physiological principles. Ryder *et al.* (1976), representing the disciplines of medicine, computer science and astronomy, respectively, compared the changes in velocities from 1900 to 1970 for all running events and extrapolated to arrive at expected world records for the years 2004 and 2028; their linear relationships were rather optimistic and none of their 2004 predictions have been realized to date. Mathematicians (Deakin, 1961; Lucy, 1958), statisticians (Blest, 1996; Smith, 1988) and psychometricians (Schutz and McBryde, 1983) have fitted world records or yearly best times with a variety of mathematical functions, and predicted future record performances.

As is readily apparent from the above examples, interest in the nature and extent of track and field performances (past, present and future) has a long and diversified history, and has attracted scholars from many walks of life – it is truly a multidisciplinary area of interest. Unfortunately, it has not been *inter*disciplinary, and many of the authors have apparently been unaware of previously published work in another discipline. Hopefully, the recently (1992) formed Statistics in Sports Section of the American Statistical Association will act as a central focus for the dissemination of much of this work, and we will see a more integrative and sequential development of ideas in this area.

9.3.2 Prediction models of men's running

The following section contains a brief description of the historical development of statistical models of running performance. We have been selective in including some papers and excluding others; however, readers wishing to see a more detailed and less statistical discussion of the study of athletic records are encouraged to read Meade (1966). All models and references to world records in this section refer only to men's events, since the scarcity of suitable data sets precluded the analyses of women's performances in most early studies. Also, this section frequently refers to the record for the mile run, even though this event is not part of the Olympics or World Championships. The mile is used as the standard to compare predictions over time, as it was the premier track event for the first part of this century and most of the early writers focused their attention on that event.

9.3.2.1   Kennelly

Over 90 years ago, when the world record for the mile was 4:15.6, Kennelly (1906) predicted that eventually man would run this distance in the unbelievably fast time of 3:58.1. Kennelly used log-log graphic plots of the time–distance relationship for distances from 100 yards to 10 miles. His predictions for future performance, which were not central to his paper, were based on subjective extrapolation. His time–distance analyses, however, were mathematically derived, yielding the equation:

$$\log(t) = 1.125 \log(d) - 1.236 \tag{9.1}$$

where t is time in seconds and d is distance in meters. This log-log function can be expressed as the power function (the more common form in the more recent literature):

$$t = 0.0588 d^{1.125} \tag{9.2}$$

9.3.2.2   Meade, Hamilton and Nurmi

Meade (1916), on examining the log-log curve for 220 yards to 5 miles, remarked that 'The probability of any marked change in the records represented by the curve is very remote' (Meade, 1916, p. 598), and thus concluded that the then-current record of 4:12.6 would likely never be broken. It lasted for eight years. Virtually no further mathematical modelling of track and field performance was done for the next 40 years.

However, a number of sports writers and athletes did offer their thoughts on the matter. For example, in 1925, Paavo Nurmi, the Finnish Olympic gold medallist, is reported to have concluded that a time of 4:04.0 was the fastest time that could ever be achieved for the mile run (Deakin, 1961). Given that Nurmi held the record (4:10.4) for the mile at that time, as well as the world records for the 1500 m, 2000 m, 3000 m, 5000 m, 10 000 m, 20 000 m and 1 hour run, his predictions were given considerable credence.

With the record at 4:06.8 (still well above Kennelly's mathematical projection of 3:58.1), Hamilton (1934) published his list of 'perfect records', values which he felt would never be surpassed. All his predicted ultimate records were broken by 1970, with his 4:01.2 'perfect record' for the mile lasting until 1954. Obviously, athletes and sportswriters did not give much credence to Kennelly's mathematical projections.

9.3.2.3 *Lucy*

Lucy's (1958) paper appears to be the earliest reported study which used mathematical modelling for purely predictive purposes. He used the nine best times per year in the mile run for years 1951–58 to predict ultimate mile performance. Lucy claimed that due to the interruptive effect of World War II one should not use data prior to 1950, a claim that others have empirically examined and refuted (Schutz and McBryde, 1983). Unlike Kennelly, who modelled the relationship between velocity and distance over all running distances, Lucy modelled performance time and chronological time for one event, allowing for mathematical predictions of future performance. He derived the exponential function:

$$t_n = t_0 + ka^n \qquad (9.3)$$

where t_n is the time in seconds for the best mile time in year $1950 + n$ and a and k are parameters to be estimated. Lucy reformulated equation (9.3) as a log-log function, solved for a and k using least squares, and then performed an iterative process to estimate the minimum value of t_n (because he used the average of nine runs at each value of n) to refine his estimates of a and k. Today, his procedures appear rather crude, but at the time it was an acceptable approach.

Unfortunately, Lucy's results are only provided in a hand-drawn graph of his predicted values from 1951 to 1962. As we would like to compare Lucy's function with subsequent work, we have estimated the parameter values for a and k. Using his reported ultimate value $t_\infty = 3{:}38$ and values for t_n (4:09 at $n = 0$, 3:52.5 at $n = 11$) estimated from the graph yields the equation:

$$t_n = 218 + 31.0(0.933)^n \qquad (9.4)$$

where t_n is the expected time (in seconds) for the mile run in year $1950 + n$. Lucy concluded that his projections were extreme and that 'it is difficult to believe that this time [3:38] could ever be achieved'. He did suggest that times below 3:50 were achievable and that 'perhaps even 3:45 is possible'. Given that the current record is 3:44.40, his ultimate projection of 3:38.0 does not seem that unreasonable today.

9.3.2.4 *Deakin*

Deakin (1961) proposed fitting a three-parameter exponential model to world record data (21 data points), using the function:

$$t(n) = \alpha e^{-\beta(n-1)} + \gamma \qquad (9.5)$$

where $t(n)$ is the nth world record ($n = 0$ to 20, representing the 21 years in which world records were set from 1911 to 1965), and α, β and γ are parameters to be estimated. It is not clear how Deakin estimated his parameters, but it appears that he may have fitted a curve to estimate γ, the asymptotic performance or ultimate record. He then used an iterative procedure to modify it and estimate α and β. The only result reported is his final estimate of $\gamma = 3{:}34$, which he notes is very close to Lucy's value of 3:38. The similarity of these results is not surprising considering that Deakin's equation (9.5) and Lucy's equation (9.3) are equivalent functional forms (derived independently and fitted to different sets of data).

Deakin (1967) tested out a different function to see how sensitive the ultimate value was to a 'different form of the fitted curve'. He felt a more valid function would permit extrapolation both forward and backward, thus providing estimates of not only ultimate performance, but also of a time which 'always has been within the bounds of some living person's capability'. This is certainly an interesting approach, but his result of 5:00.0 as the value which could have been achieved since the beginning of time is open to debate. He used the following function to model assumed world records over the course of history:

$$t(n) = \alpha - (2\beta/\pi)\arctan(\gamma(n-1)+\delta) \qquad (9.6)$$

where $t(n)$ was the world record, measured as the deviation (in tenths of seconds) from a time of four minutes, and α, β, γ and δ were parameters to be estimated. How these parameters were estimated was not apparent (to us), but Deakin describes a process of making initial estimates for α and β, plotting some values and fitting a straight line 'by eye', arriving at estimates for γ and δ, once again plotting values and fitting a straight line, and eventually arriving at final estimates of α and β. This yielded a final estimate of 3:34 for t_∞ (which Deakin lowered to 3:32 because records have been broken by as much as 2 s).

It seems surprising that Deakin arrived at exactly the same asymptotic performance using two different functions (and different data sets). We are of the opinion that the large number of estimates and assumptions inherent in his methods enabled him to arrive at identical results. Nevertheless, Deakin's efforts, along with those of Lucy, were the first serious attempts to model present and future track performances, and undoubtedly led to an increased awareness of this interesting problem.

9.3.2.5   Ryder, Carr and Herget

Ryder *et al.* (1976), using world records, examined the changes from 1910 to 1975 in velocities (m min^{-1}) for distances 100 m to the marathon. They concluded that the rate of improvement in velocities over that 65-year period was linear, with the 100 m showing the smallest change (an increase in velocity of 0.6 m min^{-1} year^{-1}) and the marathon the greatest rate of change (0.9 m min^{-1} year^{-1}). They then extrapolated these velocities to the years 2004 and 2028, yielding projections of 3:38.3 and 3:30.0, respectively, for the mile run. Their function for the mile would project a current world record of 3:41.0, 3.4 s better than the current record. These projections reveal the problem with all linear functions (even if the dependent variable is velocity rather than time), namely that they tend to overestimate future performance. More troublesome is the obvious extension that performance will never reach an asymptote and eventually the mile record would approach zero.

9.3.2.6   Chatterjee and Chatterjee

In a paper which received a considerable amount of attention, Chatterjee and Chatterjee (1982) fitted a number of models to winning times in the Olympic Games from 1900 to 1976. They then predicted winning times for the 1980 and 1984 Games. Interestingly, when I look back at the notes I made about this paper in 1982 (I started looking into world record predictions, and wrote an unpublished paper on the topic, in 1975), I see that I had trouble following some of the steps and could not replicate their results. Apparently, others have had the same problem. Morton (1984) points out the inadequacy of their data set and Blest (1996) cites a number of papers which raise questions regarding the accuracy and validity of the analyses. Consequently, Chatterjee and Chatterjee's paper will be reviewed no further here.

9.3.2.7 *Schutz and McBryde*

Schutz and McBryde (1983) reviewed the work done on the prediction of world track and field records and noted the very large discrepancies in the predicted outcomes. This was primarily due to two factors: (1) different studies utilized different prediction methods; and (2) different studies utilized different data sets. They concluded that there were problems with both. The early predictions (e.g. Hamilton, Nurmi) were primarily subjective, being based on experience and knowledge of the sport itself. The early scientists (e.g. Kennelly, Meade) used subjective extrapolations from a limited database. Lucy (1958), Deakin (1961, 1967) and others used exponential decay models in which an ultimate asymptotic performance in time was predicted from a limited series of previous performances. Lloyd (1966) and Ryder *et al.* (1976) extrapolated from linear trends over historical time in velocity for a given distance. Such a process is somewhat similar to the exponential decay functions with historical time and performance time in that a linear function in velocity produces a non-linear asymptotic type of function in time. However, these linear velocity functions asymptote to zero; thus, predictions into the distant future tend to be unacceptably low.

Schutz and McBryde concluded that the databases used in these early studies (and in most subsequent studies) have limited potential for predicting future records. Data sets consisting of world records for all running events at a particular point in time, while being useful for examining the time–distance relationship, are of very limited value for predicting future times. Data sets consisting of world records over time for each event, while of some value, have the disadvantage of being discontinuous, as records progress in step functions of varying sizes. For example, Sebastian Coe's 1981 world record for the men's 800 m was broken twice within a ten-day period in August 1997, resulting in a 16-year plateau followed by (at least) two data points in the same year. Fitting smooth curves to such data often leads to poor fits.

In an attempt to overcome some of these perceived shortcomings in earlier studies, Schutz and McBryde used continuous data (the best performance each year in each event from approximately 1900 to 1982) and compared four different models. Models were fitted with the full data sets, and with subsets (e.g. 1945–82), to examine if the war years affected the goodness of fit and/or predictions. Owing to the length of the time period (over 80 years), including or excluding certain periods of time (e.g. 1914–18, 1939–45) did not make a noticeable difference, and thus the full data sets were used for all further analyses. The four models tested were as follows.

Model A.   An exponential model relating time or distance and historical date for a specific event.

$$t_n = \alpha + \beta\,e^{\gamma n} \tag{9.7}$$

where t_n is the predicted time in year $1899 + n$, and α (the estimated asymptotic value of t_n as n approaches infinity) and β and γ are parameters (estimated using least squares with a modified Gauss–Newton algorithm). If the magnitudes of performance improvements are declining over time, then the function will be asymptotic with respect to time and γ will be negative. For running events (decreasing times), β will be positive and, for field events (increasing distances), β will be negative. This model is similar to those used by Lucy (1958) and Deakin (1961).

Model B.   The same form as Model A, but relating velocity (rather than running time) to year.

Model C.   A linear model between running velocity and historical date for a specific event.

$$t_n = \alpha + \beta n \tag{9.8}$$

where t_n is running time for a specific event, $1899 + n$ is the historical year, and α and β are estimated intercept and slope parameters (least squares estimation procedures were used). This model is similar to that used by Ryder *et al.* (1976).

Model D.   A two-parameter power function relating time and distance over all running events for a given year.

$$t_j = \alpha d_j^{\beta} \tag{9.9}$$

where t_j is the time in seconds for distance d_j (100 m to 10 000 m) and α and β are parameters to be estimated (using the same methods as for Model A). Lietzke (1954) and Riegel (1981) used this model. Riegel interpreted the parameter β as the 'fatigue factor', as it describes the rate at which the average velocity of a race decreases with an increase in the distance.

The results of Schutz and McBryde suggested that all four models provided reasonably good fits and projections for the shorter races, but Model A was the most appropriate one for longer distances. Their projection for ultimate performance in the mile was 3:28.5, which is quite close to Deakin's prediction of 3:32.0, and the projected 1500 m world record in the year 2000 was 3:37.5, very close to the current record of 3:37.4.

9.3.2.8   Smith

Smith (1988) took issue with the basic assumption underlying most of the previous models, namely that the records come from a sequence of independent and identically distributed random variables. He therefore used a method based on maximum likelihood principles, and used a form of residual analysis to test the models. He used world record data (which he treated as censored data) for the mile and the marathon from 1860 to 1985.[1] Smith fitted nine different models, all combinations of (1) an assumed distribution (normal, Gumbel, or generalized extreme value (GEV)), and (2) a 'drift' or pattern of

[1]We do not recommend using 'world records' prior to 1900 as there were no standards or verification procedures in place to validate any records prior to 1900.

record times (linear, quadratic or exponential). Smith states that the mile data suggested a linear trend from 1931 to 1985, so he tested all nine models using only this subset of the data (for reasons not obvious to us).

The normal distribution was generally superior to the Gumbel, and the GEV was superior to the normal in those cases for which the GEV algorithm converged and a proper solution was obtained. Smith presented his results in graphical form, making it difficult to arrive at an exact predictive value; however, a projection of a mile record of 3:40.0 in the year 2000 was made from close inspection of his Fig. 1. This value is considerably less than the current record of 3:44.4 – a characteristic of all studies that have assumed a linear rate of improvement. Smith arrived at an ultimate mile record of 2:58.0, considerably less than most other predicted ultimates. Smith noted that the standard error of this prediction is a very large 63 seconds, and concluded that 'it is questionable whether any useful estimate of the asymptote is possible'. Our conclusion is that Smith derived a sophisticated statistical method to model world record data, but it lacks predictive validity.

9.3.2.9 *Schutz and Liu*

Schutz and Liu (1993) replicated and extended the work of Schutz and McBryde (1983) by refitting the four models with updated data sets, comparing male and female performance trends, and introducing a new model. Comparing world record data and best performance per year data (from the time the IAAF formally recognized the event up to and including 1992), they found that the latter data resulted in considerably smaller standard errors (of parameter estimates) and higher R^2 values (ranging from 0.90 to 0.98 with a median of 0.96). Additionally, their findings supported Schutz and McBryde's conclusion that the three-parameter exponential model was the best model.

Figures 9.1 to 9.6 graphically portray their findings. The solid line is the best fit of the three-parameter exponential function as described above, and the dotted line is the best fitting linear function. The figures' inclusion of the best performances per year for 1993 to 1996 (noted with an 'X' and not used in the fit) provide a visual test of the degree to which the exponential function predicts short-term future performance. It is apparent that performance is plateauing for both males and females in most running events.

Schutz and Liu's new model, called the Random Sampling Model, was developed and tested on one event, the men's 1500 m. It assumed that the true score mean of the top performances has now stabilized and is expected to be constant for the next 50 years or more. Under this assumption, new world records can still be set even though the average ultimate performance has been reached; that is, random sampling from the population of all future world class runners will generally yield times close to the mean, but occasionally an outlier will occur, thus setting a new record. The top-50 performances per year since 1980 (the estimated first year of stabilization ascertained by trend analysis) were used to determine the population mean and variance of mean ultimate performance. Foster and Stuart's (1954) procedures were used to test the randomness and stability of the 650 (50 per year for 13 years) 1500 m top performances over the period 1980–92, with the results supporting the hypothesis of no trend ($p > 0.25$).

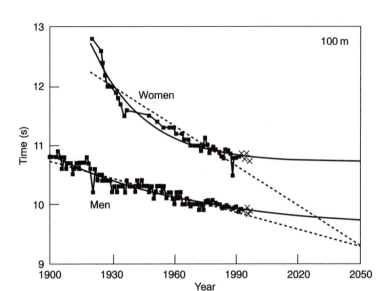

Figure 9.1   Best performance per year data and fitted curves for the 100 m.

A Monte Carlo program was written to estimate future world records and waiting times to each new record. Based on the calculated population mean and variance of ultimate performance, top-50 performances per year for each year in the future were randomly generated. Extreme value theory, utilizing

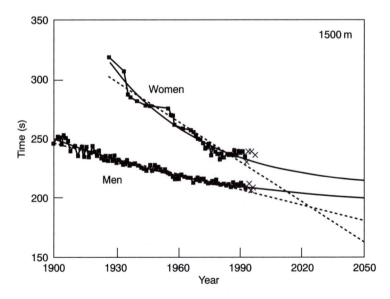

Figure 9.2   Best performance per year data and fitted curves for the 1500 m.

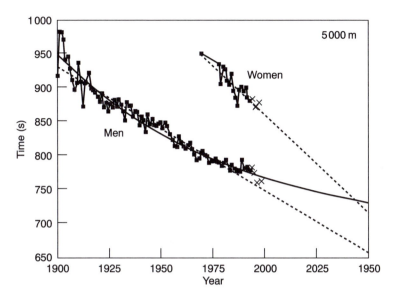

Figure 9.3 Best performance per year data and fitted curves for the 5000 m.

an exponential distribution, was used to model the ordered extreme performances and generate new world records. The dependent variable of interest for each simulation was the value of each world record and the waiting time (number of years) for it to occur. The program compared the generated best

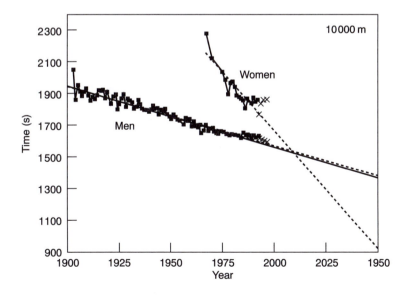

Figure 9.4 Best performance per year data and fitted curves for the 10 000 m.

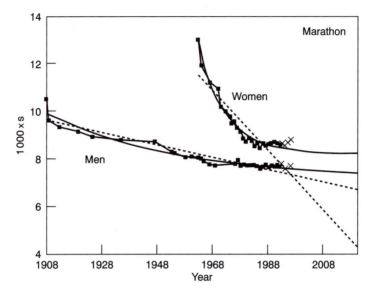

Figure 9.5   Best performance per year data and fitted curves for the marathon.

performance of each year with the previous world record, and identified and recorded any new world records. This simulation continued until a specific number (3) of world records occurred. This process was replicated 1200 times, yielding a distribution for both variables (record performance and waiting time)

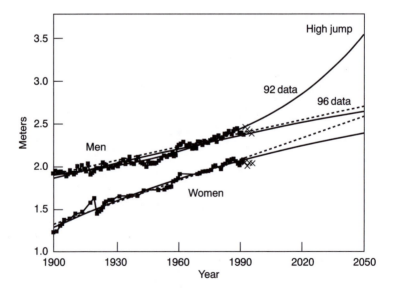

Figure 9.6   Best performance per year data and fitted curves for the high jump.

for the first new world record, second world record and third world record. As a partial validity check, the simulated waiting times to the next world record were compared with the theoretical expected waiting times calculated using Glick's (1978) recommended procedures and shown to fall within expectations.

Since the world record distributions were skewed, the medians of the 1200 simulations were used to predict the world records. The results indicated that the first world record after 1992 (at which time the world record was 3:29.46, for 1500 m, set in 1985) would be 3:28.39, with a standard error of 0.013 s and a mean waiting time of 10 years (±0.16). History has shown us to be very close with respect to both the date and the magnitude of the next world record – Morceli set the next (and current) record in exactly ten years (1995) with a time of 3:28.37, only 0.02 s off our predicted time. The second world record would be 3:27.44 (±0.048) with a waiting time of 17 years (±0.43), and the third world record would be 3:25.22 (±0.11) with a waiting time of 28.95 years (±0.85). This would result in a world record of approximately 3:25.00 by the year 2041. A comparison with the predictions given by our other models indicates that, as expected, these predicted values are considerably slower than what was predicted with the exponential model (3:19.11 in 2050). It remains to be seen if 1500 m runners have indeed reached a plateau or if in fact a downward trend will re-establish itself over the next few years.

9.3.2.10 Blest

In the most recent paper of note on world records, Blest (1996) used world running records at the time (year) of each of the 18 Olympiads 1912 to 1992, since he considered this to be 'a more extensive and well-behaved set' of data. He did not make clear as to what the comparison data sets are, but he did cite Chatterjee and Chatterjee's (1982) Olympic winning times and Deakin's (1967) world records. As pointed out previously, such data sets are incomplete and, although Blest's data are certainly preferable, they still do not reflect the actual level of performance increments over time. For example, the 400 m record set in 1968 remained as the world record until after the 1984 Games, thus suggesting that there was no improvement in 400 m running over that 16-year period. However, examination of the best-per-year performances over that 16-year period clearly indicated that there was a gradual decrease in times, but not enough to break the record. Lee Evans' 400 m record of 43.86 s, set in 1968 at the Mexico Olympics, was a full second better than the world record which existed at the time of the 1964 games (and the record was broken twice between 1964 and 1968). Using best times per year eliminates these plateaus and provides a more accurate description of year-to-year running performances.

Blest went on to suggest a power function relating time and distance (over all running distances), and stated that 'despite the apparently natural choice of this model, this author has discovered no other reference to its use'. As already pointed out in this chapter, Kennelly (1906), Lietzke (1954), Riegel (1981), Schutz and McBryde (1983) and Schutz and Liu (1993), and undoubtedly a number of others, have used this model.

Despite the fact that Blest did use world record data and a well-established power model to fit the data, his paper did add to our knowledge of predicting world records. Blest fitted the power function $t_j = \alpha d_j^{\beta}$ to each of the 18 data sets, and then computed an average value for β and refitted the data using this average value for each of the 18 data sets. He showed that there was no significant degradation of the fit to the data. More importantly, he then examined the change over time of the parameter β (the slope of the time–distance relationship if expressed in log-log form) using a simple linear model and seven non-linear functions. The three functions (all containing four estimatable parameters) which provided the best fits were a logistic, a reparameterized Gompertz, and an 'antisymmetric exponential' model. He then estimated lower bounds (ultimate world records) for each of these functions, with the antisymmetric exponential producing the lowest estimates and the reparameterized Gompertz the largest. Blest's logistic projections (e.g. a lower bound of 3:10.43 for the 1500 m) were very similar to those obtained by Schutz and Liu using the three-parameter exponential model with world-best times (e.g. an ultimate performance of 3:09.04 for the 1500 m).

9.3.3   Prediction models for field events

Relative to the amount of interest shown in the modelling of running performances, very little work has been done on modelling and predicting field events. The reasons for this are probably twofold. First, field events attract much less public attention than do running events and are analyzed only for the rare occasion related to an outstanding performance, such as Bob Beamon's 1968 long jump record which stood for 23 years. Second, the modelling of record performances would have to take into account the effects of changes in equipment (a difficult task). For example, the introduction of the fiberglass pole resulted in dramatic increases in pole vault performances, and changes in the rules regarding both the weight and the release of the javelin affected record performances in that event. Nevertheless, a few studies have looked at some of the non-running events.

Stefani (1977), using only percentage increases from one Olympic Games to the next, predicted performances in the year 2000 for all men's and women's track and field events (and swimming events too). His predictions for the running events appear reasonable – the current world records for some events are already better than his predicted Olympic times for the year 2000; in some events his predictions are unlikely to be achieved (e.g. 9.72 for the 100 m); and in some events his predictions may be right on (e.g. the 110 m hurdles where his prediction of 12.90 s is only 0.01 s below the current world record). In the high jump, long jump and triple jump his predictions are all quite close to current records, but in the throws (shot, discus, javelin and hammer), his predictions exceed the current world records by at least 10% for each event. The reason for this large error in predictions for the throwing events is that the rate of improvement in these events has slowed down since 1977, and linear projections like Stefani's tend to estimate performances considerably superior to what we are observing in the 1990s. Chapter 12 on predicting outcomes presents more details on these models.

Schutz and McBryde (1983) tried to fit some of the field events with their exponential function and met with mixed success. For some events the exponential function (Model A) could not fit the data, and for some others (e.g. high jump) the data exhibited an *increasing* rate of improvement (yielding an asymptotic value of infinity). However, for the long jump and shot put (the only men's field events for which Schutz and McBryde reported projections), their predictions for world records in the year 2000 may turn out to be surprisingly accurate. They predicted a long jump record of 8.97 m (current record is 8.95 m) and a shot record of 23.45 m (current record is 23.12 m).

Interestingly, in some recent work, Schutz and Liu (1993) were still not able to discern any non-linearity in performance in the men's high jump, with a linear model providing a better fit to the data than any of a number of non-linear functions. Given that we were not willing to assume that this trend would continue forever, we concluded that performance in the high jump has not yet reached the early stages of asymptotic performance seen in all running events, and it may be another five or ten years before realistic future performances can be predicted. However, we fit the exponential function using data through 1996 and, as can be seen in Fig. 9.6, the pattern now conforms to a negatively accelerating exponential function.

9.3.4   Comparing the predictions

Table 9.1 contains the predicted world records for the year 2000 and Table 9.2 shows the expected ultimate performance records as reported by those authors who calculated such values (and were fearless enough to put them in print). It should be noted that the year 2000 values listed for Ryder *et al.* are actually their year 2004 values (they did not predict performance in year 2000). Also, the Rumball and Coleman values are converted values (they reported ultimates for non-metric distances).

Looking at the year 2000 men's predictions, we can see that all of the Ryder *et al.* values are considerably below the current world records, supporting our

Table 9.1   Predictions of world records in the year 2000[1]

Authors[2]		Ryder et al.	Stefani	Schutz and McBryde	Peronnet and Thibault	Current world record[3]
100 m	M	9.56	9.72	**9.85**	9.74	9.84
	W		10.65		10.66	10.49
200 m	M	18.97	19.76	19.53	19.53	19.32
	W		21.09	19.71	**21.46**	21.34
400 m	M	42.49	42.65	**43.20**	**43.44**	43.29
	W			45.20	46.85	47.60
800 m	M	1:38.30	1:38.05	1:39.30	1:39.88	1:41.11
	W			**1:53.10**	1:51.16	1:53.28
1500 m	M	3:22.20	3:33.08	**3:26.80**	**3:25.45**	3:27.37
	W				3:47.93	3:50.46
5000 m	M	12:24.80	**12:44.07**	**12:46.50**	**12:42.72**	12:39.74
	W				14:19.33	14:36.45
10 000 m	M	25:44.00	26:07.85	**26:43.10**	**26:43.63**	26:27.85
	W				**29:38.41**	29:31.78
Marathon	M				2:05:23.72	2:06:50
	W				**2:18:43.34**	2:21:06

[1] Bold face values indicate that the predicted time is within ±1% of the current world record. Single under-lines indicate an underprediction, double underlines an overprediction
[2] Ryder *et al.* (1976), Stefani (1977), Schutz and McBryde (1983) and Peronnet and Thibault (1989)
[3] World record as of October 1997 (men's 800 m, 5000 m and 10 000 m not yet official)

Table 9.2   Predictions of ultimate performances

Authors[1]		Hamilton	Rumball and Coleman	Morton	Peronnet and Thibault	Schutz and Liu	Blest
100 m	M	10.06		9.15	9.37	9.56	9.47
	W				10.15	10.71	
200 m	M	20.05		18.15	18.32		18.66
	W				20.25		
400 m	M	46.20		39.33	39.60		41.82
	W				44.71		
800 m	M	1:46.70	1:26.49	1:33.00	1:30.86		1:34.88
	W				1:42.71		
1500 m	M	3:44.78	2:47.77	3:05.70	3:04.27	3:09.04	3:10.43
	W				3:26.95	3:26.96	
5000 m	M	14:02.00	10:03.77	11:22.90	11:11.61	11:20.06	12:03.26
	W				12:33.36		
10 000 m	M	29:17.70	21:07.61	24:04.60	23:36.89		25:59.40
	W				26:19.48		
Marathon	M		1:37:30.00		1:48:25.25	1:59:25.21	2:00:26.33
	W				2:00:33.22	2:17:12.26	

[1] Hamilton (1934), Rumball and Coleman (1970), Morton (1984), Peronnet and Thibault (1989), Schutz and Liu (1993), Blest (1996)

claim that their linear model is overly optimistic. Stefani's predictions have held up quite well, considering that they were based on percentage improvements and were made over 20 years ago. Three of his predicted values have already been surpassed, but it is possible that the other four could turn out to be the best predictors by the year 2000. Schutz and McBryde's predictions show the smallest absolute deviation from the current record in four of the seven events for which they reported predictions. Peronnet and Thibault (whose work is described in Section 9.4) are more accurate for two events, and both papers report the same predicted record for the 200 m. Given this advantage in accuracy, and that Schutz and McBryde's predictions were made six years before the Peronnet and Thibault paper was published, we think the model used by them (and subsequently by us) is the best predictive model.

Looking at the ultimate predictions, it is readily apparent that Hamilton underestimated the future development of athletic abilities, and Rumball and Coleman have such high expectations that it is most unlikely that their ultimate values will ever be reached. With respect to the other four sets of predictions, we will never know which one is the most valid. Morton tends to expect greater improvements in the sprints as indicated by his lower ultimate values for the 100 m, 200 m and 400 m runs. Blest is the most conservative, with his predicted ultimates being higher than the other four for all but the 100 m. Peronnet and Thibault provide the most comprehensive set of predictions, with their sprint values being higher than Morton's and their predictions from 800 m to the marathon being lower. Schutz and Liu's predictions, being based on the same model and type of data used by Schutz and McBryde, could well prove to be the most accurate (given the accuracy of the Schutz and McBryde predictions for the year 2000). Unfortunately, the Schutz and Liu predictions are incomplete (due to space limitations in publication and difficulties fitting the exponential model to some events at that time).

9.4 Physiological models

The majority of published studies dealing with the modeling of world record and future performance in track and field have been essentially time series analyses; that is, the models have been based solely on the relationship of observed past performances with historical time. However, there is another set of literature which may not be as extensive but is certainly just as interesting. In these studies performance (past, present and future) is modelled not just with chronological time, but also with physiological and/or biomechanical factors. The parameters or components of these models have a direct physiological representation, and the mathematical functions are similar to the time versus distance models in form. Projections are made by setting theoretical limits on one or more components of the physiological function.

The most cited early study is Hill's (1925) paper, in which he explains the relationship between running time and physiological (oxygen) requirements. His mathematical relationships between oxygen consumption and velocity, however, are expressed only in graphical terms, and he neither attempts to derive any mathematical functions nor makes any predictions.

Henry (1954) developed a comprehensive polynomial exponential model, relating velocity and running time, for a specific year:

$$V(t) = b_1 e^{-k_1 t} + b_2 e^{-k_2 t} + b_3 e^{-k_3 t} + b_4 e^{-k_4 t} + b_5 e^{-k_5 t} \qquad (9.10)$$

where $V(t)$ is velocity, t is time in seconds, k_i is a rate constant ($i = 1, 2, \ldots, 5$), and b_i is the velocity constant in yards per second. The five components of this model represent the energy loss, alactate O_2 debt, lactate O_2 debt, glycogen depletion, and fat metabolization. This is generally accepted as an accurate and complete model of the energy requirements for running.

Peronnet and Thibault (1989) developed a model to describe the average power output $P(t)$ ($W kg^{-1}$) sustained over the race duration T (seconds):

$$Pt = [S/T(1 - e^{-T/b_2})] + 1/T \int_0^T [BMR + B(1 - e^{-t/b_1})] \, dt \qquad (9.11)$$

where BMR is the basal metabolic rate, B ($W kg^{-1}$) is the difference between peak and BMR, b_1 is a time constant for the kinetics of aerobic metabolism at the beginning of exercise, b_2 is the time constant for the kinetics of anaerobic metabolism at the beginning of exercise, S is the energy from anaerobic metabolism actually available to the runner over T, and T is the race duration (seconds). Peronnet and Thibault are the only authors who extended their physiological-based model to formulate prediction models. They generated predictions for all men's and women's running events for the years 2000, 2028 and 2040, as well as an ultimate performance. Their year 2000 predictions are surprisingly close to those obtained by Schutz and McBryde (1983) using an exponential model relating performance and chronological time. Although these two approaches may provide equally accurate predictions, the Peronnet and Thibault model must be considered a more useful model as it contains some explanatory power to accompany the predictive capabilities.

Other works of note in this area, which include more general laws of nature (such as the first law of thermodynamics), are those by Lloyd (1966), Keller (1974), Wilkerson (1982), Ward-Smith (1985), and Behncke (1987).

9.5   Male–female comparisons

A topic of recent interest and controversy centers on the question of when, or if, women will catch up to men in track performance. Although some women's track and field events were incorporated into the Olympic Games in 1928, full participation by women in a wide range of events is a very recent phenomenon. However, given the chance, women exhibited a remarkable rate of improvement in their world records from approximately 1960 to the mid-1980s. Some investigators assumed that this rate of improvement would continue indefinitely, and predicted that women runners may surpass their male counterparts early in the 21st century. For example, Dyer (1977) has stated that, 'Equality between the sexes will be achieved shortly after the turn of the century, ... and even in the events with the slowest change equality is predicted in the year 2044', and Whipp and Ward (1992), two well-respected physiologists, wrote in *Nature* that 'Women will be running at the same velocities as men before the year 2050, and will exceed men in the marathon by 1998.' It is apparent that these authors, and a number of sports writers and broadcasters, were deceived by the very rapid advances that women made in the 1960s to 1980s. The linear rate of improvement during that time period reflected the sudden increase in participation by women in a number of events – just as some men's events exhibited a linear increase for the first 30 or 40 years of this century.

History now shows that women's performance has levelled off, and in fact it appears to have reached near-asymptotic levels very quickly. Table 9.3 presents the current world record for men's and women's track and field events, and the percentage difference in performances for the years 1970, 1980 and 1996. It is interesting to note that in the nine metric running events, 100 m to marathon, the current world record for women was set in the 1980s for five of those events. In contrast, only two records from the 1980s remain for the men (and six of the current world records were set in 1996 or 1997). In some of the field events such as the pole vault and triple jump, which were just recently included as world championship or Olympic events for women, there is still a very marked 'learning curve', and it may be a number of years before we see a plateauing of performance in these events. Overall, although the percentage difference values declined dramatically from 1970 to 1990, there has been very little further decline – in fact, the difference has increased in four events, decreased in four and remained constant in one other. It appears that the male–female difference of approximately 10–12% is a performance difference that may remain constant over time.

There were some early attempts to analyze women's running performances and predict future achievements, but most of these were based on limited data, used linear models, and did not attempt to predict beyond the next Olympics or some other relatively short time period (e.g. Craig, 1963; Frucht and Jokl, 1964; Hodgkins and Skubic, 1968). One exception is the comprehensive work of Dyer (1977, 1982) who made extensive comparisons between men's and women's performances in a number of sporting events. However, Dyer primarily relied on statistics such as percentage changes and percentage differences (between men and women), and did not utilize any non-linear model fitting procedures.

Table 9.3 World records in track and field (as of 30 December 1996), with male–female comparisons

Track events	Men Record	Year	Women Record	Year	Percentage difference[1] 1970	1980	1996
100 m	9.84	1996	10.49	1988	11.1	11.1	6.6
100 m Hurdles	–	–	12.21	1988			
110 m Hurdles	12.91	1993	–	–			
200 m	19.32	1996	21.34	1988	13.1	10.1	10.5
400 m	43.29	1988	47.60	1985	14.1	9.8	10.0
400 m Hurdles	46.78	1992	52.61	1995			12.5
800 m	1:41.73[2]	1981	1:53.28	1983	15.5	10.8	11.4
1500 m	3:27.37	1995	3:50.46	1993	17.6	10.0	11.1
1 mile	3:44.39	1993	4:12.56	1996			12.6
3000 m	7:20.67[2]	1996	8:06.11	1993	25.8	12.2	10.3
Steeplechase	7:59.18	1995	–	–			
5000 m	12:44.38[2]	1995	14:36.45	1995	19.7	14.7	14.7
10 000 m	26:38.08[2]	1996	29:31.78	1993	28.4	16.0	10.9
Marathon	2:06:50	1988	2:21:06	1985	42.3	13.3	11.2
Field events[3]							
High jump	2.45	1993	2.09	1987	16.6	14.8	14.7
Pole vault	6.14	1994	4.45	1996			27.5
Long jump	8.95	1991	7.52	1988	23.1	20.3	16.0
Triple jump	18.29	1995	15.50	1995			15.3
Shot	23.12	1990	22.63	1987			
Discus	74.08	1986	76.80	1988			
Hammer	86.74	1986	69.42	1996			
Javelin	98.48	1996	80.00	1988			

[1] Calculated as (women's record − men's record)/(men's record)
[2] New world records were set in these events in 1997, but have not yet received IAAF formal ratification
[3] All field event records are distances in meters

In the early 1980s, Schutz and McBryde (1983) were somewhat unsuccessful in fitting any type of non-linear function to most sets of women's track and field data (best performances per year). However, by the early 1990s, women's achievements had shown a distinct levelling off, and thus in 1993 we (Schutz and Liu) were able to fit the three-parameter exponential model successfully to a number of women's track and field events. An exception was the women's 100 m data, which could not be fitted successfully with this function because of the extraordinary low value of the current world record. Florence Griffiths-Joyner set the record of 10.49 s in 1988, breaking the existing record set in 1984 by over a quarter of a second (0.27 s), a remarkable feat considering that 100 m records are usually broken by hundredths of seconds, and on the rare occasion, by a tenth of a second. Since 1988, no other female runner has come remotely close to Griffiths-Joyner's time (the suggested reasons for her amazing performance are another story!), and thus the best fitting curve asymptotes shortly after 1988, and predicts a record of 10.72 s in the year 2050 (see Fig. 9.1).

Figures 9.1 to 9.6, explained previously, provide a clear graphic representation of the relative recent progress of men and women in a selection of track and field events. The linear function is obviously not a good fit in most cases and is included to support our argument that women are no longer progressing at a linear rate and, in fact, have generated data over the last 10 years which conforms well to an asymptotic function. It is clear that both women and men are approaching asymptotic levels of performance in many events, and the women's asymptote is approximately 12% above that of the men. Our results certainly do not support claims that women's performances will equal those of men sometime in the next century.

9.6   Conclusions

The focus of this chapter has been on the prediction of ultimate performance in track and field. It is obviously a fascinating, and challenging, exercise, as evidenced by the number of individuals who have expended a considerable amount of effort on this problem. To answer the question 'What are the limits of human athletic achievements?' will be a never-ending quest, but one of continuing interest and study. Undoubtedly, the future holds surprising changes, in equipment, in bio-engineering, in performance-enhancing drugs, etc, and we will never be able to predict confidently an 'ultimate performance'. However, as Deakin (1961) stated, 'it is good fun, and probably about as exact as some of the uses that mathematics is put to elsewhere'.

References

Behncke, H. (1987) Optimization models for the force and energy in competitive sports. *Mathematical Methods in the Applied Sciences*, **9**, 298–311.

Blest, D. C. (1996) Lower bounds for athletic performance. *The Statistician*, **45**, 243–253.

Chatterjee, S. and Chatterjee, S. (1982) New lamps for old: an exploratory analysis of running times in Olympic Games. *Applied Statistics*, **31**, 14–22.

Craig, A. B. (1963) Evaluation and predictions of world running and swimming records. *Journal of Sport Medicine*, **3**, 14–21.

Dawkins, B. P., Andreae, P. M. and O'Connor, P. M. (1994) Analysis of Olympic heptathlon data. *Journal of the American Statistical Association*, **89**, 1100–1106.

Deakin, B. (1961) Mathematics of athletic performance. *Matrix, Melbourne University Mathematical Society Magazine*, issue 4.

Deakin, B. (1967) Estimating bounds on athletic performance. *The Mathematical Gazette*, **51**, 100–103.

Dyer, K. F. (1977) The trend of the male–female performance differential in athletics, swimming and cycling 1948–76. *Journal of Biosocial Sciences*, **9**, 325–338.

Dyer, K. F. (1982) *Catching Up with the Men*. London: Junction Books.

Foster, F. G. and Stuart, A. (1954) Distribution-free tests in time-series based on the breaking of records. *Journal of the Royal Statistical Society*, **16**, 1–22.

Frucht, A. H. and Jokl, E. (1964) Parabolic extrapolation of Olympic performance growth since 1900. *Journal of Sports Medicine*, **4**, 142–152.

Glick, N. (1978) Breaking records and breaking boards. *The American Mathematical Monthly*, **85**, 2–26.

Hamilton, B. (1934) Table of the ultimate of human effort in track and field. *Seattle Post Intelligencer*, 24 July 1974.

Heffley, D. R. (1977) Assigning runners to a relay team. In S. P. Ladany and R. E. Machol (eds), *Optimal Strategies in Sports*, pp. 169–171. Amsterdam: North-Holland.

Henry, F. M. (1954) Theoretical rate equation for world record running speeds. *Science*, **120**, 1073–1074.

Hill, A. V. (1925) The physiological basis of athletic records. *British Association for the Advancement of Science, Report of the 93rd Meeting*, pp. 156–173.

Hodgkins, J. and Skubic, V. (1968) Women's track and field records. *Journal of Sports Medicine*, **8**, 36–42.

Keller, J. B. (1974) Optimal velocity in a race. *The American Mathematical Monthly*, **81**, 474–480.

Kennelly, A. (1906) An approximate law of fatigue in speeds of racing animals. *Proceedings of the American Academy of Arts and Sciences*, **42**, 275–331.

Kennelly, A. (1926) Changes during the last twenty years in the world's speed records of racing animals. *Proceedings of the American Academy of Arts and Sciences*, **61**, 487–523.

Lietzke, M. H. (1954) An analytical study of world and Olympic racing records. *Science*, **119**, 333–336.

Lloyd, B. B. (1966) The energetics of running: an analysis of world records. *British Association for the Advancement of Science, Report of the 133rd Meeting*, pp. 515–530.

Lucy, L. B. (1958) Future progress in the mile. *IOTA*, **1**, 8–11.

Meade, G. P. (1916) An analytic study of athletic records. *Scientific Monthly*, **2**, 596–600.

Meade, G. P. (1966) *Athletic Records: The Ways and Wherefores*. New York: Vantage Press.

Morton, R. H. (1984) New lamps for old? (Letter to the Editor). *Applied Statistics*, **33**, 317–318.

Naik, N. and Khattree, R. (1996) Revisiting Olympic track records: some practical considerations in the principal component analysis. *The American Statistician*, **50**, 140–144.

Peronnet, F. and Thibault, G. (1989) Mathematical analysis of running performance and world running records. *Journal of Applied Physiology*, **67**, 453–465.

Pritchard, W. G. and Pritchard, J. K. (1994) Mathematical models of running. *American Scientist*, **82**, 546–553.

Queracentani, R. L. (1964) *A World History of Track and Field Athletics: 1864–1964*. London: Oxford University Press.

Riegel, P. S. (1981) Athletic records and human endurance. *American Scientist*, **69**, 285–290.

Rumball, W. and Coleman, C. (1970) Analysis of running and the prediction of ultimate performance. *Nature*, **228**, 184–185.

Ryder, R. W., Carr, H. J. and Herget, P. (1976) Analysis of running and the prediction of ultimate performances. *Nature*, **228**, 184–195.

Schutz, R. W. (1994) The systematic study of 'Statistics in Sports': Do we need a framework? *American Statistical Association: Proceedings of the Section on Statistics in Sports*. Alexandria, VA: ASA, pp. 16–20.

Schutz, R. W. and McBryde, J. P. (1983) The prediction of ultimate track and field performance: past, present, future. *Proceedings of the FISU Conference Universiade '83 in Association with the Xth HISPA Congress*, pp. 498–515.

Schutz, R. and Liu, Y. (1993) Track and field world record projections. *American Statistical Association: Proceedings of the Section on Statistics in Sports*. Alexandria, VA: ASA, pp. 52–57.

Sidney, J. B. and Sidney, S. J. (1976) Intransitivity in the scoring of cross-country competitions. In R. E. Machol, S. P. Ladany and D. G. Morrison (eds), *Management Science in Sports*, pp. 145–151. Amsterdam: North-Holland.

Smith, R. L. (1988) Forecasting records by maximum likelihood. *Journal of the American Statistical Association*, **83**, 331–338.

Sphicas, G. P. and Ladany, S. P. (1976) Dynamic policies in the long jump. In R. E. Machol, S. P. Ladany and D. G. Morrison (eds), *Management Science in Sports*, pp. 111–124. Amsterdam: North-Holland.

Stefani, R. T. (1977) Trends in Olympic winning performances. *Athletic Journal*, December, 44–46.

Tibshirani, R. (1997) Who is the fastest man in the world? *The American Statistician*, **51**, 106–111.

Tidow, G. (1989) The 1985 IAAF decathlon scoring tables: an attempt at analysis. *New Studies in Athletics*, **2**, 45–62.

Townend, M. S. (1984) *Mathematics in Sport*. West Sussex: Ellis Horwood.

Wainer, H. and De Veaux, R. D. (1994) Resizing triathlons for fairness. *Chance*, **7**, 20–25.

Ward-Smith, A. J. (1985) A mathematical theory of running, based on the first law of thermodynamics, and its application to the performance of world-class athletes. *Journal of Biomechanics*, **18**, 337–349.

Whipp, B. J. and Ward, S. A. (1992) Will women soon outrun men? *Nature*, **335** (January), 25.

Wilkerson, J. (1982) Running faster than the animals: will man match a bear's speed? *Runner's World*, August, 36–40.

Part II
SPORTS THEMES

10

On the Design of Sports Tournaments

Tim McGarry

University of British Columbia, Canada

10.1 Analysing sports tournaments as a series of paired contests (comparisons)

A sporting contest invariably takes place in the context of wider competition; that is, as part of a series of sports contests organized in such a way as to allow comparison between entrants in order to identify a winner. The organization of a series of sports contests is a sports tournament. The intent of this paper is to review the validity of various tournament designs that are frequently used in sporting competition. We limit our scope, for the most part, to sports contests that consist of only two entrants.

A sports contest between two entrants can be analysed using the method of paired comparison. Here, a paired comparison refers to a binary comparison, or contest, in which a preference for one of the pair is expressed. (For an extensive review of paired comparisons, see David, 1988.) In the context of a sports contest, the preference for an entrant is expressed in the win–loss outcome. Since a sports tournament is an organized series of sports contests, a tournament can be analysed as a series of paired comparisons. This allows for the analysis of various outcomes under a variety of initial conditions; for example, the probability of accurately ranking the first n entrants, the probability that an entrant will rank in the first n places and the probability that the two strongest entrants will meet in a final contest.

We define a sports tournament as a formal system that seeks to rank entrants fairly in accord with their ability. (Note that throughout this chapter, ranking refers to an ordering of entrants after a tournament, while seeding refers to a similar ordering of entrants before a tournament.) Thus, an entrant of stronger (higher) ability should rank above an entrant of weaker (lower) ability, while, at the same time, any entrant should not rank differently from any other entrant of equal ability. In fact, the designs of some sports tournaments preclude this requirement (see later). In this light, statisticians have sought to analyse various tournament designs by quantifying their validity (Kendall, 1955; Maurice, 1958;

David, 1959; Glenn, 1960; Searls, 1963), as well as identifying equitable proce-dures for breaking ties in unbalanced designs (Andrews and David, 1990), often as an extension of the Bradley–Terry (1952) method (Bradley, 1954; Dykstra, 1960; Rao and Kupper, 1967; Davidson, 1970; David, 1987; Joe, 1990) or the Thurstone–Mosteller (Mosteller, 1951) method (Glenn and David, 1960; Henerey, 1992; Kuk, 1995).

10.2   The structure of various sports tournaments

Sports tournaments often take the form of a single elimination or knock-out (KO), a round robin (RR) or, as is often the case in professional sports, some hybrid RR–KO. For example, many sports rank their best entrants from some form of league (RR) competition in the regular season, prior to their seeded entry into an elimination tournament (i.e. playoffs) from which an overall champion is produced. Other sports, like tennis, use the results from various contests to seed the entrants. (For an example of ranking the best player in tennis from an unequal number of comparisons, see Cowden, 1974.) With this in mind, McGarry and Schutz (1997) undertook a study to assess the efficacy of the single elimination, double elimination, RR, RR–KO and RR–RR tournaments under different initial conditions. The various assessment criteria required the use of consolation contests in order to rank all entrants (Figs 10.1–10.4). (In Figs 10.1–10.4, the letters A–H represent the eight entrants and the numbers 1–8 represent the rank of the entrant that finishes at that level.) We assume, in keeping with most sports tournaments, that the structures are balanced, although some unbalanced tournaments will be considered later. (In an unbalanced tournament, each entrant does not compete the same number of times as every other like entrant. In a balanced tournament, on the other hand, each entrant competes the same number of times as every other like entrant. Thus, byes are unnecessary in elimination competition and all entrants contest each other entrant the same number of times in RR competition.) We used $t = 8$ entrants in all cases for purposes of comparison between tournaments. The value of t is not expected to affect our interpretation of a tournament's efficacy relative to that of other tournaments.

10.2.1   Knockout tournaments

The single elimination (KO) proceeds on the basis of elimination where the losing entrants in each round are eliminated from winning the championship until a sole entrant (winner) remains (Fig. 10.1). The double elimination (DKO) also proceeds on the basis of elimination, except that entrants must lose twice before elimination from the championship. (For a variant of this design that can rank any number of entrants, as well as managing entrant with-drawals during the course of tournament competition, see Rokosz, 1990.) Losing entrants proceed to a loser's bracket and are eliminated on a subsequent loss in this bracket (Fig. 10.2). The winner of the winner's bracket and the winner of the loser's bracket proceed to a final contest, which is repeated once if the winner of the loser's bracket wins the first final.

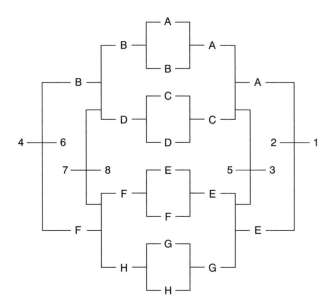

Figure 10.1   KO. Single knockout tournament design with consolation at each stage of the competition. From McGarry and Schutz (1997) *Journal of the Operational Research Society*, **48**, p. 66. Copyright 1997 by Macmillan Press Ltd. Reprinted by permission.

The KO (unseeded) tournament is used in situations where there is little or no *a priori* knowledge about the relative abilities of the entrants, or, alternatively, where blind chance is an accepted component of the tournament outcome. Soccer KO tournaments, both in Britain and Europe, use a random draw (in each round) in the latter stages of competition. A non-intuitive aspect of the KO (unseeded) tournament (random draw for the first round only) is observed in the following consideration. If the entrants are ordered *a priori* from strongest (1) to weakest (t) and the stronger entrant of the pair is presumed to always win the contest (i.e. $k - 1$ is stronger than k for $t \geqslant k > 1$), then the expected rank of the second best player is close to 3 (Wiorkowski, 1972). In other words, the second best entrant is expected to finish in third place, or thereabouts, by virtue of the tournament design. This result arises from the chance placing of the two best entrants in the same half of the draw which occurs with probability $[t - 2]/[2t - 2]$. In these cases, approximately half the time, the second best entrant ranks as a consequence of what round in the tournament the contest against the best entrant is scheduled.

This injustice is also evidenced in the DKO (unseeded) tournament which, assuming the same initial conditions (i.e. $k - 1$ is stronger than k for $t \geqslant k > 1$), accurately ranks the best two entrants but not the third best for the same reason. The expected rank of the third strongest entrant is 3.5 (Fox, 1973), a rank that we interpret as a 0.5 probability of ranking in third and fourth place respectively. The alternative possibility, that of a tie for third place, is not in fact a viable outcome for the DKO (see Fig. 10.2). While an objection might be raised that the initial conditions are never met in sports practice (Palmer, 1973), the point is that a KO and DKO using an initial

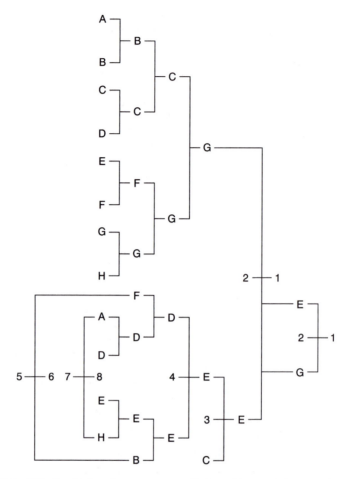

Figure 10.2   DKO. Double knockout tournament design with consolation at each stage of the competition. From McGarry and Schutz (1997) *Journal of the Operational Research Society*, **48**, p. 67. Copyright 1997 by Macmillan Press Ltd. Reprinted by permission.

random draw reduces the prospect of fairly ranking the second and third best entrants respectively. Katz (1977), on the basis of transitivity (i.e. if $A > B$ and $B > C$ then $A > C$), argued against unnecessary contests if preferences have already been established and, consequently, suggested an efficiency index for tournaments based on the expected number of contests required to establish the correct rank order of the entrants. While there is no assurance that transitivity holds in the context of sports competition, the efficiency rating of a tournament is a worthy consideration in the analysis of tournament design.

An easy way to address the vagaries of random draw in an elimination tournament is to seed the entrants beforehand. This is a common technique that is used to advance the progress of the better (stronger) entrants throughout the tournament and, in particular, to promote the prospect of a final contest between the two best entrants. This is achieved by placing the two best entrants in separate halves of the draw, the next two best players in separate halves of each half of the draw and so on. The common seeding allocation for an eight-entrant tournament, which we used to analyse the KO (seeded) and DKO (seeded) tournament, is P_1-P_8, P_5-P_4, P_3-P_6, P_7-P_2, where P_1 denotes the best entrant, P_2 the next best entrant and so on. This type of seeding is frequently used for many championship playoff competitions, including ice hockey, basketball, American football and tennis. A counter-intuitive aspect of this seeding is that, under certain conditions of strong stochastic transitivity, the best entrant in fact has a lower chance of winning the tournament than the next best entrant (Chung and Hwang, 1978; Israel, 1981). This aspect can be redressed by re-seeding at each round of the tournament, based on each entrant's performance within that tournament (Hwang, 1982), a practice that is used in the Stanley Cup (ice hockey) and the McIntyre Final Eight (Australian Rules) competitions.

The McIntyre Final Eight is a variant of the seeded eight-entrant KO and is currently used in the playoffs in the Australian (Rules) Football League (Clarke, 1996[1]). In brief, the entrants contest the first round P_1-P_8, P_5-P_4, P_3-P_6, P_7-P_2. The two highest seeded winners from the first round are awarded a bye in the second round to the preliminary final and the two lowest seeded losers are eliminated from the tournament. The remaining four entrants meanwhile contest a semi-final for entry to the preliminary final. Thus, the preliminary final consists of the two entrants awarded byes and the two winners from the semi-final. The two winners from the preliminary final contest the championship final. This unbalanced design is interesting in the sense that the route that an entrant takes through the tournament, and thus its chance of success, is contingent on the progress of other entrants. The history of the tournament (from four teams in 1931, to five teams in 1972, six teams in 1991 and eight teams in 1994) also provides an interesting commentary in that analysis shows how different unbalanced tournament designs notably alter the winning probabilities of various teams (Clarke, 1996).

Seeding is awarded on the basis of playing history. This is usually taken from the results of previous tournament competition, either the final league rankings of the regular season or some other equivalent tournament rankings. An alternative method is to use a panel of experts to seed the entrants, as is the case of basketball team members of the National Collegiate Athletic Association. Here, a panel selects 64 from 292 teams for seeded entry into four regional tournament playoffs of 16 teams each. (The winners of the four regional playoffs proceed to a Final Four from which a national champion emerges after winning in the fifth and sixth rounds respectively.) An interesting question pertains to the accuracy of the seeding and the effect of inaccurate seeding on the tournament outcome. The efficacy of the seeding method used by the National Collegiate Athletic Association was analysed from comparison of the probabilities that each entrant will win the tournament (Schwertman *et al.*, 1991), and later each contest (Schwertman *et al.*, 1996), as a function of its seed position, as well as additional criteria, including the win–loss record, the rating percentage index and the point spread (Carlin, 1996). (The rating percentage index is a measure of team strength based on some function of playing history over the course of

---

[1] I thank Stephen R. Clarke for kindly bringing this tournament design to my attention.

the preceding season. The point spread is the margin of victory with which an entrant is predicted to win over another entrant.) See Chapter 3 on basketball for further descriptions of these models and Chapter 12 for general descriptions of sports rating systems.

10.2.2   Round robin tournaments

The RR proceeds on the basis that each entrant competes against each other entrant within the structure, a process that is usually repeated in league-based competition in order to eliminate the bias of home advantage. (For a review of home advantage in various sports, see Courneya and Carron, 1992, as well as Chapter 12 on predicting outcomes.) While the RR might serve as an end in itself, it might also serve as a seeding process for entry into further competition, usually an elimination tournament. In some cases, the best ranked entrants from multiple RRs feed into an elimination structure. We analysed this type of tournament as an RR–KO hybrid, where two RRs of four entrants each were used to rank the entrants before their advance to a seeded KO (Fig. 10.3). Soccer, rugby football, American football, ice hockey, basketball, baseball and cricket all use this type of tournament design in some form. We also analysed a filtered RR (RR–RR) that consists of two successive RRs of four entrants each (Fig. 10.4)

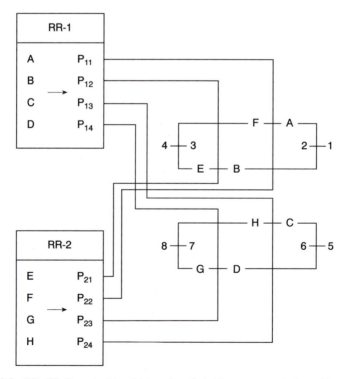

Figure 10.3   RR–KO. Round robin with knockout hybrid tournament design with consolation at each stage of the competition. From McGarry and Schutz (1997) *Journal of the Operational Research Society*, **48**, p. 68. Copyright 1997 by Macmillan Press Ltd. Reprinted by permission.

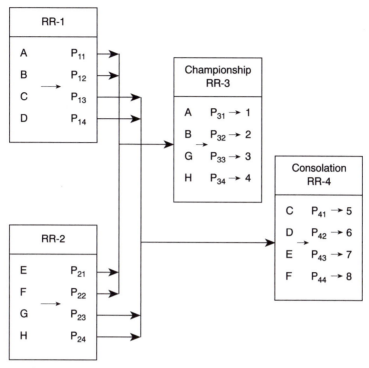

Figure 10.4 RR–RR. Filtered round robin tournament design with consolation at each stage of the competition. From McGarry and Schutz (1997) *Journal of the Operational Research Society*, **48**, p. 67. Copyright 1997 by Macmillan Press Ltd. Reprinted by permission.

as a possible alternative to the RR, in that it requires fewer contests. The first two RRs are used to seed the entrants for entry into the second two RRs. The first and second seeds from each RR gain entry to the championship RR and the third and fourth seeds gain entry to the consolation RR. This type of tournament design is not commonly used in sport although promotion and relegation between divisions in league competition at the end of the regular season proceeds on a similar basis. In soccer, for example, the tournament design typically consists of a few divisions in which each team within a division competes against each other team twice during the course of the regular season (i.e. home and away). The first *n* placed (ranked) teams after season competition are promoted to the next highest division and, likewise, the lowest *n* ranked teams are relegated to the next lowest division. Ties (i.e. equal number of games won) were resolved in all RRs with an equal-odds 'coin toss' between each of the tied entrants. Game ties were not allowed in these analyses in order to allow comparison of different tournaments under stable conditions.

10.2.3 The efficacy of KO, RR and RR–KO tournaments

The tournaments were analysed by assigning a probability matrix $i \times j$. The probability that any entrant i will win against any entrant j ($p_{i,j}$) reads row

Table 10.1   Playing matrix where $p_{i,j}$ = probability that *i* wins a contest against *j*

Entrant	1	2	3	4	5	6	7	8	Average
1	–	0.55	0.60	0.65	0.70	0.75	0.80	0.85	0.70
2	0.45	–	0.55	0.60	0.65	0.70	0.75	0.80	0.64
3	0.40	0.45	–	0.55	0.60	0.65	0.70	0.75	0.59
4	0.35	0.40	0.45	–	0.55	0.60	0.65	0.70	0.53
5	0.30	0.35	0.40	0.45	–	0.55	0.60	0.65	0.47
6	0.25	0.30	0.35	0.40	0.45	–	0.55	0.60	0.41
7	0.20	0.25	0.30	0.35	0.40	0.45	–	0.55	0.36
8	0.15	0.20	0.25	0.30	0.35	0.40	0.45	–	0.30

by column and the complementary probability that any entrant *j* will win against any entrant *i* ($p_{j,i}$) reads column by row, where $p_{i,j} + p_{j,i} = 1$. Table 10.1 reflects an example of a playing matrix that contains a linear ordering of ability, from strongest to weakest, according to the probability of winning a contest. We assume stationarity and independence within the playing matrix. We used additional matrices in order to analyse different initial conditions, but these are excluded here for reasons of brevity. For further detail see McGarry and Schutz (1997).

We consider the most valid tournament to be that which best ranks the entrants according to their standings, as defined in the playing matrix. The kappa statistic (κ) provides a satisfactory measure in this regard. κ reflects the degree of parity in the cross-diagonal cells of a two-way matrix (Fleiss, 1975), in this case the rankings from the playing matrix and the tournament. Specifically, κ is expressed as the ratio of observed to maximum improvement in tournament rankings over chance rankings. Thus, $\kappa = [p_o - p_c]/[1 - p_c]$, where p_o is the proportion of observed agreement and p_c is the proportion of chance agreement. In fact, we used a weighted kappa (κ_w), where disparity in the off-diagonal cells further from the cross-diagonal are weighted more than those nearer to it. The validity of a tournament, as inferred from κ_w, is highest for the DKO (seeded), the RR and the KO (seeded), assuming accurate seeding, and lowest for the KO (unseeded) (see Table 10.2).

An alternative measure of a tournament's validity is to maximize the expected value of each entrant winning the tournament. Monahan and Berger (1977) used the points from a season of league-based ice hockey competitions as a rating index for each entrant in the Stanley Cup playoffs of the National Hockey League. (At that time, the Stanley Cup playoff was an eight-team

Table 10.2   Validity and efficiency measures for various tournaments with respect to the playing matrix in Table 10.1

	No. of games	κ_w	W_{max}	Pr(2 Best in Final)
KO (unseeded)	12	0.078	0.597	0.128
KO (seeded)	12	0.113	0.611	0.258
DKO (unseeded)	16 or 17	0.096	0.615	0.289
DKO (seeded)	16 or 17	0.121	0.625	0.369
RR–KO	20	0.089	0.602	0.158
RR–RR	24	0.095	0.604	–
RR	28	0.119	0.613	–

KO with a seven-game series in each round.) Here, the maximum expected value of the tournament, provided in the formula

$$W_{max} = \sum p(i \text{ wins Cup}) [i \text{ points rating}] \text{ where } i = 1 \text{ to } 8$$

is the maximized expected points of the Cup winner summed over each entrant. We used the $p_{i,j}$ from the playing matrix, averaged across rows, as the rating index instead of a points measure. (Note that $p_{i,j}$ was taken to be the probability of winning the seven-game series and not the probability of winning each game in the series.) The validity of a tournament using this measure is highest for the DKO (seeded), the DKO (unseeded), the RR and the KO (seeded). Taking both κ_w and the W_{max} into account, the DKO (seeded) tournament is the best tournament to discriminate accurately between entrants of different (linearly ordered) abilities (Table 10.2). The DKO (unseeded) well approximates the RR and would seem an expedient alternative, if a fewer number of games are required. (In practice, professional sports may prefer more games rather than less games to determine a champion, in order to increase revenue generation from ticket sales.)

An important aim of many tournaments is to maximize the probability that the two best entrants meet in a single final contest. The double elimination tournaments produce notably higher probabilities than the single elimination tournaments in this regard (see Table 10.2), although the possibility of two final contests might not be attractive to a tournament organizer. For example, the change in the college baseball world series in 1988, from a double elimination of eight teams to two double eliminations of four teams from which the two winners meet in a final contest, was undertaken to negate the possibility of a repeat final and so satisfy television requirements for a single championship contest. Interestingly, this change occurred at little cost to the winning probabilities of each of the eight entrants (Ladwig and Schwertman, 1992). Of note is that the DKO (unseeded) is better than the KO (seeded) in producing a final between the two best entrants, a finding that was repeated in varying amounts for each of the playing matrices that were used by McGarry and Schutz (1997). Thus, we recommend a double elimination tournament if maximizing the likelihood that the two best entrants meet in a final contest is a tournament objective. Seeding of course further enhances this prospect. On the other hand, the KO (unseeded) and RR–KO yield much lower probabilities and should, in our view, only be used if chance is recognised as an acceptable facet of the tournament outcome. (The RR and RR–RR tournaments are excluded since these designs do not produce a final contest.) We conclude, from measures of validity, probability of the two best entrants meeting in the final, and efficiency (i.e. the number of contests required) that the DKO (seeded), the DKO (unseeded) and the KO (seeded) tournaments offer the best designs for most sports competition purposes. Also, the extra four (or five) games required for the DKO, from that of the KO, would seem to be a worthwhile addition.

The validity of a tournament as a function of the number of contests raises the question of a tournament's efficiency (cf. Katz, 1977). We note that this measure of efficiency tends to favour a tournament with few contests since it is unlikely that more contests will contribute information in equal part. If a KO is indeed a more efficient method of comparison than an RR, then an alternative KO tournament of approximately the same number of contests as that of an RR should prove a more valid tournament. The number of contests in a KO can be increased by introducing an n-game series in each round (see later) or, alternatively, by using a KO–KO hybrid like a draw and process tournament (Appleton, 1995). This tournament, sometimes used in croquet, consists of two separate KOs where the second (process) is determined from the first (draw). Entrants that meet in the first round of the draw ($t = 8$) cannot meet those same entrants until the final of the process (and vice versa for the first round of the process). The draw and process tournament can be unseeded or seeded and, as expected, the latter is an effective tournament design for identifying the best entrant (Appleton, 1995). A variant of this hybrid would be to use the first KO to seed the second KO. Whether these types of tournaments are more or less valid than an equivalent RR is an open question.

These comments are provided from analysis of tournaments under certain initial conditions. Other initial conditions are possible (Israel, 1981; Hwang, 1982) where the RRs, as well as the random KOs, are better than the seeded KOs in ranking the entrants in accord with their abilities. In this light, the RR is the most valid (or safe) tournament with which to rank entrants given the limitations of the random KOs listed above. This view also holds if, as expected, the contest probabilities vary over the course of competition. In this event, the RR is expected to distinguish itself from the KO for two reasons. First, the RR is even-handed in that each entrant contests against each other entrant, often taking home advantage into account through repeat contests. The variation in contest probabilities is thus expected to even out in the long run average over the course of competition. This is not the case in elimination competition because of the consequence of early losses. Second, the increased number of contests in an RR affords better discrimination between entrants whose win probabilities vary over the course of competition, an argument that also applies as the ability of entrants tends to homogeneity. The effects of transient probabilities on the outcome of various tournaments, at least to our knowledge, are an unexplored area of enquiry to date. We offer some brief considerations below.

10.3 Stationarity and independence of contests

We assume stationarity and independence within the probability matrix. Stationarity assumes that the probabilities do not change as a function of time, while independence assumes that the probabilities do not change as a function of history. In other words, $p_{i,j}$ is constant (stationarity) and not contingent on any other $p_{i,j}$ (independent). It is unlikely that these assumptions hold in practice and the question begs as to how well these assumptions approximate the real situation. (See Mosteller, 1952, for a considered analysis of stationarity and independence from data taken from the Major League Baseball World Series, 1903–51.) Transient probabilities have been considered in some paired comparison analyses of sports competition in the National League (baseball) (Barry and Hartigan, 1993) and the German National Soccer League (Bundesliga) (Fahrmeir and Tutz, 1994). The focus of these studies was to identify a valid model that captures the changes in team ability over the course of time, as indicated from the contest outcomes. These models can be applied to analyse the behaviour of sports tournaments. In fact, the validity of these models as they relate to empirical observation is not a limitation, since the temporal update of the paired comparisons in line with any *a priori* function will suffice for the analysis of tournaments, as long as the changes in abilities are sufficient to cover the real situation. If changing strengths are allowed, then the prediction of division winners at various stages of the season yields more conservative probabilities of the best team winning (Barry and Hartigan, 1993). The team with the best win–loss record has a lower probability of winning the division under conditions of changing strength because they are likely to have lower strength in the future than their record to date would suggest. We expect that these findings would be mirrored in the analysis of tournaments for the same reason.

It seems reasonable to suggest that violating the assumptions of stationarity and independence will have a larger impact on those sports tournaments that decide an outcome on the basis of less information. For example, transient probabilities would be expected to affect the ranking probabilities of a KO more so than an RR on three counts. First, the KO contains fewer contests (comparisons) and so identifies the winner from less information; second, as noted earlier, the KO contests are not even-handed; and third, contests in a tournament need not contribute the same information.

This last reason raises the question of the importance of a contest in a sports tournament. This issue, first raised by Morris (1977) with respect to the importance of a point at various stages of a tennis match, has been defined as the probability of winning the contest given the point is won minus the probability of winning the contest given the point is lost. (See Chapter 8 for more details on importance of points in tennis.) If this definition is accepted, then a contest in elimination competition assumes more importance the further the tournament progresses. In fact, the importance of a contest changes not only with the stage of the tournament, but also with the results of other contests. In the McIntyre Final Eight playoff, a game's importance is a direct consequence of the outcome of the other games. In this design, a loss might or might not result in elimination from the tournament, a consequence that plainly affects the importance of that contest. Incidentally, if it is known that a loss will not result in elimination from the tournament, then the possibility arises that each team might try to lose that contest in order to avoid entry into one half of the draw. Fortunately, this absurdity is corrected by scheduling the contest so that the win–loss consequence of the contest is not known ahead of time (Clarke, 1996).

The scheduling of a contest is an important consideration in the design of a tournament for two reasons. First, as we have already seen, a strategic advantage is awarded to entrants of a later contest if the outcome from an earlier contest provides pertinent information to the later contest. Horse jumping provides another example in this regard where later entrants can optimize their decision strategies based on the performances of earlier entrants. Second, the schedule order is important when the performance is scored in a subjective way. For example, judges are more likely to rank entrants scheduled early in a music competition lower than entrants that are scheduled later (Flores and Ginsburgh, 1996).

There is no reason that this finding should not extend to sports competitions. Sports like gymnastics, diving and figure skating all use judges' ratings to assess athletic performance according to various criteria. The subjectivity of these ratings is assuaged somewhat by using a panel of experts, although the various methods of scoring these ratings speak against a widely accepted statistical practice of robust estimation. The scores for diving performance are produced from the judges' mean ratings after discarding the two highest and two lowest ratings (Bruno, 1986), while the scores for figure skating are produced from the judges' median ratings (Bassett and Persky, 1994). Objectivity in scoring is further compromised in the grading of performances of different levels of difficulty, both in terms of judges' ratings (Bruno, 1986) and in the strategic advantage awarded to entrants that compete later in the competition (Henig and O'Neill, 1992). These findings all speak for the importance of the contest order on the outcome of some sports tournaments.

10.4   The inequities of some unbalanced tournament designs

The outcome of a sports tournament is influenced not only by the scheduling of a sporting contest, but also by how these contests are compared, especially in unbalanced tournaments. The bias of an unbalanced tournament is evidenced, at least under certain initial conditions, in the World Cup Soccer Finals (McGarry and Schutz, 1994). In the 1986, 1990 and 1994 World Cups, six RRs of four teams each competed for entry to a 16-team seeded KO, consisting of the top two teams from each RR, as well as four wild card entries (the best four from six third-place teams). The promotion of some teams occurs at the expense of other teams by virtue of allocation to the different RRs. In the 1998 World Cup Soccer Finals, eight RRs of four teams each competed for entry to a 16-team seeded KO consisting of the top two teams from each RR, a change that promises a more equitable design. We note that Major League Baseball uses a similar unbalanced design for entry to their playoffs in both the National League and the American League. In each league, the three division winners and a wild card entry (the best second-place team from the three divisions) compete in a semi-final from which a winner is produced. This tournament design is further confounded in that each division does not contain the same number of teams. Once again, some teams are rewarded and other teams penalized by virtue of the inequities of the tournament design.

10.5   Equitable ranking procedures in the event of tied outcomes

Franks and McGarry (1996) categorized a sporting contest (of two entrants) as being either score dependent or time dependent. In a score-dependent contest (e.g. tennis, volleyball), a contest outcome (win–loss) is produced in unspecified time from the first entrant to reach a fixed score. In a time-dependent contest (e.g. ice hockey, soccer), a contest outcome (win–loss or tie) is produced with an unspecified score in a fixed period of time. A tied result in these sports is a frequent occurrence and is usually accepted in league-based (RR) competition. How a contest tie is broken in elimination competition, invariably through the use of scheduled overtime (at least in the first instance), is a consequence of the sport and is not of concern here.

A contest outcome in an RR is usually converted to points that reflect the value attached to each outcome (win, loss or tie). These points are tallied over the course of the competition and the winner is the entrant with the highest points tally. The points tally does not distinguish between win–tie–loss records, nor between wins against different opponents. Ties on points are broken through a secondary measure, such as goal differential (i.e. goals for minus goals against). Goal differential credits (debits) an entrant with goals for (against) in equal measure and rewards (penalizes) entrants on the margin of the win (loss). It is a fair measure for breaking ties providing each entrant plays each like entrant over the course of the competition. European soccer KO competitions use goal difference to break a tied outcome over a two-game (two-leg) series. In the event of a tie on goal difference, ties are further resolved by a weak ranking procedure that halves the worth of home goals (or doubles the worth of away goals) on the basis of home advantage. Further ties are resolved through penalties (i.e. a penalty shoot-out).

The use of unbalanced RRs necessitates a different method of breaking ties. The Swiss tournament, used in sports like chess and bridge, is an example of this type of design. Successive rounds are played without elimination, but the pairings in each round are determined by the outcomes of the previous rounds. Thus, the better players tend to compete against each other as the tournament progresses. The final score is a function of the number of wins and ties that a player has attained but, importantly, wins and ties against stronger opponents are more highly valued than wins and ties against weaker opponents. This scoring method can also be used to break ties (i.e. equal number of games won) in sports like baseball. Groeneveld (1990) showed, from an analysis of the National League (baseball), that a suitable index to break ties in an unbalanced design – an extension of Kendall's (1955) method – is to assign a positive (negative) point for each win (loss), as well as a positive (negative) point for each win (loss) of i against j. This index compares favourably with the Bradley–Terry (1952) ranking procedure and has the advantage of ease of computation. This consideration is important since complex analyses for breaking ties and determining tournament winners are not well received in most sports. For further consideration of ranking tied outcomes, see Bradley (1954), Dykstra (1960), Ali *et al.* (1986), David (1987), Cook *et al.* (1988), Cook and Kress (1990) and Kaykobad *et al.* (1995).

10.6 Efficacy of a sporting contest

The efficacy of a sporting contest (i.e. the likelihood that the stronger entrant will win the contest) is facilitated by lengthening the single contest to an n-game series. Likewise, the efficacy of a sports tournament (i.e. the likelihood that the stronger entrant will win the tournament) is increased if each contest is lengthened to an n-game series. For example, Searls (1963) showed that increasing each contest in KO and DKO competition from a one-game series to a three-game series increases the probability that the best entrant will win the tournament under any initial conditions of seeding. Various sports – for example, ice hockey, baseball and basketball – make use of an n-game series in each round of KO competition to this effect. Snooker and darts also use an odd integer *n*-game series in order to decide a contest outcome in their elimination tournaments. The nature of these latter sports of course allows for a much lengthened series in each round of competition.

European soccer playoffs use a two-leg series, where a single contest can result in a win, loss or tie. Here, the preferred *n* is an even integer since it provides for a fairer scheduling of home (H) and away (A) games, while the possibility of ties in a single contest means that neither an even nor an odd integer series guarantees an outright result. In sports where only a win–loss outcome in a single contest is allowed, the preferred *n* is an odd integer because this ensures a win from the series. Odd integer series are common features of playoffs in North American sports. Major League Baseball uses a five-game series in the first round of playoffs to determine the National League and American League champions. From the perspective of the team with home field advantage, the first round schedule is AAHHH. Since the playoff schedule is decided by the leagues before the start of the season, the home field advantage is effectively random. Ice hockey and basketball have both used a seven-game series (AHHAAHH) in their playoff (single elimination) competitions. Home advantage is awarded on the basis of rankings in league (RR) competition from the regular season.

The Major League Baseball World Series, contested between the champions of the National League and the American League, uses a different game schedule, namely HHAAAHH, and home advantage alternates between the two leagues each year. If we assume stationarity and independence of home wins, then the probability that the team with home advantage wins the series is not, in fact, contingent on the schedule. Hurley (1993) provided various probabilities that the home team wins the World Series as a consequence of home field advantage and offers two alternatives to eliminate this bias. The first is a first-to-four, win-by-two series (analogous to a tennis tie break) which produces a series of indefinite length. The second is to schedule (in any order) the extra home game at a neutral venue, which keeps the series at a fixed length.

The likelihood that the stronger entrant will win a lengthened series is affirmed through a random walk of the possible win sequences. Mosteller (1952) reported the win probability of the stronger entrant for a one-game, three-game, five-game, seven-game and nine-game series, as well as the win probability and the tie probability (i.e. split series) for a two-game, four-game, six-game and eight-game series. (Ties in single contests are not allowed in these analyses.) While the increase in the length of a series increases the probability that the stronger entrant will win the series, it does so at a smaller return per two-game increment. That is, the difference in win probability between a one-game and a three-game series is larger than the difference between a three-game and a five-game series and so on. This result suggests that increasing the contest from a one-game series to a three-game series is the most expedient (efficient) option in that it yields the largest win dividend per unit cost. On the other hand, the extra contests in a lengthened series might be justified if the two entrants are closely matched (i.e. $p \approx q \approx 0.5$). This view was confirmed by Maisel (1966) who, in an extension of Mosteller's (1952) work, reports the optimum series length as a function of the win probability for a single contest. Maisel suggested that it is better to over-estimate (rather than under-estimate) the optimized *k* in a series of $2k - 1$ contests.

These comments pertain to sports that force a win–loss outcome for a single contest. Maisel (1966) also analysed an *n*-game series in which ties in single contests are allowed. These analyses show that, for odd integers of *n*, the likelihood of a tied series decreases as *n* increases while, at the same time, the win probability for one of the entrants increases and the win probability for the other entrant increases or decreases depending on the initial conditions.[2] We conducted a preliminary analysis for a two-game and a four-game series and found similar results. The probability of a tied series tends to decrease as the series lengthens, although this is not always the case. Likewise, the probability of winning and of losing the series increases or decreases in a non-systematic way as a consequence of the initial values of *p*, *q* and *r* (where $p + q + r = 1$; *p* is the probability of winning a single contest, *q* is the probability of losing a single contest and *r* is the probability of tying a single contest).

10.7 Overview of some other sports tournament designs

We have restricted our focus to tournament designs that use the KO, RR or some variation thereof. We will now briefly consider some other tournament designs that extend beyond this scope. These comments are an abridged form of those of McGarry and Schutz (1997).

In many individual performance sports, the entrants compete against some external performance measure and the individual with the highest (or lowest) score is declared the winner. Golf and archery are common examples of this type of tournament. Some sports, like Alpine (downhill) skiing, measure performance by the timepiece where the fastest (i.e. shortest) time yields the best performance. An advantage of these types of sports is that a tournament can host a large number of entrants, each of whom can be rank ordered in a short period of time. Because the entrants compete solo rather than head to head (i.e. in a paired contest), these sports afford a more efficient scheduling of events and, also, reduce the number of comparisons required to establish a winner. A minimum of one (multiple) comparison, or contest (e.g. round, stage, heat), is required to determine a winner in these sports, whereas a minimum of $\log N / \log 2$ and $N - 1$ (multiple) comparisons are required in a single elimination and RR respectively. This is not to say that individual performance sports are limited to single events. Sports like Formula 1 racing, for example, use a series of contests, or stages, in order to derive a composite performance score, thereby increasing both the reliability and validity of the final result.

Olympic track sports use an elimination tournament. Here, the entrants form heats of typically four or eight and the top placed entrants (usually the top two or three) from each heat advance to the next heat. (This design is analogous to the RR–RR design, except that each entrant is ranked as a result of one comparison per heat rather than $N - 1$ comparisons per RR.) The remaining available places are then assigned to the near qualifying finishers from the best performance scores from all the heats combined. A limitation of this type of design is that the performance measures are only comparable across

---

[2] Maisel (1966) actually reports data for even integers of *k* for series of $2k - 1$ (or *n*) contests.

entrants because of sensitivity to varying external conditions. In sports such as rowing, where performance comparisons between heats may not be suitable because of changing conditions, the near qualifying finishers participate in a repechage, or loser's heat, with the winner of that heat advancing to the next heat to join the earlier placed qualifiers. The repechage increases the validity of the tournament structure in determining the final placings, since it compensates for the possibility that the better entrants are matched against each other by a random draw in an earlier heat. While the repechage is not necessary in seeded tournament structures (if the seeding is accurate), it nevertheless guards against an early upset and consequently is an attractive feature for tournament organizers.

10.8   Summary

Various tournament designs exist to determine a winner and to rank entrants in accord with their abilities. The tournaments considered in this chapter have different ranking properties depending on, amongst other factors, the initial starting conditions. To date, analysis of tournaments has assumed stationarity and independence in the contest probabilities, and the inclusion of transient probabilities would be expected to yield important information to their overall ratings. (An efficiency metric for tournaments would also be a valuable addition in this regard.) Since it is unlikely that the usual assumptions of stationarity, independence and transitivity hold in sports practice, it begs the question as to how well these assumptions approximate the real situation (cf. Mosteller, 1952). We suggest that the evolution of statistical models that breach these assumptions is a useful step in the future analysis of tournaments, if their bearing on sports outcomes is to be fully realized.

Acknowledgments

I would like to thank Jay Bennett for inviting me to contribute a chapter to this book and for his editorial comments throughout the write-up process.

References

Ali, I., Cook, W. D. and Kress, M. (1986) On the minimum violations ranking of a tournament. *Management Science*, **32**, 660–674.

Andrews, D. M. and David, H. A. (1990) Nonparametric analysis of unbalanced paired comparisons or ranked data. *Journal of the American Statistical Association*, **85**, 1140–1146.

Appleton, D. R. (1995) May the best man win? *The Statistician*, **44**, 529–538.

Barry, D. and Hartigan, S. A. (1993) Choice models for predicting divisional winners in major league baseball. *Journal of the American Statistical Association*, **88**, 766–774.

Bassett, Jr., G. W. and Persky, J. (1994) Rating skating. *Journal of the American Statistical Association*, **89**, 1075–1079.

Bradley, R. A. (1954) Incomplete block rank analysis: on the appropriateness of the model for a method of paired comparisons. *Biometrics*, **10**, 375–390.

Bradley, R. A. and Terry, M. E. (1952) The rank analysis of incomplete block designs. I. The method of paired comparisons. *Biometrika*, **39**, 324–345.

Bruno, J. E. (1986) Scoring bias in springboard and platform diving: a Bayesian approach to revision of DD tables. *Research Quarterly for Exercise and Sport*, **57**, 16–22.

Carlin, B. P. (1996) Improved NCAA basketball tournament modeling via point spread and team strength information. *The American Statistician*, **50**, 39–43.

Chung, F. R. K. and Hwang, F. K. (1978) Do stronger players win more knock-out tournaments? *Journal of the American Statistical Association*, **73**, 593–596.

Clarke, S. R. (1996) Calculating premiership odds by computer: an analysis of the AFL final eight playoff system. *Asia–Pacific Journal of Operational Research*, **13**, 89–104.

Cook, W. D., Golan, I. and Kress, M. (1988) Heuristics for ranking players in a round-robin tournament. *Computers and Operations Research*, **15**, 135–144.

Cook, W. D. and Kress, M. (1990) An *m*th generation model for weak ranking of players in a tournament. *Journal of the Operational Research Society*, **41**, 1111–1119.

Courneya, K. S. and Carron, A. V. (1992) The home court advantage in sport competitions: a literature review. *Journal of Sport and Exercise Psychology*, **14**, 13–27.

Cowden, D. J. (1974) A method of evaluating contestants. *The American Statistician*, **29**, 82–84.

David, H. A. (1959) Tournaments and paired comparisons. *Biometrika*, **46**, 139–149.

David, H. A. (1987) Ranking from unbalanced comparison data. *Biometrika*, **74**, 432–436.

David, H. A. (1988) *The Method of Paired Comparisons*. New York: Oxford University Press.

Davidson, R. R. (1970) On extending the Bradley–Terry model to accommodate ties in paired comparison experiments. *Journal of the American Statistical Association*, **65**, 317–328.

Dykstra, O. (1960) Rank analysis of incomplete block designs: a method of paired comparisons employing unequal repetitions on pairs. *Biometrics*, **16**, 176–188.

Fahrmeir, L. and Tutz, G. (1994) Dynamic stochastic models for time-dependent ordered paired comparison systems. *Journal of the American Statistical Association*, **89**, 1438–1449.

Fleiss, J. L. (1975) Measuring agreement between two judges on the presence or absence of a trait. *Biometrics*, **31**, 651–659.

Flores, R. G. and Ginsburgh, V. A. (1996) The Queen Elizabeth musical competition: how fair is the final ranking? *The Statistician*, **45**, 97–104.

Fox, M. (1973) Double-elimination tournaments [Letter to the editor]. *The American Statistician*, **27**, 90–91.

Franks, I. M. and McGarry, T. (1996) The science of match analysis. In T. Reilly (ed.), *Science and Soccer*, pp. 363–375. London: E. & F. N. Spon.

Glenn, W. A. (1960) A comparison of the effectiveness of tournaments. *Biometrika*, **47**, 253–262.

Glenn, W. A. and David, H. A. (1960) Ties in paired comparison experiments using a modified Thurstone–Mosteller model. *Biometrics*, **16**, 86–109.

Groeneveld, R. A. (1990) Ranking teams in a league with two divisions of *t* teams. *The American Statistician*, **44**, 277–281.

Henerey, R. J. (1992) An extension of the Thurstone–Mosteller model for chess. *The Statistician*, **41**, 559–567.

Henig, M. and O'Neill, B. (1992) Games of boldness, where the player performing the hardest task wins. *Operations Research*, **40**, 76–86.

Hurley, W. (1993) What sort of tournament should the World Series be? *Chance: New Directions for Statistics and Computing*, **6**, 31–33.

Hwang, F. K. (1982) New concepts in seeding knockout tournaments. *American Mathematical Monthly*, **89**, 235–239.

Israel, R. B. (1981) Stronger players need not win more knockout tournaments. *Journal of the American Statistical Association*, **76**, 950–951.

Joe, H. (1990) Extended use of paired comparison models, with application to chess rankings. *Applied Statistics*, **39**, 85–93.

Katz, L. (1977) An efficient sequential ranking procedure. *Journal of the American Statistical Association*, **72**, 841–844.

Kaykobad, M., Ahmed, Q. N. U., Kahid, A. T. M. S. and Bakhtiar, R. (1995) A new algorithm for ranking players of a round-robin tournament. *Computers and Operations Research*, **22**, 221–226.

Kendall, M. G. (1955) Further contributions to the theory of paired comparisons. *Biometrics*, **11**, 43–62.

Kuk, A. Y. C. (1995) Modelling paired comparisons data with large numbers of draws and large variability of draw percentages among players. *The Statistician*, **44**, 523–528.

Ladwig, J. A. and Schwertman, N. C. (1992) Using probability and statistics to analyse tournament competitions. *Chance: New Directions for Statistics and Computing*, **5**, 49–53.

Maisel, H. (1966) Best *k* of 2*k* − 1 comparisons. *Journal of the American Statistical Association*, **61**, 329–344.

Maurice, R. J. (1958) Selection of the population with the largest mean when comparisons can be made only in pairs. *Biometrika*, **45**, 581–586.

McGarry, T. and Schutz, R. W. (1994) Analysis of the 1986 and 1994 World Cup soccer tournament. *Proceedings of the 154th Annual Meeting of the American Statistical Association: Section on Statistics in Sport*, pp. 61–65.

McGarry, T. and Schutz, R. W. (1997) Efficacy of traditional sports tournament structures. *Journal of the Operational Research Society*, **48**, 65–74.

Monahan, J. P. and Berger, P. D. (1977) Playoff structures in the national hockey league. In S. P. Ladany and R. E. Machol (eds), *Optimal Strategies in Sports*, pp. 123–128. Amsterdam: North-Holland.

Morris, C. (1977) The most important points in tennis. In S. P. Ladany and R. E. Machol (eds), *Optimal Strategies in Sports*, pp. 131–140. Amsterdam: North-Holland.

Mosteller, F. (1951) Remarks on the method of paired comparisons: I. The least squares solution assuming equal standard deviations and equal correlations. *Psychometrika*, **16**, 3–9.

Mosteller, F. (1952) The World Series competition. *Journal of the American Statistical Association*, **47**, 355–380.

Palmer, B. Z. (1973) Knock-out tournaments [Letter to the editor]. *The American Statistician*, **27**, 40.

Rao, P. V. and Kupper, L. L. (1967) Ties in paired comparison experiments: a generalization of the Bradley–Terry model. *Journal of the American Statistical Association*, **62**, 192–204.

Rokosz, F. (1990) The wave – a new tournament structure. *Journal of Physical Education, Recreation and Dance*, **61**, 92–94.

Schwertman, N. C., McCready, T. A. and Howard, L. (1991) Probability models for the NCAA regional basketball tournaments. *The American Statistician*, **45**, 35–38.

Schwertman, N. C., Schenk, K. L. and Hollbrook, B. C. (1996) More probability models for the NCAA regional basketball tournaments. *The American Statistician*, **50**, 34–38.

Searls, D. T. (1963) On the probability of winning with different tournament procedures. *Journal of the American Statistical Association*, **58**, 1064–1081.

Wiorkowski, J. J. (1972) A curious aspect of knockout tournaments of size 2^m. *The American Statistician*, **26**, 28–30.

11

Statistical Data Graphics in Sports

John A. Flueck

University of Nevada-Las Vegas, USA

11.1 Introduction

The world is becoming more visually oriented, whether it be in international signage, advertising layouts, newspaper and magazine pictures, television, movies, and even computer software. Thus, it should be no surprise that the graphic revolution has begun to make its presence felt in sports. The origin of graphical displays dates back at least to 2000 BC, when the Sumerians and Babylonian tribes lived in the valley between the Tigris and Euphrates rivers and wrote their historical records on bricks of baked clay. Over the centuries, this statistical activity increased steadily until 1786 when William Playfair, younger brother of the well-known scientist John Playfair and draftsman to James Watt, the inventor of the steam engine, published his *Commercial and Political Atlas* in London (Playfair, 1786). Playfair designed, promoted, and experimented with different graphical forms in order to enhance the messages encoded in his displays (Wainer, 1996), and the use of data graphics began to increase at an exponential rate.

Recent applications of the graphical display of data and results have been advanced by Tukey (1977), Tufte (1983), Cleveland (1985) and others. In sports, the use of statistical data displays dates back to at least the early days of professional baseball, and the 1871–75 National Association Register of baseball player's performance statistics (Reichler, 1982). The concept of spatial data graphics can be traced back to the pixel characterization of Ted Williams' strike zone in his book *The Science of Hitting* (Williams and Underwood, 1971). It is the forerunner of the hot/cold zone graphs used by the Fox Sports channel in its television coverage of the 1997 baseball playoffs. Davenport's (1979) *Baseball Graphics* appears to be the first book solely dedicated to graphical display of sports data, and *USA Today* has led the way in the print media. Television also appears to be using more graphical displays of sports as seen in the use of rotating three-dimensional spatial displays of pitches in NBC's coverage of the 1997 World Series. Even Internet web sites have been producing and displaying more graphs of sports performances (e.g. Fig. 4.3 in the chapter on cricket).

This chapter will examine data-based visual displays used to investigate, present, and understand the activities of a sport. Graphs, plots, panels, tables and other visual forms belong to this subject, and hence it is an important component of Exploratory Data Analysis (Tukey, 1977). Displays of sports data and results will be reviewed and classified according to type, content and graphical message. Lessons learned will be judged in terms of Cleveland's four main principles of graphic construction: (1) clear vision; (2) clear understanding; (3) scales; and (4) general strategy (Cleveland, 1985).

11.2 Tables

To classify the types of display properly one needs guidelines, and thus the following definitions will be used (Webster's Ninth New Collegiate Dictionary, 1988).

> Table: 'A systematic arrangement of data usually in rows and columns for ready use'
>
> Graph: 'A diagram (as a series of one or more points, lines, line segments, curves, areas) that represents variation of a variable in comparison with that of one or more other variables'

In North America, the sport of baseball appears to be one of the first users of statistical data graphics, and Table 11.1 presents an example of the National Association's Player Register, 1871–75, for two outstanding players from the Boston Red Stockings: pitcher Al Spalding and shortstop George Wright. This display of yearly and five-year performance totals shows total games played (G), as well as games by position (G by POS). Four measures of hitting performance (at-bats (AB), hits (H), runs (R), batting average (BA)) and, for pitchers, three additional performance summary measures (wins (W), losses (L), and winning percentage (PCT)) are presented. All nine measures are counts or related fractions.

This display clearly is a table, with its rows representing years and its columns measuring individual performance. The Player Register uses bold type for some summary statistics to highlight yearly record performances, but then it clutters the data region with biographical information which reduces the focus on performance statistics and interrupts the column designations. This information should be condensed and reported at the side of each player's record.

A year later, 1876, the National League was formed, and more performance measures could be extracted from the improved record keeping of each game. Fast forward 100 years to 1976 and the items tracked had increased slightly but the box scores were largely unchanged. Clearly, baseball statisticians still saw themselves as accountants recording the numbers of thousands of box scores and millions of data, with no real effort to investigate or understand why a game evolved as it did. The strategy still was to collect storehouses of data presented in tables, leaving the message to the readers' imaginations.

Jumping ahead 16 years, Fig. 11.1 presents a 1992 scouting report on both pitcher Roger Clemens and shortstop Cal Ripken. The data and information have become dense with player information, overall statistics, situational stats

Table 11.1 Example of National Association's Player Register, 1871–75, for Al Spalding and George Wright (from Reichler, 1982)

		G	AB	H	R	BA	W	L	PCT	G by POS
Al Spalding										BR TR 6'1" 170 lbs.
	SPALDING, ALBERT GOODWILL									
	B. Sept. 2, 1850, Byron, Ill. D. Sept. 9, 1915, Point Loma, Calif.									
	Hall of Fame 1939.									
1871	Boston Red Stockings	31	151	40	43	.265	**20**	10	.667	P-31, OF-2
1872		48	248	84	59	.339	**37**	8	**.822**	P-48, OF-2
1873		60	331	105	85	.317	**41**	15	**.732**	P-57, OF-3
1874		71	**363**	121	80	.333	**52**	18	**.743**	P-71
1875		74	352	112	67	.318	**57**	5	**.919**	P-66, OF-10, 1B-2
5 yrs.		284	1445	462	334	.320	207	56	.787	P-273, OF-17, 1B-2
George Wright										BR TR 5'9½" 150 lbs.
	WRIGHT, GEORGE									
	Brother of Harry Wright. Brother of Sam Wright.									
	B. Jan. 28, 1847, Yonkers, N.Y. D. Aug. 21, 1937, Boston, Mass.									
	Hall of Fame 1937.									
1871	Boston Red Stockings	16	88	36	35	.409				SS-15, 1B-1
1872		48	253	85	86	.336				SS-48
1873		59	333	126	98	.378				SS-59
1874		60	319	110	75	.345				SS-60
1875		79	**407**	137	105	.337				SS-79
5 yrs.		262	1400	494	399	.353				SS-261, 1B-1

ROGER CLEMENS

Position: SP
Bats: R **Throws:** R
Ht: 6′ 4″ **Wt:** 220

Opening Day Age: 29
Born: 8/4/62 in Dayton, OH

ML Seasons: 8

Overall Statistics

	W	L	ERA	G	GS	Sv	IP	H	R	BB	SO	HR
1991	18	10	2.62	35	35	0	271.1	219	93	65	241	15
Career	134	61	2.85	241	240	0	1784.1	1500	628	490	1665	117

How Often He Throws Strikes

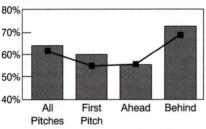

1991 Situational Stats

	W	L	ERA	Sv	IP		AB	H	HR	RBI	AVG
Home	8	5	2.59	0	142.2	LHB	563	123	3	45	.218
Road	10	5	2.66	0	128.2	RHB	430	96	12	37	.223
Day	6	2	2.22	0	73.0	Sc Pos	199	46	3	60	.231
Night	12	8	2.77	0	198.1	Clutch	125	28	0	8	.224

1991 Rankings (American League)
→ 1st in ERA (2.62), games started (35), shutouts (4), innings (271.1), batters faced (1,077), strikeouts (241), pitches thrown (4,025) and runners caught stealing (16)
→ 2nd in complete games (13) and lowest on-base average allowed (.270)
→ 3rd in strikeout/walk ratio (3.7), least home runs per 9 innings (.50) and most strikeouts per 9 innings (8.0)
→ Led the Red Sox in ERA, wins (18), games started, complete games, shutouts, innings, batters faced, hit batsmen (5), strikeouts, pitches thrown and strikeout/walk ratio

CAL RIPKEN

Position: SS
Bats: R **Throws:** R
Ht: 6′ 4″ **Wt:** 225

Opening Day Age: 31
Born: 8/24/60 in Havre de Grace, MD

ML Seasons: 11

Overall Statistics

	G	AB	R	H	D	T	HR	RBI	SB	BB	SO	AVG
1991	162	650	99	210	46	5	34	114	6	53	46	.323
Career	1638	6305	970	1762	340	33	259	942	28	688	747	.279

Where He Hits the Ball

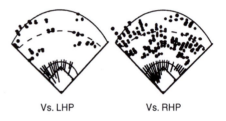

Vs. LHP   Vs. RHP

1991 Situational Stats

	AB	H	HR	RBI	AVG		AB	H	HR	RBI	AVG
Home	315	90	16	52	.286	LHP	164	57	12	31	.348
Road	335	120	18	62	.358	RHP	486	153	22	83	.315
Day	162	49	10	34	.302	Sc Pos	149	47	5	70	.315
Night	488	161	24	80	.330	Clutch	101	35	4	13	.347

1991 Rankings (American League)
→ 1st in total bases (368), games (162) and highest batting average on the road (.358)
→ 2nd in hits (210), doubles (46) and slugging percentage (.566)
→ 3rd in home runs (34) and sacrifice flies (9)
→ Led the Orioles in batting average (.323), home runs, at bats (650), runs (99), hits, doubles, total bases, RBI (114), games, slugging percentage, on-base percentage (.374), HR frequency (19.1 ABs per HR), lowest percentage of swings that missed (10.6%) and highest percentage of swings put into play (49.7%)
→ Led AL shortstops in fielding percentage (.986)

Figure 11.1   1992 scouting report on both pitcher Roger Clemens and shortstop Cal Ripken.

(home versus road games, day versus night games), rankings, and a graphical display of both pitching and hitting. The spatial graphical display superimposed on a baseball field presents the locations of Ripken's hits and their type: on the ground (lines) or in the air (dots). The display is split in two to present a comparative situational breakdown (facing a left- or right-handed pitcher). This clearly provides considerable multivariate information. The graphical data display for Clemens presents further insight into the traditional pitcher–hitter battle in baseball. The display shows the percentage of pitches Clemens threw for strikes as contrasted with the league average. The data are shown overall and at different stages of the count: first pitch, ahead (more strikes than balls), and behind (more balls than strikes) (Dewan and Zminda, 1992). The graph demonstrates Clemens' superiority over the rest of the league in control pitching. What a difference 16 years makes. Now one might classify this type of display as 'Tables becoming graphic'.

Ted Williams, who played for the Boston Red Sox from 1939 to 1960, is among the most outstanding and analytical of all baseball players. Figure 11.2 presents a spatial display of his batting average, conditional on the location of the pitched ball with respect to Williams' strike zone. This 11×7 matrix of batting averages takes advantage of a tabular structure to capture the strike zone in a coordinate-like representation. Evidently, low and outside pitches were Williams' only weakness. What data were used for this table and how they were processed is not well known (Williams and Underwood, 1971). A modification of this concept has been used by Fox Sports in its television coverage of the 1997 baseball playoffs.

A masterpiece of pitching efficiency in baseball is shown in Fig. 11.3 using two alternative data graphic displays of Greg Maddux's inning-by-inning pitch count of strikes and balls in the Atlanta Braves 4-1 win over the Chicago Cubs on 22 July 1997 at Chicago's Wrigley Field (Fossum, 1997). Figure 11.3(a) presents these data in a symbolic form, with commentary on the Cubs' longest at-bat (five pitches) and the closeness of the performance to the current record of 76 pitches (63 strikes and 15 balls for a remarkable total of 78 pitches and a game strike percentage of 81%). Still, this display really is a table with an inning time sequence. The use of panels for innings and baseballs and circles for total pitches, strikes, and balls is a poor selection of symbols, and the display loses the time series and multivariate character of the data. In short, this display strategy is defective. Alternatively, a stacked bar chart would be a better choice (Fig. 11.3(b)). This display draws attention to how few pitches Maddux threw per inning, at most 11 in the early innings and as few as 7 in the later innings. The reader immediately discovers he threw 100% strikes in the third inning. The graph also indicates how he got stronger over the innings with a median of 10 pitches in the first four innings and 8 in the final five innings. If one desired to stay with a table, a better display would be to place the inning-by-inning data in columns (e.g. 5 strikes, 2 balls, 7 pitches total for the first inning) with a game summary column. This would also allow annotation at the bottom of each column (e.g. the inning in which the Cubs scored).

Another form of a table used in sports is the tabular report which presents current standings and game results. Figure 11.4 provides an example of this type of table using the Green Bay Packers football team, the 1997 Super Bowl winners. The tabular report typically starts with divisional standing (e.g. the Packers are in first place in their division and have already qualified for conference playoffs, as indicated by the '×'). The report proceeds to the latest game, citing its key events (Tampa Bay's quarterback sprained his ankle late in the first half and the offense became ineffective), the outcome of

Figure 11.2   Matrix of Ted Williams' batting averages conditional on the location of the pitched ball in the strike zone (redrawn from Williams and Underwood, 1971).

the game (Tampa Bay lost 17 to 6), both team's statistics (e.g. Green Bay had 20 first downs and Tampa Bay 8), each team's individual statistics, and the quarter-by-quarter scoring events (e.g. Tampa Bay kicked a 24-yard field goal with 8 min 42 s left in the first quarter). Needless to say, this type of data is important for filling the football databases, but one gets the feeling of data

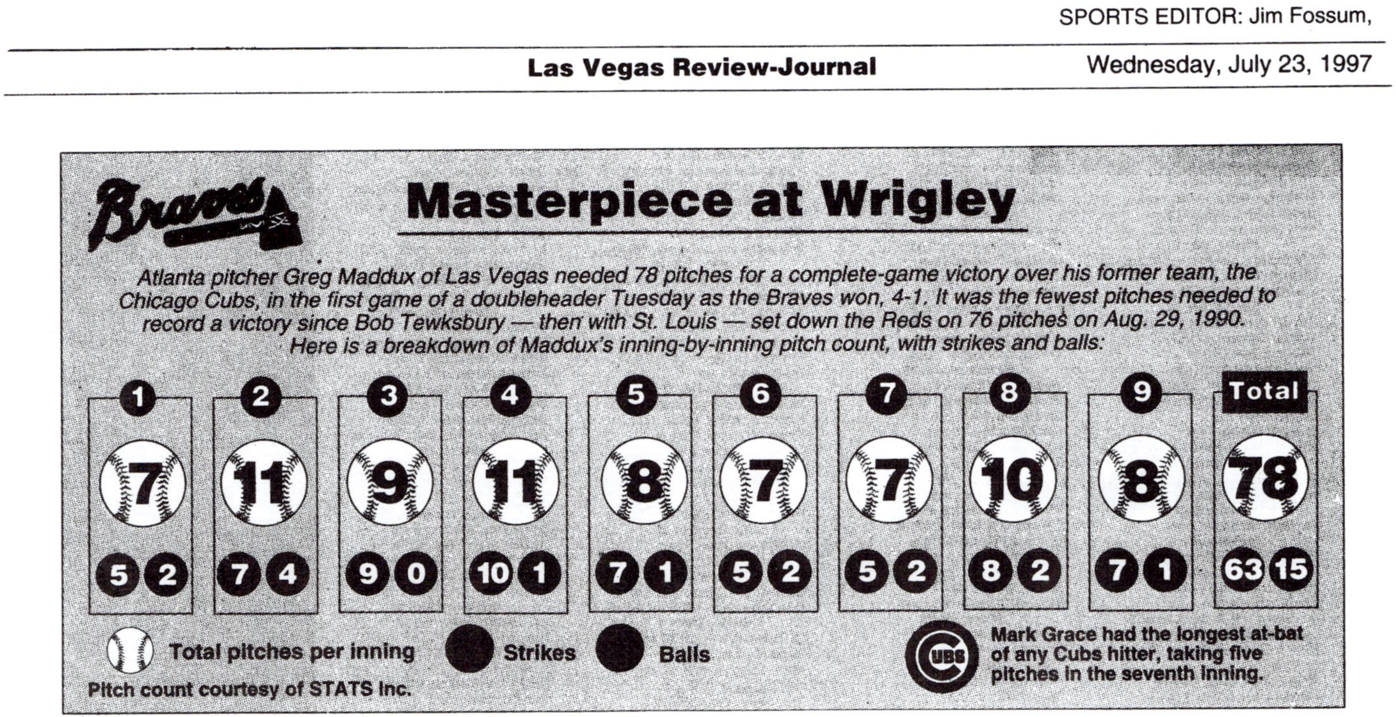

Figure 11.3 Greg Maddux's inning-by-inning pitch count of strikes and balls on 22 July 1997. (a) Symbolic form (source: *Las Vegas Review-Journal*, Wednesday, July 23, 1977);

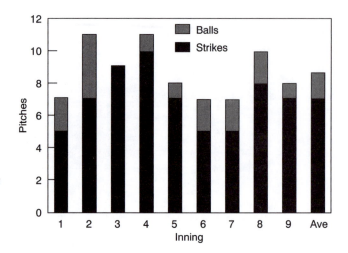

Figure 11.3   (b) stacked bar chart.

overload rather quickly. What are the key performance measures and how do they relate to each other and the time sequence of the game? The presentation neither invites nor aids analysis.

11.3   Tables to graphs

Figure 11.5 presents another approach to the display of pitching using Ron Guidry's extraordinary 1978 season for the New York Yankees. Guidry's 35 starts are ordered in a five-by-seven matrix of panels, each panel representing a different game. The panels are ordered from game 1 on 8 April at Texas, through game 35 on 2 October at Boston. The vertical line inside each panel marks the number of innings pitched (IP) using the horizontal scale of 0 to 9. Each panel is split into two parts.

(1)   The upper half displays the number of hits (H) and earned-runs (ER), using the upward scale of 0 to 9 on the left-hand side of the panel. The dark shaded triangle emphasizes the ER given up.
(2)   The lower half displays the number of strikeouts (K) and walks (W), using the downward scale of 0 to 9 also on the left of the panel. The lighter shaded triangle emphasizes the difference between strikeouts and walks.

Additional information on the pitching performance is given inside the box, including innings pitched (IP), win–loss record to date, complete games (CG), shutout number (ShO), and number of hits in the shutout. Thus, on 28 May, Guidry defeated the Toronto Blue Jays at Yankee Stadium to improve his record to 7 wins without a loss. He allowed 6 hits, 3 earned runs, 0 walks, 6 strikeouts, and went 9 innings for his third complete game. The display cleverly combines some of the features of a table with the actions of a time sequence to produce 35 tabular displays which effectively draw the reader's attention to the pitcher's classic battle.

Central

x-Green Bay	11	3	0	.786	360	251	7-0-0	4-3-0	9-2-0	2-1-0	7-1-0
Tampa Bay	9	5	0	.643	268	217	4-3-0	5-2-0	6-5-0	3-0-0	2-5-0
Minnesota	8	6	0	.571	302	317	4-2-0	4-4-0	6-5-0	2-1-0	3-4-0
Detroit	7	7	0	.500	352	283	5-2-0	2-5-0	6-5-0	1-2-0	5-2-0
Chicago	3	11	0	.214	235	380	2-6-0	1-5-0	1-9-0	2-2-0	1-6-0

Green Bay 7	0	7	3	- 17
Tampa Bay 3	3	0	0	- 6

Green Bay   **Tampa Bay**

17   6

Key: With the Packers leading 7-6 late in the first half, Tampa Bay quarterback Trent Dilfer suffered a sprained right ankle when he was sacked by Reggie White. The Bucs' offense sputtered the rest of the day.

Next: Packers (-7) at Panthers; Buccaneers (-2) at Jets.

First Quarter
TB—FG Husted 24, 8:42. Drive: 4 plays, 7 yards, 1:39. Key play: Jones recovery of Freeman fumble on Packers' 14. **Tampa Bay 3, Green Bay 0.**
GB—R.Brooks 43 pass from Favre (Longwell kick), 4:46. Drive: 2 plays, 36 yards, 1:05. Key play: Robinson recovery of Alstott fumble on Bucs' 36. **Green Bay 7, Tampa Bay 3.**
Second Quarter
TB—FG Husted 48, 1:52. Drive: 4 plays, minus 1 yard, 1:27. Key play: Lynch interception and return to Packers' 30. **Green Bay 7, Tampa Bay 6.**
Third Quarter
GB—Levens 8 pass from Favre (Longwell kick), 6:03. Drive: 10 plays, 73 yards, 5:13. Key plays: Favre 18 pass to Freeman; Favre 20 pass to Chmura; Favre 6 pass to Levens. **Green Bay 14, Tampa Bay 6.**
Fourth Quarter
GB—FG Longwell 27, 6:24. Drive: 16 plays, 88 yards, 10:31. Key plays: Favre 14 pass to Henderson; Favre 10, 12 and 8 passes to Levens. **Green Bay 17, Tampa Bay 6.**
A—73,523. No-shows—319.

INDIVIDUAL STATISTICS
RUSHING—Green Bay, Levens 22-54, Favre 6-13, Henderson 2-11, R.Brooks 1-4. Tampa Bay, Alstott 10-34, Dunn 12-33.
PASSING—Green Bay, Favre 25-33-1 280. Tampa Bay, Dilfer 6-17-0 67. Walsh 4-9-1 50.
RECEIVING—Green Bay, Levens 8-64, Freeman 5-73, Chmura 5-37, R.Brooks 3-71, Mayes 3-22, Henderson 1-13. Tampa Bay, K.Williams 5-87, Thomas 2-19, Dunn 2-9, Harris 1-2.
PUNT RETURNS—Green Bay, Mayes 3-16. Tampa Bay, K.Williams 1-13.
KICKOFF RETURNS—Green Bay, Beebe 2-60. Tampa Bay, Anthony 2-45, K.Wiliams 1-22, Rhett 1-16.
TACKLES-ASSISTS-SACKS—Green Bay, Robinson 8-0-1, B.Harris 6-1-0, Butler 5-1-0, S.Dotson 3-2-0, B.Williams 2-2-0, Joyner 2-0-1, Evans 2-0-0, Sharper 3-0-0, Wilkins 2-0-1, Brown 1-1-0, White 1-1-1, Koonce 0-3-0, Prior 1-0-0, T.Williams 1-0-0, R.Brooks 1-0-0, Chmura 1-0-0, Darkins 1-0-0, Davis 1-0-0, Henderson 1-0-0, Hollinquest 1-0-0.
Tampa Bay, Nickerson 10-2-0, D.Brooks 9-1-0, Lynch 7-0-0, Culpepper 6-1-0, Abraham 5-1-0, Mincy 5-1-0, Parker 5-0-0, Ahanotu 3-0-0, Gooch 2-1-0, Young 2-0-0, Johnson 1-1-0, Ellison 1-0-0, Harris 1-0-0, Husted 1-0-0, Legette 1-0-0, Mayberry 1-0-0, Moore 0-1-0, Quarles 0-1-0.
INTERCEPTIONS—Green Bay, Prior 1-0. Tampa Bay, Lynch 1-28.
MISSED FIELD GOALS—Green Bay, Longwell 32 (BK). Tampa Bay, None.

OFFICIALS—Referee Phil Luckett, Ump Ron Botchan, HL Gary Slaughter, LJ Tom Barnes, BJ Tony Corrente, SJ Doug Toole, FJ Don Dorkowski.
Time: 2:50.

	GB	TB
FIRST DOWNS	20	8
Rushing	3	2
Passing	15	5
Penalty	2	1
THIRD DOWN EFF	5-11	6-16
FOURTH DOWN EFF	0-0	0-1
TOTAL NET YARDS	362	161
Total Plays	64	52
Avg Gain	5.7	3.1
NET YARDS RUSHING	82	67
Rushes	31	22
Avg per rush	2.6	3.0
NET YARDS PASSING	280	94
Sacked-Yds lost	0-0	4-23
Gross-Yds passing	280	117
Completed-Att.	25-33	10-26
Had Intercepted	1	1
Yards-Pass Play	8.5	3.1
KICKOFFS-EndZone-TB	4-0-0	3-1-1
PUNTS-Avg.	4-39.5	6-43.0
Punts blocked	0	0
FGs-PATs blocked	1-0	0-0
TOTAL RETURN YARDAGE	76	124
Punts Returns	3-16	1-13
Kickoffs Returns	2-60	4-83
Interceptions	1-0	1-28
PENALTIES-Yds	3-21	6-33
FUMBLES-Lost	2-2	1-1
TIME OF POSSESSION	34:41	25:19

Figure 11.4   Example of current standings and game results (*Las Vegas Review-Journal*, Monday, 8 December 1997).

RON GUIDRY -- 1978 -- GAME BY GAME

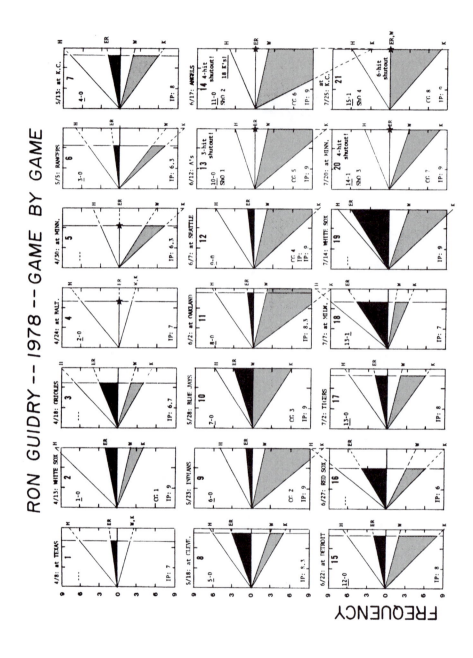

FREQUENCY

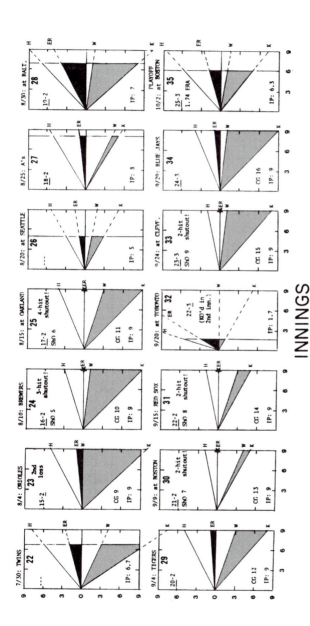

INNINGS

Figure 11.5 Display of pitching: Ron Guidry, 1978, game by game.

The 35 game pitches by Ron Guidry in 1978 through the 2 October playoff game, as summarized in 'pitching plots'. For hits (H) and earned runs (ER), read upward from zero, and for walks (W) and strikeouts (K), read downward from zero, on the scale at the left of each panel. The horizontal scale represents innings in each panel. The black shading emphasizes the earned runs given up, and the lighter shading represents the difference between strikeouts and walks (K–W). The vertical line where the shading ends shows: the number of innings pitched (IP). Other abbreviations: CG = complete game, ShO = shutout. Games identified by opposing team's nickname were home games at Yankee Stadium, those with 'at' were road games.

(1)   The ultimate goal is to minimize opponent's runs.
(2)   A technique to achieve this goal is control (to keep strike-outs ahead of walks).
(3)   The style is to maximize innings pitched.

Each panel of the matrix is rich in information compared to its size (data dense), and hence it requires time and effort to see all the relations of this multivariate plot. There are two time sequences throughout the display: one within each plot and one across the panels. However, even this informative data graphic display can be improved by following Cleveland's principles. First, keep the tick marks and scale outside the panel, as are the innings numbers. Second, extend the right-hand side of the panel downward when needed to report the number of strikeouts (e.g. panel 22). Third, remove the dark shading of the ER triangle and embolden the line to emphasize the slope that is the ER run rate. Finally, bring the symbols (e.g. ER, W and K) inside the panel, because this is where the action is reported. Yes, even good statistical data graphic displays can be improved.

11.4   Clearly graphs

In 1994, Sasakawa Sports Foundation's National Sports–Life Survey investigated the participation in a sport and/or physical activity as an important dimension of health maintenance, prevention of disease, and improvement of the quality of life. The specific goal was to 'assess the physical activity patterns and levels of the adult Japanese population' (Ikeda, 1996). A national representative sample of 2000 adults, 20 years old or more, was randomly selected, and trained interviewers visited the sampled homes. The response rate was 79.8%. As shown in Fig. 11.6, 13 activities were plotted using the two dimensions of frequency of participation in an activity (i.e. number of days in past 12 months) and the duration of time participating in the activity (minutes per activity). The annotated plot reveals three clusters of activity in the adult Japanese population: (1) exercise break and jogging were the most popular forms of activity and done almost every second day for about 30 minutes; (2) at the other end of the spectrum were skiing and golfing, done about 10 days a year for about 4 hours each time; and (3) finally, all the other sports which were done about 30 days per year for about 120 minutes each time. In short, one could classify the above clustered participation results as daily, weekends and seasonal activities.

A simple scatterplot can provide substantially more information when annotated by additional categories. In addition, due to the proper scientific design and the careful implementation of this study, estimates of the activities of various subsets of the surveyed population (e.g. activities by age groups) are also possible.

The United States Golf Association (USGA) has been keeping track of the winning score of its US Open Championship since at least 1909. Figure 11.7(a) presents a scatterplot of each year's winning total score versus the year, from 1909 to 1962. The graph has been annotated with names of some clubs which have held the US Open and the winning entrant that year. Three guidelines appear to have been subjectively drawn (Reddy, 1990), but they

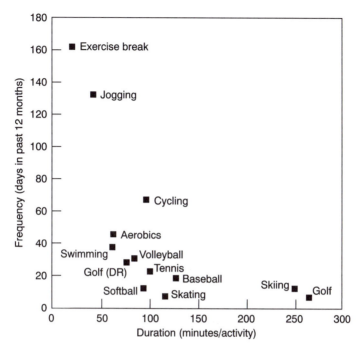

Figure 11.6 Sports/physical activities by frequency and duration (redrawn from Ikeda, 1996).

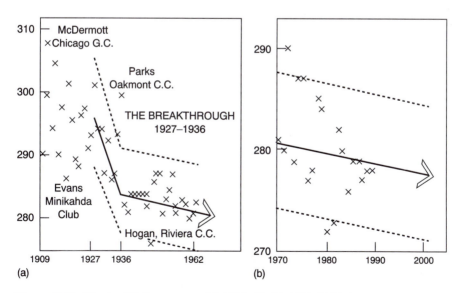

Figure 11.7 Winning US Open scores. (a) 1909–62; (b) 1970–2000 (redrawn from Reddy, 1990).

presumably are intended to show the trend of the scores over the last 50 or so years. In the base period, 1909 to 1927, the mean winning score is reported to be 295, by 1936 it was 284, and in the most recent decade, 1990 to 2000, it is predicted to fall to a mean of 277 (Fig. 11.7(b)). The break in the slope of the centerline of the plot indicates the end of the 'breakthrough' or change period, when hickory shafts were being replaced by steel shafted clubs. Other changes were also occurring in this period including standardizing the golf ball and improving the maintenance of golf courses. Lastly, there is the belief that more outstanding athletes are picking golf for their profession. While these data are multivariate and a number of factors may be influential, the graphs do not provide much information to aid in their analysis. The plotted scores in Fig. 11.7(a) appear to be too few for the time period, and neither Fig. 11.7(a) nor 11.7(b) has trend lines which follow the plotted data. There certainly appears to be a need for some statistical trending of the data. A control chart may be appropriate after the scoring process stabilizes.

The previous examples of statistical data graphics in sports focused on a single sport, but one is often interested in comparisons between sports. The following example of statistical data graphics begins to address this situation. The National Collegiate Athletic Association (NCAA) attempts to monitor injury rates of men's and women's sports through its Injury Surveillance System. This system is an entirely voluntary effort by NCAA colleges and universities. In 1991–92, the system attempted to get participation from about 10% of the teams in each of the nine men's (i.e. American football, spring football, wrestling, soccer, gymnastics, ice hockey, lacrosse, baseball and basketball) and seven women's (i.e. gymnastics, soccer, basketball, field hockey, lacrosse, volleyball and softball) sports competing in that period. Teams were requested to report weekly on injuries to its athletes. At the end of the season, the data were analyzed and a report was published for each of the 16 sports. The reports were made available to all NCAA members and other interested parties.

Figure 11.8 is a back-to-back stem-and-leaf plot of injury rates, injuries per one thousand athlete exposures, for the nine men's and seven women's sports, partitioned by whether the injury occurred in practice or in a game (Flueck *et al.*, 1992). In this stem-and-leaf plot, the vertical stem represents the ones digit and each leaf represents the tenths digit of an injury rate. For example, the injury rate in men's soccer during a game is 19 injuries per thousand athlete exposures. This rate is represented by a '9' leaf on the '1e' stem annotated by an 's' for soccer to its right. (See Tukey, 1977, for further description of stem-and-leaf plots.) If the sport is played by both men and women, then the women's injury rate annotation is prefixed by the letter 'w' (e.g. 'ws' for women's soccer). Hence, the women's injury rate for soccer games is 17 injuries per thousand athlete exposures.

This visual display provides information at a number of levels. At the macro level, one quickly sees injury rates in practice are considerably lower than in a game. As an example, football games (f) have an injury rate of 36 per 1000 athlete exposures, whereas football practices have an injury rate of 4; hence football games are nine times more likely to produce an injury than football practices. Looking only at the practice side, some sports have the same, or nearly the same, injury rates for both men and women (e.g. men's and women's soccer). Looking at the game side of the display indicates three clusters of injury

	Practice			Game	
			4*		
			3e		
			3s	6	f
			3f		
			3t	2	spf
			3*	1	w
			2e		
			2s		
			2f		
			2t		
			2*	1	wg
			1e	9	s
			1s	6777	l,ih,ws,g
			1f		
			1t		
			1*	0	b
	spf	9	0e	99	wfh,wb
	wg,w	77	0s	67	bb,wl
ws,s,wv,b,g,wb,f,wfh,l		555444444	0f	55	wsb,wv
	wl,wsb,ic,bb	3322	0t		
			0*		

Notation: w_ = womens_, spf = spring football, g = gymnastics, w = wrestling, s = soccer,
v = volleyball, b = basketball, f = football, fh = field hockey, l = lacrosse, sb = softball,
ih = ice hockey, bb = baseball

Figure 11.8 Total game and practice injury rates for nine men's and seven women's sports (injuries per 1000 athlete exposures) (redrawn from Flueck *et al.*, 1992).

rates: (1) football, spring football, and wrestling, all men's direct contact sports; (2) women's gymnastics through men's lacrosse; and (3) men's baseball through women's softball. The surprise in this side of the data graphic display is the high injury rate in women's gymnastics (which also has a relatively high rate in practice). Perhaps women's gymnastics needs a review of its efforts to prevent injuries in competition and practice.

The annotated back-to-back stem-and-leaf data graphic display (indicating sport, practice versus game, and gender) is capable of providing multivariate macro, micro, comparative, and clustering information. One can also partition the data further and generate a new stem-and-leaf plot conditional on a subset of the original data (e.g. body parts injured in women's compared with men's sports).

11.5 Time sequences

The plotting of performance versus time or distance is fundamental to many sports. Figure 11.9 presents a two-panel vertical display of time sequence plots which simultaneously track both the 1969 National League East and West Division baseball title races. The performance indicator is games over or under a .500 winning percentage (i.e. wins minus losses). An alternative could track the situation with a relative net plot, or games behind the leader. At the start of the season, all teams in the league had a value of .500 (wins

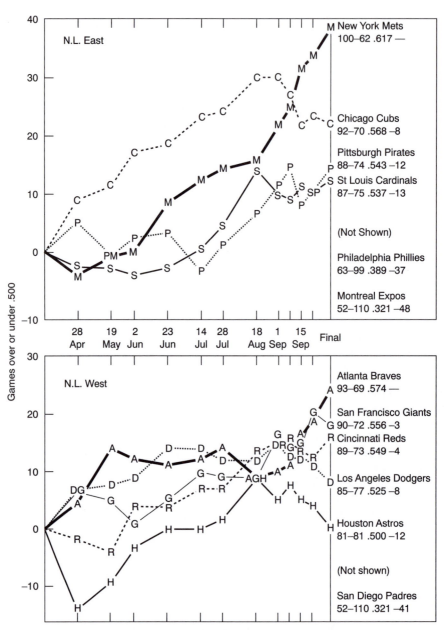

Figure 11.9 Two-panel display of the 1969 National League East and West Division baseball title races (redrawn from Davenport, 1979).

equal losses). As the season progressed, teams won more or less than 50% and, at the end of the season, the teams with the highest winning percentage in each division moved on to the National League championship playoffs with a chance to proceed to the World Series.

The National League's East Division title race appeared to be decided by 1 September with the Cubs nine games ahead of their nearest competitor, the New York Mets. True to form, the perennial losing Cubs collapsed and finished eight games short of first place. Note how the Mets overcame a mediocre start and matched the Cubs performance (i.e. same slope) throughout midseason, and overtook the Cubs in the last 30 days. In the West division of the National League, Atlanta was in the middle of the pack on 1 September and managed to beat out the Giants in the last week of the season. These dramatic changes in division rankings in so short a time were truly amazing, and the two time series plots vividly tell the story of the season ending 'cold and hot' streaks.

The use of obvious symbols (C for Cubs, M for Mets, etc.) for each team helps to keep track of the rankings over time. Also, teams out of the race are only summarized in the margin of the display (e.g. the Phillies' record of 63 wins and 99 losses placed them 37 games behind the division-winning Mets), making it easier to follow the remaining teams. In addition, the dark line marking the winning team in each division also aids in following the race. However, there is a missed opportunity in not annotating parts of each team's time series record with events that might provide reasons for the ups and downs a team encounters over a season.

Westfall (1990) extended this concept to plotting the difference in scores between two competitors in basketball. His score-difference (SD) plot allows one to assimilate rapidly the difference in basketball scores versus continuous elapsed-time. It is applicable to both real-time use, as well as post-game evaluations. This type of plot emphasizes:

(1) the length of time a team holds the lead;
(2) the difference in scores at any time;
(3) the number and rate of lead changes; and
(4) the streakiness of scoring.

The SD plot for the 1989 NCAA Championship Game, Michigan versus Seton Hall (Fig 11.10(a)), well illustrates the tempo of the game, with very little scoring in the opening minutes of the game, a trading of streaks, and Michigan only ahead by 3 points at the close of the first half. Note how narrow spikes indicate rapid scoring by both teams while wide blocks indicate periods of no scoring. In the second half, Michigan rapidly built-up a lead of 12 points in the first five minutes, and then proceeded to surrender the entire lead over the balance of the half to end the regulation time in a tie. In the five-minute overtime period (Fig. 11.10(b)), Seton Hall maintained a lead of 1 to 3 points for virtually the entire period, only to lose at the last moment.

An SD plot of the 1989 National Basketball Association (NBA) Championship game between the Pistons and the Lakers illustrates how an injury to a key player may have affected the outcome of the game (Fig. 11.10(c)). The Lakers led over the first 30 minutes of the game, at times by as much as 10 points. After an injury to their star player Magic Johnson, the Lakers built a large temporary lead which the Pistons eventually overcame to win by 3 points. The graphic display of the score-difference time series plot forces one to recognize the streaky nature of a basketball game, how large leads can evaporate with time, and how important it is to be prepared for the final minutes of a game. Annotations of events which might have an effect on a game

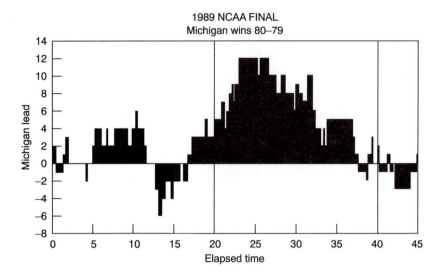

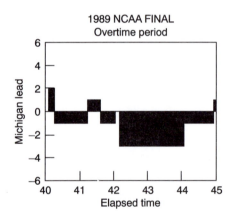

Figure 11.10 (a) Score-difference (SD) plot for the 1989 NCAA Championship Game, Michigan versus Seton Hall. (b) Five minute overtime period of the game in (a) (redrawn from Westfall, 1990).

should be noted on the SD time series plot (as the injury to Johnson is shown), so post-game evaluations can be made using this information. Finally, the headings and contents of each data graphics display should be carefully checked for accuracy, so there is a clear understanding of the message. Cleveland's advice to 'Proof read graphics' is important for all displays.

A variation of the use of the SD concept in baseball is presented in Fig. 11.11 (Bennett, 1994). The graph displays the probability of a Toronto Blue Jays victory versus time (as measured by number of batters sent to the plate) for Game 6 of the 1993 World Series. It shows how Toronto established an early lead, lost it late in the game, and won on the final play with Joe Carter's dramatic home run. The number annotations on the graph indicate Toronto's

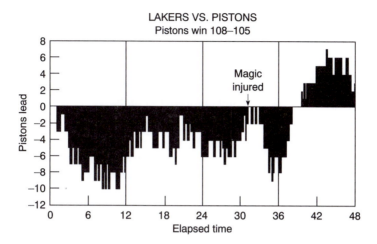

Figure 11.10   (c) SD plot of 1989 National Basketball Association Championship Game 2 between the Pistons and the Lakers (redrawn from Westfall, 1990).

lead in runs at that point in the game, and the letter annotations (e.g. 3B, FO and HR) represent Paul Molitor's performances at bat. Molitor was chosen as the Most Valuable Player of the Series, and the graph highlights some of his contributions, such as his single (1B), which set up Joe Carter's home run to win the game on the final play. The plot also shows how graphic techniques for direct measures of situations (e.g. point differentials in basketball and net wins in baseball) can also be applied to modeled measures (e.g. probability

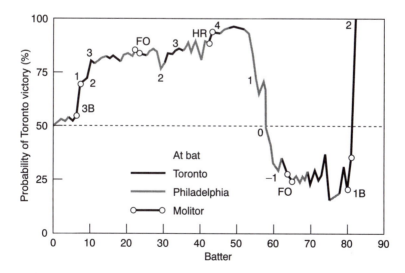

Figure 11.11   Probability of a Toronto Blue Jays victory versus time (number of batters) for Game 6 of the 1993 World Series.

of victory in the Player–Game Percentage model described in Chapter 2 on baseball).

11.6   Spatial playing fields

Figure 11.1 provided an introduction to spatial data graphical applications in showing where Cal Ripken hit the ball. These types of data graphic displays are further examples of a growing effort to add spatial dimensions to the use of sport performance displays, in order to provide additional dimensions and information. These graphs take advantage of the fan's and coach's familiarity with the playing field to establish an instantly recognizable context for the graph.

Shot charts are the analogous application in basketball. These charts show where shots are made and missed from different areas of the court. Two team shot charts from the 1997 NCAA Championship Game, Arizona versus Kentucky, are presented in Figs 11.12(a) and 11.12(b) (National Collegiate Athletic Association, 1997). The shot chart for one of the Kentucky players, Scott Padgett, is also presented in Fig. 11.12(c) (National Collegiate Athletic Association, 1997). The white circles represent shots missed and the dark circles represent shots made. The use of both halves of the court to represent the halves of the game allows comparison of play between halves. As an example, Arizona spread its shot selections in the first half, but in the second half, it concentrated them inside near the basket. Alternatively, Kentucky concentrated its shots in the free-throw lane in the first half and spread its shots between the free-throw lane and the three-point line in the second half. In terms of scoring, Arizona only made three of eight attempts at the three-point circle in the first half (38%), but improved to three of five attempts (60%) in the second half. Correspondingly, Kentucky changed its shot selection from 11 attempts at the three-point shot in the first half to 19 in the second half. Of the 19 three-point attempts by Kentucky in the second half, Scott Padgett took ten and made two for a 20% success rate. Perhaps Padgett's brief 50% success rate at three-point range in the first half encouraged further attempts, with little success, in the second half.

The addition of shot charts, with their strong emphasis on the spatial dimension, is very important in basketball. Scouting reports all emphasize the locations from which a player will typically take his shots, the type of shot, and the shooting percentage given different game situations. In fact, the game plan typically is based on establishing areas of the court which a team believes it can control, and now more information is available to annotate the spatial displays with additional information such as assists and rebounds. Video replay is widely used by NBA teams: just ask coaches what they do late at night – they watch tapes of past games. Most recently, data mining tools have been developed to aid coaches in identifying interesting competitive situations using video replays (Colet and Bhandari, 1997).

An old adage of golf is 'Drive for show, but putt for dough'. Figure 11.13 presents two examples of the former, based on a research study conducted in the early 1960s by the United States Golf Association (USGA) (Reddy, 1963). The figure displays the final locations of the golf ball from initial

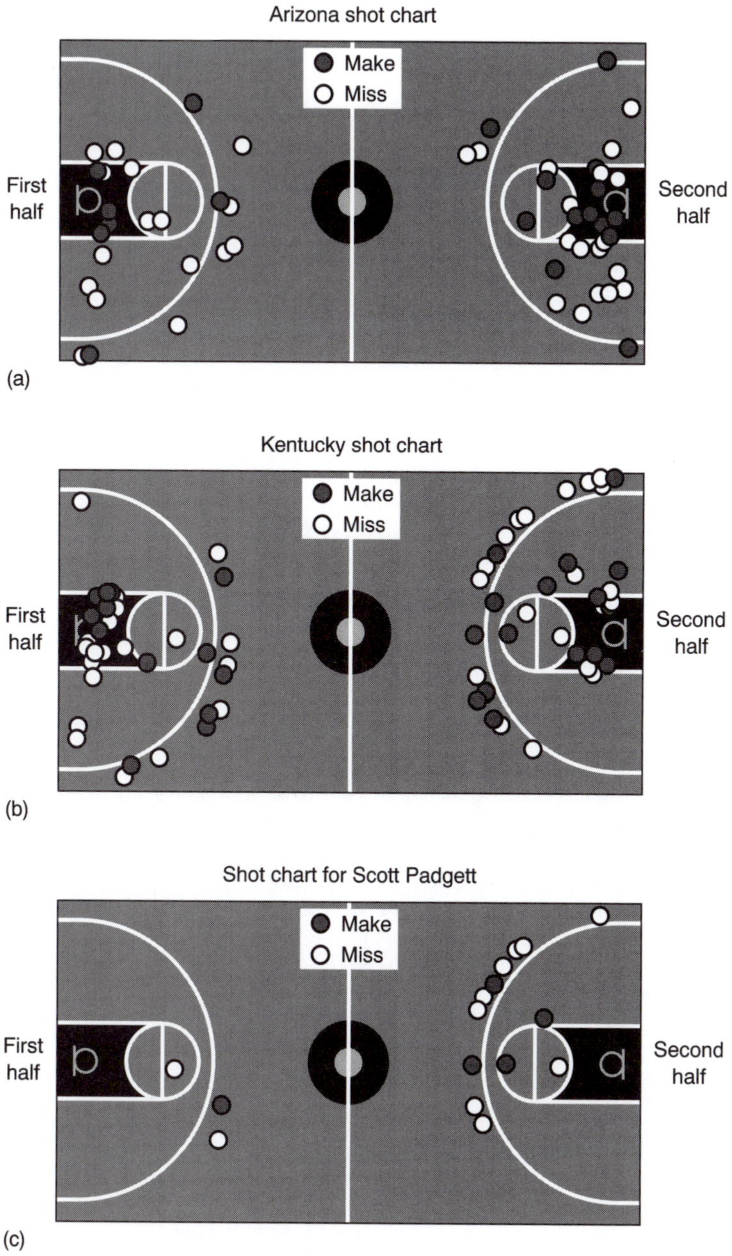

Figure 11.12 Team shot charts from the 1997 NCAA Championship Game, Arizona versus Kentucky. (a) Arizona shot chart; (b) Kentucky shot chart; (c) shot chart for Scott Padgett, one of the Kentucky players.

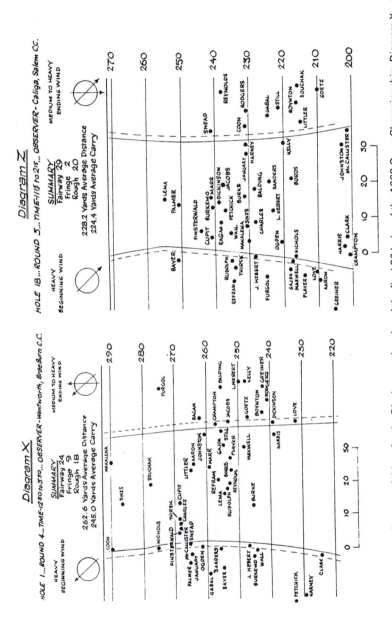

Figure 11.13 The final locations of the golf ball from initial drives of the 51 players who made the cut at the 1963 US Open championship (reproduced from Reddy, 1963).

The diagrams show the total distance of drives for the 51 players who played the final 36 holes of the 1963 Open Championship: Diagram X – Hole 1 in Round 4; Diagram Z – Hole 18 in Round 3. The wind assisted players on No. 1 and was against them on No. 18.

drives off the tee on holes 1 and 18 for the 51 players who made the cut at the 1963 United States Open championship.

Tony Lema and Arnold Palmer had the two longest drives on hole 18, each approximately 250 yards down the middle of the fairway. Only 29 (57%) of the drives stopped in the fairway, which at its narrowest point was about 30 yards wide; about equal numbers missed the fairway on the left side (20%) as on the right side (20%), with two settling in the fringe. Figure 11.13(b) suggests that players were aiming their drives for the middle of the fairway at about the 220 to 240 yard mark. Although the players on this hole had a headwind, classified as medium to heavy, it was reported to have had little effect on landing and remaining within the fairway. However, when playing hole 1, which is parallel to hole 18, the wind again was medium to heavy, but at the players' backs. Now only 45% of the player's balls stopped in the fairway and the percentage missing the fairway on the left was 20%, on the right 18%, and in the fringe 18%. Although the wind added about 20 yards to most drives the variability of the drive locations was greater with the wind at the players' backs than in their face (e.g. Palmer's drive was in the rough on the left side at 265 yards). Thus, increasing relative wind speed apparently results in less control over a player's drive and more chances for problem lies. Interestingly, virtually all players preferred hitting into the wind and getting the shorter drive. Note that the figure shifts the distance scale 20 yards between holes 1 and 18; this is not good scaling practice.

The 1965 USGA Open championship recorded all shots taken by the 50 qualifiers on the two par-three holes, 3 and 16. Both greens were over 10 000 square-feet (rather large). Hole 3 is 164 yards with a water hazard, and hole 16 is 218 yards across a valley to a plateau green with three defending sand traps. Figure 11.14 presents the spatial outline of the two greens with the day-by-day placements of the pin and each day's mean score for the holes (Reddy, 1965). The focus of the research was to assess the effect of the pin location on the scores of the qualifiers for the four days of the tournament.

The graphic display is accompanied by two tables which provide aggregate and daily scoring information for each green on each of the four days of the tournament. One quickly sees that the farther the pin is placed toward the back of each green, and better defended, the higher is the average score (e.g. 3.22 for day 2 versus 3.02 for day 1 on hole 3). In addition, the percentage of bogeys on hole 16 is substantially larger than on hole 3 (24% versus 17%). Perhaps water focuses the concentration of the golfer more than sand, which does not exact a stroke penalty. The interaction of both spatial and table information provides more than the sum of its parts.

One application of spatial graphs is of particular interest to the fan as a spectator. It is a tradition in baseball that any ball hit or thrown into the seating area can be kept by the fan who catches it. Even given the large number of balls hit into the stands, it is a relative rarity to catch a ball in a Major League Baseball game and any fan who catches a ball is thrilled (unless it is a home run by the visiting team). So, it is of interest to the fan to identify the best seats in a stadium for catching a ball. Figure 11.15 shows the distribution of balls batted into the stands of the Houston Astrodome over the course of 57 games (Frohlich and Scott, 1981). The graph shows the seating area of all three seating levels. Clearly, the upper decks are poor choices for catching balls. The seating in the home run

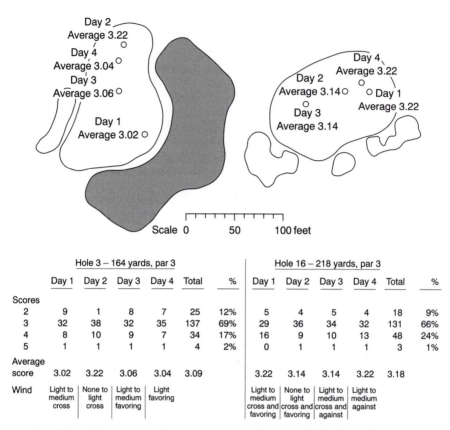

	Hole 3 – 164 yards, par 3						Hole 16 – 218 yards, par 3					
	Day 1	Day 2	Day 3	Day 4	Total	%	Day 1	Day 2	Day 3	Day 4	Total	%
Scores												
2	9	1	8	7	25	12%	5	4	5	4	18	9%
3	32	38	32	35	137	69%	29	36	34	32	131	66%
4	8	10	9	7	34	17%	16	9	10	13	48	24%
5	1	1	1	1	4	2%	0	1	1	1	3	1%
Average score	3.02	3.22	3.06	3.04	3.09		3.22	3.14	3.14	3.22	3.18	
Wind	Light to medium cross	None to light cross	Light to medium favoring	Light favoring			Light to medium cross and favoring	None to light cross and favoring	Light to medium cross and against	Light to medium against		

Figure 11.14 Effect of pin placement on scoring in the 1965 USGA Open championship.

area in the outfield is also a poor choice; however, this is somewhat redeemed by the value of catching a home run as opposed to a foul ball. The lower level near home plate provides the best chance.

11.7 Illustration

In American football, the sequence of offensive plays in which a team attempts to score within a single possession is called a 'drive'. A drive can be characterized by the number of plays executed, the number of yards gained, and the result. Figure 11.16 shows all drives from Super Bowl XXVI (e.g. the Washington Redskins' opening drive of 4 plays, 6 yards, and punt (Coddington, 1992)). This attempt at a dynamic display provides an annotated and illustrated summary display of the game. A visual comparison of Washington's first half with that of Buffalo quickly conveys the message that Washington was dominant. It also shows how two of Washington's first half touchdowns resulted from a good field position at midfield to start its drives. The drive display

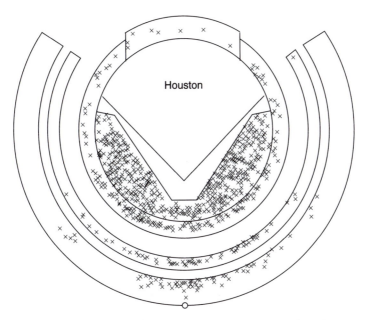

Figure 11.15 Where the grandstand balls landed in 57 games at the Astrodome.

uses a tabular approach, with rows representing drives and yard-lines representing columns. There even is a sequential time component measured in time to execute a play. However, the illustrated drive display seems caught in the middle of the hierarchy of sport information, for it is neither a strong storehouse of data nor a detailed picture of the action of the game. Coaches do not look at drive displays for evaluating offense and defense performances, but they do look long and hard at the video recording of the football game. The home for this display probably is with the fans.

Figure 11.17 provides an illustrated and annotated display of the 100 m sprint at the 1987 World Track and Field Championships in Rome (Wildbur, 1989; Pritchard and Pritchard, 1994). This illustrated multi-window time sequence display focuses on comparing two great sprinters, the Olympic champion Carl Lewis and the Canadian champion Ben Johnson, who established a new world record in winning this race. This illustrated and annotated display combines both tabular and graphical information in providing initial status (e.g. reaction time at 0 m), split time sequences (e.g. Johnson's lead in seconds), interval speeds (e.g. 12.9 mph over the first 10 m), window-by-window comments (e.g. Johnson reaches peak speed), and a sequence of illustrations focusing on the race within a race. This illustrated statistical data graphic display meets most of Cleveland's principles by providing a clear picture, clear understanding, an appropriate time scale, and an excellent presentation strategy. In addition, the display contains information useful for both reviewing the race (why was Lewis late in getting started?) and improving the athlete's future performances (too much variability in Lewis's split times).

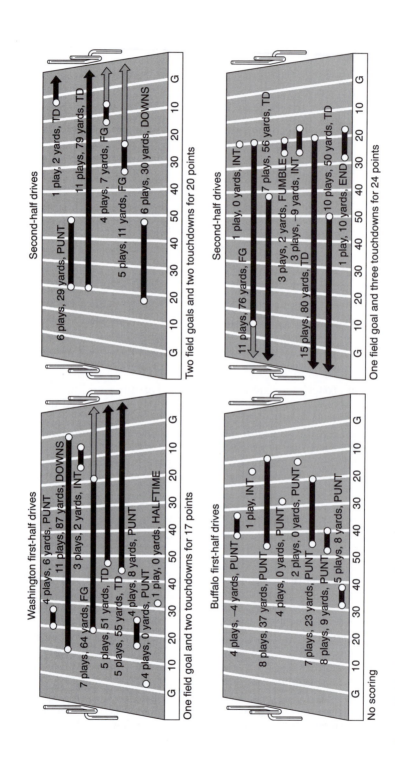

Figure 11.16 All drives for Super Bowl XXVI (redrawn from *USA Today*, January 27, 1992).

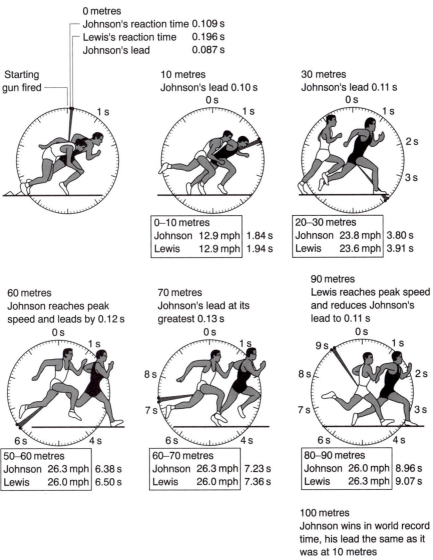

0 metres
Johnson's reaction time 0.109 s
Lewis's reaction time   0.196 s
Johnson's lead   0.087 s

Starting
gun fired

1 s

10 metres
Johnson's lead 0.10 s
0 s

1 s

0–10 metres		
Johnson	12.9 mph	1.84 s
Lewis	12.9 mph	1.94 s

30 metres
Johnson's lead 0.11 s
0 s

1 s
2 s
3 s

20–30 metres		
Johnson	23.8 mph	3.80 s
Lewis	23.6 mph	3.91 s

60 metres
Johnson reaches peak
speed and leads by 0.12 s
0 s

1 s
8 s
7 s
6 s   4 s

50–60 metres		
Johnson	26.3 mph	6.38 s
Lewis	26.0 mph	6.50 s

70 metres
Johnson's lead at its
greatest 0.13 s
0 s

1 s
8 s
7 s
6 s   4 s

60–70 metres		
Johnson	26.3 mph	7.23 s
Lewis	26.0 mph	7.36 s

90 metres
Lewis reaches peak speed
and reduces Johnson's
lead to 0.11 s
0 s

9 s   1 s
8 s   2 s
7 s   3 s
6 s   4 s

80–90 metres		
Johnson	26.0 mph	8.96 s
Lewis	26.3 mph	9.07 s

100 metres
Johnson wins in world record
time, his lead the same as it
was at 10 metres

9 s   1 s
8 s   2 s
7 s   3 s
6 s   4 s

90–100 metres		
Johnson	25.7 mph	9.83 s
Lewis	26.0 mph	9.93 s

Figure 11.17   Display of the 100 m sprint at the 1987 World Track and Field Championships
(redrawn from Wildbur, 1989).

11.8   Concluding remarks

Twenty-four statistical data graphic displays, with over 30 sports ranging from football, basketball and baseball, to skiing, swimming, golf and track, were presented, examined, and evaluated in their ability to aid in telling a sports story. Six categories of display were used to capture the essence of the 24 displays: (1) tables; (2) tables to graphs; (3) clearly graphs; (4) time sequences; (5) spatial sequences; and (6) illustrations. Each data graphic display was evaluated using Cleveland's general criteria of clear vision, clear understanding, scales, and general strategy. Results were mixed with some displays clearly in need of annotation and improved scales, and others attempting to focus too broadly. Many of the displays could profit from more attention to construction strategy. There appears to be a need for more creativity in merging both tabular and graphic components in a display.

There is clear evidence of an effort to add dynamic dimensions to a graph and even to a table. The static table is trying to become a dynamic graph, the dynamic graph is attempting to become a multi-dimensional display, and the spatial component seems to be materializing everywhere. Hardware and software people are improving the display process daily; in short, the force of the old claim that a picture is worth a thousand words now depends on who are the designing herds.

References

Bennett, J. M. (1994) MVP, LVP, and PGP: a statistical analysis of Toronto in the World Series. *1994 Proceedings of the Section on Statistics in Sports.* Alexandria, VA: American Statistical Association, pp. 66–71.

Cleveland, W. S. (1985) *The Elements of Graphing Data.* Monterey, CA: Wadsworth.

Coddington, R. (1992) Super Bowl XXVI: Washington Redskins 37, Buffalo Bills 24. *USA Today*, 27 January 1992, 5C.

Colet, E. V. and Bhandari, I. S. (1997) Advanced Scout: an application of data mining to the NBA using attribute focusing. *Presented at 1997 Joint Statistical Meetings*, 11 August 1997.

Davenport, J. W. (1979) *Baseball Graphics.* Madison, WI: First Impressions.

Dewan, J. and Zminda, D. (eds) (1992) *The Scouting Report: 1992.* New York: Harper Perennial.

Flueck, J. A., Strauss, R. H. and Dick, R. W. (1992) A study of the frequency, type, and severity of injuries in intercollegiate football. *1992 Joint Statistical Meetings, Statistics in Sports Section*, 9–13 August 1992, Boston, MA.

Fossum, J. (1997) Masterpiece at Wrigley. *Las Vegas Journal*, Sport Section, 23 July 1997.

Frohlich, C. and Scott, G. R. (1981) Where spectators sit to catch baseballs. *Baseball Research Journal*, **10**, 132–138.

Ikeda, M. (1996) Participation in sports and physical activity among Japanese adults: an international comparison. *Challenges Ahead for Improving Sports Statistics*, Italian National Olympic Committee, Document and Information Division, Roma.

National Collegiate Athletic Association (1997) Kentucky vs. Arizona. NCAA Online, www.finalfour.net.

Playfair, W. (1786) *Commercial and Political Atlas*. 1786: London.

Pritchard, W. G. and Pritchard, J. K. (1994) Mathematical models of running. *American Scientist*, **82**, 546–553.

Reddy, J. (1963) Putting and driving in the open. *United States Golf Association*, September, p. 7.

Reddy, J. (1965) Scoring analysis of the U.S. open: the short game and par-3 play. *United States Golf Association*, 12–13 November.

Reddy, J. (1990) USGA Open scores predicted for 1970–1990. *United States Golf Association*, May–June, 1990.

Reichler, J. L. (ed.) (1982) *The Baseball Encyclopedia*. New York: Macmillan.

Tufte, E. R. (1983) *The Visual Display of Quantitative Information*. Cheshire: Graphics Press.

Tukey, J. W. (1977) *Exploratory Data Analysis*. Boston: Addison-Wesley.

Wainer, H. (1996) Visual revelations. *Chance*, **9**, 43–52.

Westfall, P. H. (1990) Graphical presentation of a basketball game. *The American Statistician*, **44**, 305–307.

Wildbur, P. (1989) *Information Graphics*. New York: Van Nostrand Reinhold.

Williams, T. and Underwood, J. (1971) *The Science of Hitting*. New York: Simon and Schuster.

12

Predicting Outcomes

Ray Stefani

California State University, Long Beach, USA

12.1 Overview

After each sports competition, the media report the results by ranking competitors in running events first, second, third and so on, and showing win–draw–loss standings for teams. There are a number of purposes for ranking competitors after several competitions. It may be desirable simply to list which teams and individuals are best. Also, rankings (and generally the ratings upon which the ordinal rankings are based) may be used for gambling purposes or for seeding tournaments and competitions. A good sports rating system (SRS) adjusts for differing numbers of competitions, competitive conditions and competitions with opponents of varying skills. A simple merging of ordinal placements or tabulation of win–draw–loss fractions is not sufficient.

In this chapter, gambling schemes are covered first, including *pari-mutuel* and *fixed-odds* schemes, to facilitate an understanding of the utility of each SRS for gambling purposes.

In order to provide a systematic study of SRSs, this chapter proposes a taxonomy of sports and SRSs. Eighty three major international and regional sports are classified as: 'combat' sports where each competitor tries to control the opponent (e.g. boxing and fencing), 'object' sports where each competitor tries to control an object in direct competition with the opponent (e.g. soccer and chess) and 'independent' sports where each competitor is unimpeded by the opponent. Each SRS operates in three phases: evaluation of sports performances to provide results, weighting of the results to provide competition points and creation of a rating for each competitor using the competition points, summarized over one or more seasons. Each sports performance is evaluated and ranked by judging, measuring and/or scoring. There are two combinations of operations that classify SRSs: accumulative (using some combination of the operator's best, summation and aging) and adjustive (using some combination of averaging, recursive adjustment and aging). For example, best-based accumulative SRSs are more forgiving of future poor performance and more rewarding of improved performance, while averaging adjustive SRSs rewards future improved performance and penalizes future poor performance about equally.

The outcome of a future event can be predicted based on an SRS. Some SRSs provide a quantitative estimate of future performance difference while others simply rank competitors, providing a predicted order of finish for future competition.

12.2   Gambling schemes

In many countries, a government agency organizes the various legalized betting schemes. For convenience, the term *betting agency* will be used to define whatever organization handles the bets and distributes the winnings.

There are two basic gambling schemes: *pari-mutuel* and *fixed-odds*. In pari-mutuel gambling the betting agency takes a constant fraction of the gross bet and returns the rest in equal shares to each successful bettor. Each bettor is thus competing against the other bettors. In fixed-odds gambling, each successful bettor receives a pre-determined return while the return to the agency varies with the number of winners. Here each bettor competes with each gambling agency.

12.2.1   Pari-mutuel gambling

Table 12.1 illustrates pari-mutuel betting. Suppose each bettor has four choices (for example, horses in a race or tennis players in a tournament). Each bettor may wager any number of betting units (in some currency) on one or more of the four choices. Let n_i represent the number of units bet on choice i, where i varies from one to four, and let N represent the total units bet, which is 2000 here.

Then q_i equal to n_i/N is the fraction of the units bet on choice i and $1/q_i$ is the payoff if choice i is correct to return the entire pool of N units.

For example, 1000 units are bet on choice 1. If that choice is correct, then 2.0 units should be returned for each of the 1000 units bet on choice 1, clearing the pool of 2000 units. Naturally, the betting agency needs a profit on the gross bet to cover overheads and a return on capital. Suppose 20% of N is to be retained; thus the units returned must be discounted. Let D be that discount (0.8 here). Then if each return r_i is chosen to be D/q_i or $0.8/q_i$, then 0.8 units are returned for each unit bet regardless of which choice is correct. Odds are quoted in the form $r_i - 1:1$ which is usually expressed as a ratio of integers. The odds for choice 1 would be 0.6:1, usually quoted as 3:5. Thus, the odds indicate profit:bet. For example, a successful bet of 1 unit on choice 1 produces a

Table 12.1   Pari-mutuel example

Choice i	n_i	$q_i = n_i/N$	$1/q_i$	$r_i = 0.8/q_i$	Profit $r_i - 1$	Odds $r_i - 1:1$	Integer odds	$f_i = 1/r_i$
1	1000	0.50	2.0	1.6	0.6	0.6:1	3:5	0.6250
2	500	0.25	4.0	3.2	2.2	2.2:1	11:5	0.3125
3	400	0.20	5.0	4.0	3.0	3:1	3:1	0.2500
4	100	0.05	20.0	16.0	15.0	15:1	15:1	0.0625
	2000	1.00						1.2500

Table 12.2   5 February 1997, fourth race Santa Anita, Los Angeles, California, USA (*Los Angeles Times*, 6 February 1997)

Choice	Odds	r_i	f_1
1	2.3:1	3.3	0.303
2	9.4:1	10.4	0.096
3	22.3:1	23.3	0.043
4	30.5:1	31.5	0.032
5	89.6:1	90.6	0.011
6	1.4:1	2.4	0.417
7	18.4:1	19.4	0.052
8	7.2:1	8.2	0.122
9	142:1	143	0.007
10	7.5:1	8.5	0.118
			1.201

profit of 0.6 unit. If choice 1 is correct, 1.6 units are be returned for each unit bet so 1000×1.6 is 1600, exactly 80% of the pool. In general

$$\text{Total return to bettors} = n_i r_i = DN$$

for any correct choice. Thus, a random bet would receive the average return of 0.8 units (a loss of 20%). To help the bettor make a profit, consider the value f_i defined as $1/r_i$ which is q_i/D or $q_i/0.8$ here.

Let p_i be the probability that choice i will be correct. To make a profit, the expected return to the bettor should exceed one unit for each unit bet so

$$p_i r_i \geqslant 1 \quad p_i \geqslant (1/r_i) \quad p_i \geqslant f_i$$

For choice 1 in the example, f_1 is $1/1.6$ or 0.625; so to make a profit, the predicted probability for choice 1 should be at least 0.625 for a bettor to back that choice.

There is a simple way to determine the value of D, given odds from a betting agency. Notice that $1/D$ is $1/0.8$ or 1.25 here, exactly the sum of the f_i. In general, if $\sum_i$ means the sum over the values of i

$$\sum_i f_i = \sum_i \frac{q_i}{D} = \frac{1}{D} \sum_i q_i = \frac{1}{D}$$

$$D = \left[\sum_i f_i \right]^{-1}$$

Table 12.2 shows actual data from a thoroughbred horse race where D is $1/1.201$ or 0.83, indicating that 17% of the gross bet is retained. To make a profit, a discerning bettor must be at least 20.1% more accurate than the collective wisdom reflected in the betting volume. In the fourth race, the favorite is horse 6. That horse must be at least 0.417 likely to win for a bettor to break even.

12.2.2   Fixed-odds gambling

In fixed-odds gambling, the odds (and therefore the r_i) are set before bets are taken. For example, predicted score difference may be given by the betting

Table 12.3   Fixed-odds example for one match

Choice i	Odds	r_i	f_i	q_i
1	10:11	1.909	0.524	0.500
2	10:11	1.909	0.524	0.500
			1.048	1.000

agency, such as an offer of San Diego over Chicago by 7 in US professional (NFL) football. The bettor can support San Diego and give 7 points to Chicago, winning that bet if San Diego wins by more than 7. The alternative is to support Chicago and take the 7 points, winning that bet if Chicago wins, draws or loses by less than 7. The bet is cancelled if the difference is exactly 7. Suppose the odds for that bet are as in Table 12.3. Since $x_i : y_i$ is the same as $(x_i/y_i):1$, then $r_i - 1$ is the same as x_i/y_i so that

$$r_i = (x_i + y_i)/y_i \qquad f_i = y_i/(x_i + y_i)$$

If both odds are 10:11, then both of the f_i are 11/21 or 0.524. The total of the two f_i is 1.048 so D is 1/1.048 or 0.95. The betting agency is attempting to have a 5% profit for that bet; however, that is only a goal that is met if the betting volume follows the target q_i, which are 0.50 here. The betting agency and each potential bettor are both interested in the f_i. The bettor must decide if either option is at least 0.524 before risking a bet, whereas the betting agency can lose money if more than 0.524 bettors are successful. Conversely, by maintaining the larger of the two betting volumes to less than 0.524 of the gross, the betting agency will make a profit no matter which choice is successful. If the limit is exceeded, the betting agency may 'lay-off' some of its volume by betting on the other option with another betting agency or allow unbalanced betting if the more popular bet is deemed to be less likely to occur (but if wrong, the betting agency can lose money).

There are many forms of fixed-odds betting requiring a selection on two or more matches, all of which must be successfully predicted to win the bet. For example, a bettor must select *trebles*, three to six of about 80 games from various sports offered by the British Columbia (Canada) provincial sports lottery. Table 12.4 shows an example of r_i for three US professional baseball games offered by that lottery. A win by one run is considered a draw. For each game there are three choices, home win, draw or away win. The bettor must decide if the likelihood of at least one choice in a given game is high enough to expect a net profit for that game. Then three to six such games would be chosen. Here D is 1/1.3 or 0.77, so the betting agency has set the goal of a

Table 12.4   Three baseball games offered by the British Columbia Lottery, 11–17 July 1996

Choice i	Game 2			Game 4			Game 24		
	r_i	f_i	q_i	r_i	f_i	q_i	r_i	f_i	q_i
Home win	2.00	0.500	0.385	3.35	0.299	0.230	1.60	0.625	0.482
Draw	3.00	0.333	0.257	3.00	0.333	0.256	2.95	0.339	0.261
Away win	2.15	0.465	0.358	1.50	0.667	0.514	3.00	0.333	0.257
		1.298	1.000		1.299	1.000		1.297	1.000

23% profit. The q_is are the fraction of betting volume that the betting agency is attempting to generate. Suppose a bettor is convinced that a home win in game 2, a draw in game 4 and an away win in game 24 are each more likely than the corresponding f_i. (In general, to bet, the product of the p_i should exceed the product of the f_i, but a more conservative approach is to require each p_i to exceed each f_i.) If all three choices are correct, the return for the three-game selection would be the product of the three returns r_i, which amounts to $2 \times 3 \times 3$ or 18 times the units bet.

12.3 Taxonomy of sports rating systems

It is surprising that little appears in the formal literature regarding SRSs. Apparently simple systems are deemed unworthy of attention, some systems are published in the popular media with minimal information and some are proprietary, providing services such as information for gambling (with little incentive for publication). To analyze coherently the entire field of SRSs, the fundamental nature of the sports should also be understood. For example, do boxing and diving have anything in common? Similarly what (if anything) is common between chess and soccer or between swimming and golf?

Motivated by a desire to unify the various SRSs into a coherent whole, this chapter suggests a taxonomy for a comprehensive group of sports and the SRSs operating on those sports. To select that comprehensive group of sports, it is useful to note that the International Olympic Committee (IOC) recognizes two sets of sports federations (*Olympic Review*, 1995): those that organize sports included in the Olympic Games and those that organize major regional competition in sports that were previously included in Olympic competition or which may eventually be so recognized. The sports organized by those two sets of federations form an appropriate roster of major international and regional sports (see Tables 12.5 and 12.6).

The definition of a sport is somewhat subjective. The IOC states that 26 sports were included in the 1996 Atlanta Olympics. More precisely, 26 sports federations were recognized to organize competitions. For example, the International Swimming Federation (FINA) organized aquatic competition (officially one sport) including swimming, diving, water polo and synchronized swimming (more commonly considered to be four sports). The sports in Tables 12.5 and 12.6 follow the more common usage.

There are many other sports contested at the regional and local level. To delimit this study, Table 12.7 contains other sports and widely played intellectual games found in major sports references, *The Information Please Sports Almanac* (1994), *The World Almanac and Book of Facts* (1995) and *The Official World Encyclopedia of Sports and Games* (1979).

The 83 sports included in Tables 12.5–12.7 may be classified into three groups: combat sports, object sports, and independent sports. A combat sport, C, is one in which each competitor tries to control the opponent by direct confrontation (e.g. boxing, wrestling and fencing). An object sport, O, is one in which each competitor tries to control an object while the other competitor is in direct confrontation (e.g. soccer, baseball and chess). An independent sport, I, is one in which one competitor may not interfere with the other

Table 12.5 Taxonomy of Olympic sports organized by IOC recognized federations

Sport	Performance	Evaluation
boxing	C	J
fencing	C	J, S
judo	C	J
wrestling	C	J
alpine skiing	I	M
archery	I	S
athletics (track and field)	I	M, S
biathlon	I	S
bobsled	I	M
canoe/kayak	I	M, S
cross-country skiing	I	M
cycling	I	M, S
diving	I	J
equestrian	I	S, M, J
figure skating	I	J
gymnastics	I	J
luge	I	M
modern pentathlon	I, C	S
rhythmic gymnastics	I	J
rowing	I	M
shooting	I	S
ski jumping/nordic comb.	I	J, S
speed skating	I	M, S
swimming	I	M
synchronized swimming	I	J
weightlifting	I	M
yachting	I	S
badminton	O	S
baseball	O	S
basketball	O	S
curling	O	S
field hockey	O	S
ice hockey	O	S
soccer	O	S
softball	O	S
table tennis	O	S
team handball	O	S
tennis	O	S
volleyball	O	S
water polo	O	S

C = combat sport; I = independent sport; O = object sport; J = judged; S = scored; M = measured

competitor (e.g. swimming, shooting and golf). In an I sport, the competitors may perform at different times and even in different places and it might be said that each competitor tries to control him(her)self. In Tables 12.5–12.7 the 83 sports are classified: 8 C, 37 O, 37 I and the five-element modern pentathlon which includes both C (fencing) and I (riding, shooting, running and swimming) elements.

There are three ways in which performance is evaluated in the sports surveyed: performance is judged, J (e.g. all combat sports, diving and gymnastics),

Table 12.6 Taxonomy of regional sports organized by IOC recognized federations

Sport	Performance	Evaluation
karate	C	J
taekwondo	C	J
acrobatics	I	J
bowling	I	S
bowls	I	S
dance sport	I	J
golf	I	S
mountaineering	I	M
orienteering	I	M
roller skating	I	M
sky diving	I	J
surfing	I	J
trampoline	I	J
triathlon	I	M
water skiing	I	J, M
korfball	O	S
netball	O	S
pelota vasca (jai alai)	O	S
raquetball	O	S
rugby	O	S
squash	O	S
underwater hockey	O	S

C = combat sport; I = independent sport; O = object sport; J = judged; S = scored; M = measured

measured objectively, M (by time, length or weight as in weightlifting, swimming and bobsled), or scored objectively, S (as in baseball, archery and golf).

Thus, the taxonomy of the 83 sports in Tables 12.5–12.7 includes classification due to type of performance (C, I, O) and mode of evaluation (J, M, S). To answer the questions posed earlier, what boxing and diving have in common is that each is evaluated by J (while one is a C sport and one is I). Chess and soccer are both O sports and both are evaluated by S. Swimming and golf are both I sports, while one is evaluated by M and the other by S. Notice that all C sports are judged (except one fencing element, the epee, which is scored entirely electronically), all O sports are scored and the I sports include all three modes of evaluation.

At the conclusion of each competition, competitors are usually ranked based on the results. More than two competitors are usually ranked first, second, third and so on. For a combat sport or object sport, wins, losses and draws may be tabulated. In time, competitors may engage in various numbers of events of various importances. What is then needed is an SRS that allows ranking over the many varied competitions. Some SRSs operate on the final ordinal ranking after each competition (based on judging, scoring and/or measuring), while others operate on the actual result.

In all SRSs surveyed with regard to the sports in Tables 12.5–12.7, there are three phases (see Table 12.8). In the *evaluation* phase, each sports performance is evaluated to provide a result, using judging, measurement and scoring as already discussed. Next, in the *weighting* phase, the results are weighted to

Table 12.7 Taxonomy of other regional sports and widely played intellectual games

Other regional sports		
Sport	Performance	Evaluation
aikido	C	J
kendo	C	J
darts	I	S
horseshoes	I	S
American football	O	S
Australian Rules football	O	S
bandy	O	S
Canadian football	O	S
court handball	O	S
cricket	O	S
croquet	O	S
Gaelic football	O	S
hurling	O	S
lacrosse	O	S
paddleball	O	S
pool/billiards	O	S
roller hockey	O	S
shinty	O	S
Widely played intellectual games		
Sport	Performance	Evaluation
chess	O	S
checkers	O	S
bridge	O	S

C = combat sport; I = independent sport; O = object sport; J = judged; S = scored

provide competition points. Finally, in the *rating* phase, some operations are applied to the competition points to create a rating for each competitor (perhaps covering one season or covering several seasons).

To represent the evaluation phase, with n competitors (individuals or teams), k competitions (during some of which a competitor may be inactive) and n_d results for each competition, the data for all of competitor i's results can be stored in a $k \times n_d$ matrix Z_i in which row m is designated z_i^m, containing the n_d results for competition m. Then, the evaluated performances for all n competitors can be stored in an $n_k \times n_d$ matrix Z, partitioned with n submatrices, one for each competitor:

$$Z = \begin{bmatrix} Z_1 \\ Z_2 \\ \vdots \\ Z_n \end{bmatrix} \tag{12.1}$$

To represent the weighting phase, the results in Z are weighted by parameters in a matrix P using operations f_w and stored in a matrix Y, containing the competition points

$$Y = f_w(Z, P) \tag{12.2}$$

Table 12.8 Taxonomy of sports rating systems

Phase	Input	Operations	Output
Evaluation	Sports performance Combat sport Object sport Independent sport	Judge Measure Score	Result Z
Weighting	Z	Weight elements of Z and combine	Competition score Y
Rating	Y	Type of operation Accumulative Best Best n_b Best n_b of previous n_p Summing Aging Adjustive Averaging Recursive Aging Derivation Ad hoc Theoretical Implementation Simultaneous Recursive	Rating r

The rating phase may be represented by applying operations f_r to Y. Let r be an $n \times 1$ column vector where r_i is the rating for competitor i. The ratings are found by

$$r = f_r(Y) \tag{12.3}$$

Any such SRS may be completely characterized by

$$r = \text{SRS}(Z, P, f_w, f_r) \tag{12.4}$$

Suppose a skiing organization allocates competition points to the first five finishers assigning 10, 5, 3, 2 and 1 points respectively, where the order of finish is established by measured time. Suppose skier i places second, first, third, fourth, sixth and second in six successive races. Each competition result z_i^m may be represented by a 1×5 row vector

[first   second   third   fourth   fifth]

where a one identifies that the competitor finished at the indicated position and a zero otherwise. Here, there are $k = 6$ competitions, so Z_i is 6×5. Then

$$Z_i = \begin{bmatrix} 0 & 1 & 0 & 0 & 0 \\ 1 & 0 & 0 & 0 & 0 \\ 0 & 0 & 1 & 0 & 0 \\ 0 & 0 & 0 & 1 & 0 \\ 0 & 0 & 0 & 0 & 0 \\ 0 & 1 & 0 & 0 & 0 \end{bmatrix}$$

and each P is $[10\ 5\ 3\ 2\ 1]^T$ for weighting. The competition points are contained in Y equal to Z_iP or $[5\ 10\ 3\ 2\ 0\ 5]^T$. If competition points are summed to provide a rating, then r_i is 25.

In Table 12.8 there are two types of operations in the rating phase by which f_r may be classified: *accumulative* and *adjustive*. An accumulative SRS is one in which (except for data aging) each rating improves or (at worst) remains the same after each competition. Usually, competition points are positive for each successful competition and zero for each unsuccessful competition. For example, positive points are accumulated by the skier for each fifth place finish or better, but no points accrue for sixth place or worse.

In contrast, an adjustive SRS is one in which each rating adjustment may be positive, zero or negative, whether or not data is aged. For example, an average of all competitions changes as the new result exceeds, is the same as or is less than the previous average. The operations used by accumulative and adjustive SRSs are discussed later in this chapter.

Other classifications may distinguish operations in the third phase of an SRS. For example, f_r may be derived by ad hoc methods that are deemed acceptable by some organization or by theoretical methods optimizing some performance measure. In some cases, the ratings must be implemented by simultaneously calculating all values of r using Y, while in other cases the ratings are implemented recursively; that is, using only the previous value of r and the most recent weighted results. Then (12.4) becomes

$$r^k = f_r(r^{k-1}, Y^k) \tag{12.5}$$

where r^k is the rating vector for all competitors at the end of competition period k and Y^k contains the competition points y_i^k for competition period k only. In many recursive cases, each competitor's rating can be found separately by

$$r_i^k = f_r(r_i^{k-1}, Y_i^k) \tag{12.6}$$

For adjustive SRSs that can be recursively implemented, (12.6) usually becomes

$$r_i^k = r_i^{k-1} + f_{r1}(\text{actual value} - \text{reference value}) \tag{12.7}$$

Thus (12.7) has the predictor–corrector form where the predictor is the reference value and corrector action occurs as the actual value differs from the predictor value.

12.4 Combat sports

There are nine combat sports (including the modern pentathlon) listed in Tables 12.5–12.7. Fencing and the fencing element of the modern pentathlon have electronic evaluation while the others are evaluated by judging. In most cases, ratings depend on the number of wins, losses and draws. The most widely published ratings are in professional boxing, where the win–draw–loss record is augmented by subjective judgment, including the quality of opponents and recent performance. Those ratings are used to schedule championship bouts and presumably a higher-rated boxer should defeat a lower-rated boxer. The ratings do not indicate the quantitative ability gap between boxers; hence, the ratings do not provide a direct means for deciding whether a betting agency's

fixed odds are worthy of accepting a bet (whether a p_i exceeds an f_i). Lee *et al.*
(1997) examined 91 Korean boxing matches. An array of physical and psychological factors were evaluated from questionnaires, and regression coefficients
fitted to the data correctly classified 81% of the winning boxers.

12.5 Independent sports

In Tables 12.5–12.7 there are 38 I sports, including the modern pentathlon
which has both I and C elements. The evaluation phase of the I sports includes
scoring, judging and measuring. Golf and track and field (athletics) are I sports
covered in other chapters of this book. There are two common ways to rate
competitors in I sports. One way is to choose the best performances from a
subset of important contests, resulting in a list of local, national or world-
best performances (often for the calendar year) as in bobsled, rowing, swimming
and athletics. Presumably, higher-rated athlete would defeat a lower-rated
athlete. Such a system allows a new competitor to move rapidly up the list
after one outstanding performance. Conversely, poor subsequent performances
do not cause a rating drop, a feature reducing the sensitivity of the predictor.

A more robust rating procedure uses many competitions. For example, an
average of some subset of competition points (usually for the calendar year
or season) is used for bowling and golf. That rating is a more reliable measure
of future performance. Such an SRS is adjustive. Thus, the rating for competitor
i after k competitions is

$$r_i^k = (1/k) \sum_{m=1}^{k} y_i^m = r_i^{k-1} + (1/k)(y_i^k - r_i^{k-1}) \qquad (12.8)$$

Then (12.8) has the predictor-corrector form of (12.7), where the actual value
y_i^k is compared to r_i^{k-1} and the rating increases or decreases based on that difference. Many sports fans probably view an average as just another ad hoc rating
tool, while, of course, an average is optimal in a least squares sense.

12.5.1 World golf ratings

The Royal and Ancient Golf Club of St Andrews and Sony sponsor an SRS
(Greer, 1995; Stefani, 1997), for the world's male professional golfers: a
three-year average, weighted toward the most recent year. Competition points
are acquired for each tournament based on the strength of field and the golfer's
finishing position. The strength of field is measured by points based on the entry
by each golfer rated among the top 100. The total of these entry points establishes the number of points w for the winner. Other golfers receive approximately $1.2w/position$ points.

In the rating phase, the total competition points for the most recent 52-week
period are multiplied by four. Added to this total are points for the previous 52-
week period multiplied by two and points for the oldest 52-week period multiplied by one. This total is divided by the total number of tournaments entered
during the three-year period, with a minimum divisor of 20 for each period. This
SRS is adjustive. Gambling on golf is covered in Larkey (1990).

12.5.2   FIS World Cup skiing ratings

The International Skiing Federation (FIS) sponsors individual competition for men and women in a series of important international alpine and cross-country races, referred to as World Cup (WC) competition. Alpine WC competition began with the 1966–67 season as described in FIS (1995/96a) and Stefani (1997). Cross-country WC competition began with the 1981–82 season as described in FIS (1995/96b) and Stefani (1997). Currently, a common point system is used for all WC races, with 100 points allocated to first place down to one point for 30th place.

The period from November to March, inclusive, comprises the WC season. The WC rating is the sum of competition points acquired during a given season (points are set to zero at the start of the next season). The WC ratings increase monotonically during each season, so this SRS is accumulative.

The WC ratings include competition points for relatively few races early in the season, so the rankings are not as good a predictor of future order of finish as would result from using more races. Realizing this, the FIS employs a second system (not as widely publicized) called the World Cup Start List (WCSL), used for seeding starting order. Let r^k_{iWC} represent the current WC rating including k races, where K races are eventually to be run for the entire current season. Let the superscript 'Last' denote the cumulative WC points at the end of the last season. Then the WCSL through week k is given by

$$r^k_{iWCSL} = r^{Last}_{iWC} - (k/K)[r^{Last}_{iWC} - r^k_{iWC}]$$

Prior to the first race of the current season, $k = 0$ and the WCSL rating is the WC rating from the end of the last season. At the end of the current season, $k = K$ and the WCSL is the same as the end-of-season WC points (which will be used for WCSL purposes at the start of the next season). As the current season progresses, the WCSL approximates the WC points for the last 52 weeks. This simple approximation avoids storing the results from each past race.

12.5.3   Olympics

It has been an interest of mine (Stefani, 1977b, 1982, 1985, 1989a, 1989b, 1992, 1994a,b,c) to examine trends in Olympic winning performances (providing insight into the rates of improvement and the effect of historical events upon performances) and then to project future winning performances. A linear regression model is applied to the Olympic winning result in event i during Olympics numbered m and $m - 1$

$$z^m_i = c_i z^{m-1}_i + e^m_i \qquad (12.9)$$

for a random error e^m_i. The normal four-year span was halved between 1992 and 1994 for the winter Olympics to change the former pattern where both summer and winter games were staged in the same calendar year. After 1994, the span returned to four years with both competitions no longer in the same year. To apply linear regression to a halved competitive span, it is necessary to provide a linear estimate of the square root of c_i. Knowing that c_i is close to one, a

Table 12.9   Percentage improvement per Olympiad

Sport	1952–72		1972–94/96	
	M	W	M	W
Athletics				
Running	0.79	1.11	0.37	0.28
Jumping	1.58	2.14	0.77	0.91
Throwing	3.86	5.61	0.93	0.29
Speed skating	1.79	1.58	1.63	2.32
Swimming	2.41	2.48	0.88	1.23
Weightlifting	2.95		2.53	

Newton–Raphson approach yields

$$z_i^m = 0.5(1 + c_i)z_i^{m-1} + e_i^m \tag{12.10}$$

For example, if c_i is 1.1, the square root is 1.049 while the estimate $0.5(1 + c_i)$ is 1.050, a close approximation. Each value of c_i can be found to minimize the sum of the squared errors over a sequence of successive competitions of two- or four-year intervals. As a uniform measure of improvement, I define each event rating r_i to be the percentage improvement per Olympiad ($\%I/O$) given by $100 \times (1 - c_i)$ for a timed event and $100 \times (c_i - 1)$ for an event determined by height or distance. Table 12.9 shows average $\%I/O$ over ensembles of Olympic summer and winter events, starting with 1952 when the former Soviet Union re-entered Olympic competition and conditions had returned to something approaching pre-war conditions.

From 1952 to 1972, women improved more than men in events other than speed skating (there are no women's weightlifting events). From 1972 to 1994/96, the rate of improvement dropped significantly in all events except speed skating. During that period, of course, the summer Games were affected by boycotts (African nations in 1976, Western-bloc nations in 1980 and Communist-bloc nations in 1984) and by the recent break up of sports systems in the Soviet Union, East Germany and Yugoslavia. By contrast, the Winter Games have not suffered from boycotting and the nations winning the most speed skating medals have generally been politically stable. From 1972 to 1994/96, women improved more than men in speed skating and swimming but, in athletics, men and women improved at about the same rate. Over both periods, improvement is least in running (where leg and lower body strength is required). Improvement appears to increase in events requiring progressively more upper body flexibility, arm strength and reliance on technique.

Each c_i provides a means for predicting future Olympic performance. Let ˆ imply prediction and assume that all k past Olympic games have been used in calculating c_i, so the next games are numbered $k + 1$. Then

$$\hat{z}_i^{k+1} = c_i z_i^k \tag{12.11}$$

Table 12.10 contains the average absolute percentage error (AAPeE) between true and predicted winning performance. Predictions are calculated one Olympics ahead using data starting in 1952. AAPeE is about 2% for athletics and swimming, 3% for speed skating and 6% for weightlifting.

Table 12.10   Average absolute percentage error of predictions

Sport	1976	1980	1984	1988	1992	1994	1996	Avg.
Athletics	1.64	1.75	2.85	2.00	3.23		1.77	2.22
Speed skating	1.21	1.83	2.56	2.81	5.53	2.55		2.80
Swimming	1.89	2.77	2.23	1.11	1.34		1.82	1.86
Weightlifting	3.87	4.49	9.73	8.14	6.64		3.22	6.10

12.5.4   Horse racing

Horse racing is an equestrian independent sport with a great deal of pari-mutuel betting. The distribution of bets, discounted for the track's take of the proceeds, establishes the payoffs and the f_i as discussed earlier. For a race to be profitable, a bettor must believe that a horse is more likely to win than the f_i for that horse.

For predicting the winning horse, it is necessary to create a rating system that weights results such as speed, pace (consistency of speed), jockey skill, finish, post position (starting position in the gate), and class (a way of separating horses related to past performance and earnings), as suggested by Bernstein (1976) and Quinn (1987). It is difficult to make a profit by simply betting on the horse most likely to win each race, because the pari-mutuel system assigns less return for the horse deemed best by the gamblers. For example, the *Los Angeles Times* published selections by three predictors and the consensus of those three predictors for 1926 races contested at Santa Anita, Hollywood Park and Del Mar from 1 January 1997 to 3 November 1997. Since 17% was retained by each track, then random selection would lose 17%. Only one of the selectors (Bob Mieszerski) did better than random chance, losing 16% of the money while selecting the winner in 30% of the races. The selected winner fraction is about 18% better than random chance (assuming about eight horses per race), but making a profit is thwarted by the pari-mutuel payoff system. For the other two predictors and the consensus, the money lost was 21% (8% winners, where long shots were picked), 23% (24% winners) and 21% (26% winners) respectively. Thus, the average predictor lost more money than for random selection.

Instead, the probability for profit must be calculated for each horse, and the odds change right up until post time (except for fixed-odds betting, which is available through some on-track and off-track betting agencies) as in Ali (1994) and Bolton and Chapman (1986). Because there may be few playable races or because of low probability 'exotic' bets (on the outcome of consecutive races or the exact order of finish of two horses in one race) payoffs may occur infrequently, making money management a crucial element of a betting strategy (Passer, 1981).

12.6   Object sports

In an object sport, each competitor's performance (and rating) is influenced by the performance of the opponent attempting to control the object, making the task of calculating ratings that much more difficult. It follows that the most sophisticated rating systems tend to be those associated with the 37 object sports listed in Tables 12.5–12.7, all of which are evaluated by scoring. A number of object sports are covered in other chapters of this book: American football (Chapter 1), baseball (Chapter 2), basketball (Chapter 3), cricket (Chapter 4), soccer (Chapter 5), ice hockey (Chapter 7) and tennis (Chapter 8). There are widely published rating systems for soccer, chess and tennis.

12.6.1 FIFA/Coca Cola world soccer ratings

The international soccer football federation (FIFA) (FIFA, 1995; Stefani, 1997) sponsors the FIFA/Coca Cola SRS for the world's national soccer football teams. It is surprising that a rather complicated SRS is used for a sport that prides itself on simplicity. The period from 1985–92 was used to initialize the ratings, which were first published in August 1993.

As part of the weighting phase for each game, FIFA continues to allocate two points based on the outcome (win, lose or draw) depending on the relative strengths of the teams, where two points are assigned when defeating a team of the same strength, more than two points for defeating a stronger team and less than two points for defeating a weaker team, with each team of equal strength receiving one point for a draw. The two-for-a-win and one-for-a-draw assignment is surprising since FIFA now mandates three points for a win and one for a draw for all other FIFA-sponsored competitions. Additional points are added for goals scored (with the average goal counting 0.3 points) where the number of these goal-related points depends on the relative strengths of the teams (a goal against a stronger team counts more than against a weaker team). The visiting team receives a 0.3 point bonus (equivalent to one goal). The sum of the outcome points and goal points is multiplied by two factors: one depending on the continent of the competing teams and one depending on the importance of the match. The continental factor varies from 0.8 for Asia and Oceania to 1.0 for Europe and South America. Cross-continental matches and World Cup matches are assigned multiplying factors of 1. The importance factor varies from 1.0 for a friendly up to 1.5 for a World Cup final. It seems ill-conceived that a 'friendly' (essentially an exhibition game for player development) counts 2/3 as much as a World Cup final. Thus:

$$\text{Competition points} = (\text{outcome points} + \text{goal points})$$
$$\times \text{continental factor} \times \text{importance factor}$$

In the rating phase, competition points accumulate during each of six 52-week periods. For each 52-week period, the total for the best eight matches is averaged with $8/k$ times the total for all k games. The total for the current 52-week period is then added to a similarly-calculated total for each of the five previous 52-week periods, each multiplied by 5/6, 4/6, 3/6, 2/6 and 1/6 respectively.

While a higher-rated team in this accumulative SRS should defeat a lower-rated team, the rating difference does not easily predict the likelihood of that win or the score difference.

12.6.2 FIDE world chess ratings

In 1970, the International Chess Federation (FIDE) implemented an SRS devised by Arpad Elo (1978) based on probability theory. The Elo SRS strives to create ratings that are normally distributed $N(2000, 200^2)$ with a mean of 2000 and a standard deviation of 200. If a player with rating r_i competes against one or more players with average rating $\bar{r}$, then the distribution of the differential performance $N(r_i - \bar{r}, 283^2)$ can be used to predict the fraction of matches that player i should win. Let each y_i^k indicate the fraction of matches won by player i in competition k and let $\hat{y}_i^k$ be the predicted fraction of matches (based on the rating after competition $k - 1$ and the Gaussian assumption). The Elo SRS updates ratings using the predictor–corrector form:

$$r_i^k = r_i^{k-1} + c_i(y_i^k - \hat{y}_i^k) \tag{12.12}$$

where c_i is between 10 and 45, depending on the number of previous matches by player i. This SRS has been applied retroactively to generate an all-time ranking of chess champions. Inherent in this adjustive SRS is the ability to predict the fraction of matches that should be won in the next match.

12.6.3   World tennis ratings (ATP and WTA)

The Association of (men's) Tennis Professionals (ATP, 1995; Stefani, 1997) sponsors the IBM/ATP world SRS first published in 1 January 1990, replacing a previous average-based SRS started in 1973. The Women's Tennis Association (WTA) (WTA, 1997; Stefani, 1997) sponsors the Corel/WTA world SRS, introduced in late December 1996, also replacing an average-based system started in 1984. The ratings are used to seed players in each subsequent tournament and to exempt top players from qualifying, so a higher-rated player should defeat a lower-rated player.

In both systems, competition points are accumulated as each player progresses through the succeeding rounds of each tennis tournament and bonus points are accumulated based on the rating of each defeated opponent. Tournament points are assigned roughly in proportion to the prize money offered by a tournament. For the ATP, there are 43 tournament grades and a player receives 660 points for winning the top grade tournament, 60 points for winning the bottom grade tournament, 50 bonus points for defeating the top ranked player and one bonus point for defeating a player ranked 151–200. For the WTA, there are 26 tournament grades and a player receives 520 points for winning the top grade tournament, 5 points for winning the bottom grade tournament, 100 bonus points for defeating the top ranked player and one bonus point for defeating a player ranked 251–500. Bonus points are somewhat more important (compared to tournament points) for the WTA.

Each ATP rating is the sum of the best 14 competition points over the most recent 52 weeks, while each WTA rating is the sum of all competition points over the same (most recent) 52-week period. At the end of each calendar year, the best players are honored by both organizations.

In the current systems, a rating drops only if aged (discarded) competition points are higher than the newly earned competition points. Previously, when an average-based system was used, ATP noted that high-ranked players would decline to play some tournaments (especially if a player's fitness was in question) to avoid having a rating drop due to a below-average performance. Since moving to the new system, both organizations contend that players are entering more tournaments and that each tournament has somewhat higher-ranked entrants.

There is a theoretical explanation for the downward movement of the averages of top-ranked players as more tournaments are entered. Using mean square considerations with certain assumptions, James and Stein (1961) and Efron and Morris (1973, 1975, 1977) showed that early-season averages tend consistently to over- or under-estimate future averages (taken when more data are available) in that those future averages are expected to tend toward the grand average (the average of all averages in a class). Thus, by theory, the ratings of the top players would be expected to move downward toward the grand average as more tournaments are entered – exactly the observed trend.

12.6.4 Predictive team rating systems

It is often useful to create an SRS in which the rating difference can be used to predict the margin of victory of an object sport during team competition. In most sports, playing at home has a significant impact on the performance of the home team; therefore, an accurate SRS (one that may be used for gambling) must correct the margin of victory by removing any home advantage. Obviously, the home advantage must be quantitatively understood.

12.6.4.1 Home advantage

Table 12.11 shows the home advantage for 31 112 games played under a variety of sports competitions. Data include the fraction of home wins (HW), draws (D, often called ties) and away wins (AW). Home advantage is indicated by HW-AW. The top part of the table covers regular season competition, while the rest of the table covers playoff (post-season) competition among the best teams based generally on regular season play. The National Basketball Association (NBA) data cover the 1992–93, 1993–94, 1995–96 and 1996–97 seasons. The Four Nations soccer data are for the top divisions in England, Italy and

Table 12.11 Home advantage for 31 112 games

Sport	S	G	HW	D	AW	HW-AW
Regular season						
Four Nations soccer	3	4346	0.480	0.262	0.258	0.222
NBA	4	4593	0.600	0.000	0.400	0.200
Australian Rules Football	15	2282	0.590	0.008	0.402	0.188
US college football	3	1669	0.574	0.017	0.409	0.165
US pro football	3	671	0.574	0.003	0.423	0.151
NHL	3	3223	0.485	0.131	0.384	0.101
Major League Baseball	5	10923	0.540	0.000	0.460	0.080
Playoff						
NBA	4	287	0.686	0.000	0.314	0.372
US pro football	11	72	0.681	0.000	0.319	0.362
Three European cups (soccer)	12	2654	0.546	0.214	0.240	0.306
NHL	3	258	0.550	0.000	0.450	0.100
Major League Baseball	5	134	0.515	0.000	0.485	0.030

S = seasons, G = games, HW = fraction of home wins, D = fraction of draws (ties), AW = fraction of away wins

Germany and the top three divisions in Norway for the 1994–95 to 1996–97 seasons. The three European cups refer to the Champions Cup, Cup Winners Cup and UEFA Cup for international European soccer club competition during the 1985–86 to 1996–97 seasons. Other seasons covered are Australian Rules Football (1980–94), US college and professional football regular seasons (1978–79 to 1980–81), US professional football playoffs (1970–71 to 1980–81), National Hockey League (NHL, 1993–94, 1995–96 and 1996–97) and Major League Baseball (USA, 1992, 1993 and 1995–97).

Pollard (1986) and Harville and Smith (1994) suggest causes of home advantage (visiting disadvantage) such as travel fatigue acting on the visiting team, tactical advantage for the home team due to familiarity with playing conditions and crowd intimidation of the visiting team by followers of the home team. Data for regular season competition may be explained by those causes. For regular season competition, each team generally plays each opponent an equal number of times both at home and away, thus canceling any bias due to a higher skill level of some teams. The highest values of HW-AW (22% and 20%) are for domestic (Four Nations) soccer and NBA competitions which feature substantial crowd intimidation of the away team.

Both sports are physically demanding so travel fatigue would have a significant effect on the visitors and familiarity with playing conditions aids in making accurate passes and shots. Australian Rules Football has a somewhat lower home advantage and that sport will be discussed shortly. US college and pro football have a lower home advantage (16 and 15%) while every play is a restart, diminishing the effect of travel fatigue and the tactical advantage of familiarity. NHL hockey and baseball (10 and 8%) have the lowest home advantages. For the NHL, frequent personnel change (during play) reduces fatigue (as in the US football sports), tactical conditions are reasonably consistent for the various rinks (reducing familiarity advantage) and the protective glass around each rink tends to separate the players from the crowd. In fact, given the violent nature of the game, the glass probably protects the fans from player intimidation.

As a sport, baseball is far less physically demanding than ice hockey and expenses are reduced for the 81 away games by playing a three-games series, which reduces travel fatigue. The rather deliberate nature of the game tends to mitigate most of the advantage due to familiarity, so it is logical that baseball has the lowest regular season home advantage.

No playoff (post-season) data are shown for US college football and Australian Rules Football since most games are played on neutral grounds. For playoff competition (compared with the regular season) the home advantage is larger in NBA, US pro football and European cup competitions, the same in NHL hockey competition and less in baseball. In US pro football, each round consists of one game played at the home field of the team with the better regular-season record. Much of the added home advantage of 21% is probably due to the skill level of the home team, but the home advantage is probably increased somewhat by weather effects and additional crowd interest. For NBA, NHL and Major League Baseball, each round consists of a best-of-five or seven series, with each team playing about equally at home and away. The European cup rounds consist of home–away pairs (competition is in the following season and is international rather than domestic).

Table 12.12   Home advantage versus time: Australian Rules Football and three European cups

Sport	S	G	HW	D	AW	HW-AW
Australian Rules Football	5	690	0.563	0.005	0.432	0.131
	5	756	0.596	0.009	0.395	0.201
	5	836	0.610	0.009	0.381	0.229
Three European cups (soccer)	4	629	0.604	0.198	0.198	0.406
	4	1027	0.530	0.226	0.244	0.286
	4	998	0.527	0.212	0.261	0.266

S = seasons, G = games, HW = fraction of home wins, D = fraction of draws (ties), AW = fraction of away wins

Additional home advantages in the NBA and European cups (17% and 8%) are probably due to increased crowd intimidation and accentuated tactical advantage aiding shot making. The factors mentioned above for NHL home advantage are unchanged so home advantage is also unchanged. If anything, baseball competition becomes even more deliberate and the relatively long series cumulatively further reduces travel fatigue so, predictably, there is only a 3% home advantage, 5% less than for the regular season.

Home advantage can change in time as shown in Table 12.12 for the Australian Football League (AFL) and the three European cup soccer competitions. The 15 AFL seasons are divided chronologically into three sets of five and the 12 seasons of soccer competition are divided into three sets of four. The home advantage is increasing for the AFL. In 1980 all 12 teams were located around Melbourne in the State of Victoria (the league was then called the Victorian Football League). By 1984 there were 11 teams in Victoria and four teams in other states. In 1997 there were ten teams in Victoria and six teams in other states. The expansion into other states has increased travel fatigue. At the same time, more clusters of teams share a common ground so 32% of the games have no home advantage, increasing the differential performance in the rest of the games.

International soccer club competition exhibits progressively less home advantage probably due to diminished crowd intimidation. FIFA requires more reserved seating and less standing room attendance, while the demise of the Iron Curtain has reduced the tensions of east–west competition.

12.6.4.2   Making predictions

Of course, any number of ad hoc SRSs are possible. Many of these SRSs have been used to create predictions and rankings published in the media, while many others may be subscribed to. Applied to American football, P. B. Williamsen published nationally an SRS based on least squares in the 1920s and 1930s, E. E. Litkenhaus published an SRS in the US Midwest and South from 1941 until his death in 1984, and Richard Dunkel published nationally the ad hoc Dunkel Index from 1955 until his death in 1975. Advertisements for SRSs applied to college and professional football appear in many sports magazines; however, accuracies are usually unsubstantiated and probably inflated. Rarely do any such SRSs appear in the literature for proprietary reasons.

12.6.4.3   *Least squares-Gaussian prediction system*

A few methods appear in the formal literature and are based on theory. For example, I have proposed a least squares-Gaussian (LSG) prediction system in Stefani (1977a, 1980, 1987, 1992) and Stefani and Clarke (1992) using three steps:

Step 1   a rating is found for each team using win margin (score difference) corrected for home advantage;

Step 2   the win margin is predicted for the next game using rating difference adjusted by a shrinking factor plus the home advantage; and

Step 3   the predicted win margin is used to estimate the probability of a home win, draw and away win using a Gaussian distribution.

In the first step, using the nomenclature of (12.1)–(12.4), from the perspective of the home team, the evaluated performance (home score minus visitor score) may be modeled by

$$z_i^m = r_i - r_j + h + e_i^m \tag{12.13}$$

where z_i^m represents the win margin for home team i during competition period m, often a one week period, h is an estimate of the home advantage (one value for all teams), r_i is the estimated rating for team i, r_j is the estimated rating for team j, team i's opponent, and e_i^m is a zero-mean random error due to errors in estimating the ratings and home advantage. The LS home advantage and ratings can be found to minimize the sum of squared errors.

Since LS causes the average error to be zero, the LS value of h is the average margin of victory for the home team minus the average rating difference $r_i - r_j$. For a large number of games it can be assumed that the average rating difference is zero, so h is simply the average win margin for past home games. The adjusted margin of victory (y) follows by subtracting h from z.

After each game, as shown in the Stefani references above, the LS rating of each pair of opponents may be adjusted independently from all other teams. The [.] term in (12.14) can be limited to two standard deviations to reduce large rating changes due to unusually large errors:

$$r_i^k = r_i^{k-1} + \frac{n_j - 1}{n_i n_j - 1}[y_i^k - (r_i^{k-1} - r_j^{k-1})] \tag{12.14}$$

where n_i is the number of games completed by team i through competition k. Note that (12.14) has the form of (12.7).

For Step 2, future margins of victory tend to exceed the differences between LS ratings as related to the work of James and Stein discussed earlier, so that a model for the margin of victory in the next game is

$$z_i^{k+1} = L(r_i^k - r_j^k) + h + e_i^{k+1} \tag{12.15}$$

where L is a shrinking factor between zero and one which reduces each subsequent prediction compared to the rating difference. The sum of squared prediction errors can be minimized by selecting L based on past predictions. Compactly, L is

$$L = \frac{\text{covar}(y_i^{k+1}, r_i^k - r_j^k)}{\text{var}(r_i^k - r_j^k)} \tag{12.16}$$

which leads to the prediction

$$\hat{z}_i^{k+1} = L(r_i^k - r_i^k) + h \tag{12.17}$$

A Gaussian distribution is assumed in Step 3 for the predicted z_i. To estimate the probability of a home win, draw and away win, thresholds T_1 and $-T_2$ are used where

$$P_{\text{home win}} = P(z_i^{k+1} \geq T_1)$$

$$P_{\text{draw}} = P(-T_2 \leq z_i^{k+1} \leq T_1) \tag{12.18}$$

$$P_{\text{away win}} = P(z_i^{k+1} \leq -T_2)$$

12.6.4.4   *Other prediction systems*

Harville (1977, 1980) uses maximum likelihood to create ratings intended to provide an unbiased estimate of the outcome. Clarke (1993) reports an ad hoc system for rating Australian Rules football teams using a 0.75 power to provide exponential smoothing:

$$r_i^k = r_i^{k-1} + 0.2[|y_i^k|^{0.75} \, \text{sign}(.) - |(r_i^{k-1} - r_j^{k-1})|^{0.75} \, \text{sign}\,(.)] \tag{12.19}$$

where sign (.) restores the sign of each term whose magnitude is raised to the 0.75 power. The rating difference of future opponents provides an essentially unbiased estimate of the outcome.

The formal literature contains few soccer prediction methods, possibly because such methods may be considered proprietary. One published method by Maher (1982) employs four parameters for each team: home attack (ha), home defense (hd), away attack (aa) and away defense (ad). Attack relates to scoring goals while defense relates to allowing goals. The predicted win margin for home team i against visiting team j after the next game equals predicted home goals minus predicted away goals:

$$\hat{z}_i^{k+1} = ha_i^k \, ad_j^k - aa_j^k \, hd_i^k \tag{12.20}$$

Goals scored and goals allowed are assumed to follow a Poisson distribution. Piotr Janmaat implemented Maher's method for soccer competition in the Netherlands, updating parameters in a manner equivalent to exponential smoothing. To update ha, for example,

$$ha_i^k = ha_i^{k-1}(1 + (\text{factor}/2) \, \text{correction}^k) \tag{12.21}$$

The correction in (12.21) equals predicted minus actual home goals divided by predicted home goals. A similar equation updates the other parameters. The home team's home parameters and the away team's away parameters are adjusted using an average factor of 0.16. The home team's away parameters and the away team's home parameters are adjusted with an average factor 0.055, selected empirically. He obtained predicted values for HW, D and AW from a joint Poisson distribution using predicted home and away goals. Janmaat also applied the Maher–Poisson (MP) procedure to competition in other nations. Chapter 5 on soccer provides a further description of Maher's work and home advantage in soccer.

Table 12.13   Predictions

Sport	Method	S	G	L	h	C	C–R	AAE	AAE/TS
US college basketball	Stefani	2	1926			0.690	0.190	9.8	0.06
Australian Rules Football		15	2282	0.66	10.4	0.680	0.180	33.4	0.16
US pro football		8	1449			0.676	0.176	12.1	0.30
US pro football		3	672	0.66	3.27	0.634	0.134	10.4	0.26
US College football		8	4346			0.715	0.215	14.3	0.33
US College football		3	1763	0.87	3.71	0.722	0.222	12.8	0.30
Four Nations soccer		3	4346	0.54	0.52	0.497	0.164	1.36	0.47
Australian Rules Football	Stefani	15	2282	0.66	10.4	0.680	0.180	33.4	0.162
	Clarke	15	2282	1.0	10.4	0.674	0.174	33.0	0.160
Three Nations soccer	Stefani	3	1984	0.59	0.49	0.472	0.139	1.25	0.46
	Maher–Janmaat	3	1984	1.0	0.49	0.465	0.132	1.30	0.48

S = season, G = games, L = shrinking factor, h = home advantage, R = random chance (0.5 or 0.333), C = fraction of correctly predicted outcomes, TS = total score per game, AAE = average absolute error

12.6.4.5   Comparison of prediction systems

The top panel in Table 12.13 shows results for 16 784 predictions using the LSG method from Stefani (1977a, 1980, 1987, 1992) and Stefani and Clarke (1992). Basketball predictions cover 1972–73 and 1973–74, where the shrinking factor and home advantage were not used. US college and professional football predictions cover 1970–71 to 1980–81, where the shrinking factor and home advantage were not used for the first eight seasons but AAE was reduced for the last three seasons when L and h were used. Australian Rules Football and soccer predictions cover the same seasons as in Table 12.11.

Excluding US college basketball for which L was not used, the fraction of correctly chosen outcomes minus random chance $(C - R)$ is highest for US college football (which has the highest L) and $C - R$ is about the same for the other sports which have about the same L values. The fraction of correct winners using random chance (R) is 0.5 for sports with few draws and 0.333 for soccer. Overall, the correct outcome is predicted about 19% more often than using random chance. The ratio of AAE divided by TS increases for US college basketball (lowest), Australian Rules Football, US football and soccer (highest), indicating increased difficulty in predicting the score difference, given skill differences for the competing teams. It is interesting that basketball has no offside rule and frequent restarts, so scoring is easier than in the other sports. Australian Rules Football has no offside rule and the length of the field makes scoring more difficult than in basketball. US football has an offside rule but every play is a restart, making scoring harder than in Australian Rules Football but easier than in soccer, which has an offside rule and few restarts. Predicting the score difference accurately appears to be strongly influenced by the rules of each sport.

The middle section of Table 12.13 compares the LSG method of Stefani and the 0.75 power method of Clarke, using 2882 games played over 15 seasons. The LSG method picked the correct outcome about 1% more often with an AAE about 1% higher. Thus, there was no significant difference.

The bottom panel compares LSG with predictions made by Janmaat using the Maher–Poisson (MP) method for 1984 games over three seasons (1994–95 to 1996–97) of the top divisions in England, Germany and Italy. LSG has

Table 12.14 Aggregated results for 4346 Four Nations soccer games

Country	HW		D		AW		E	AAE	TS	AAE	h
	Pr	Ac	Pr	Ac	Pr	Ac				/TS	
England	0.44	0.45	0.27	0.29	0.29	0.26	−0.08	1.20	2.58	0.47	0.17
Germany	0.48	0.46	0.27	0.29	0.25	0.25	0.05	1.37	2.90	0.47	0.17
Italy	0.48	0.51	0.31	0.28	0.21	0.21	−0.02	1.21	2.60	0.47	0.23
Norway	0.52	0.50	0.18	0.20	0.30	0.30	0.02	1.62	3.41	0.48	0.17
	0.48	0.48	0.25	0.26	0.27	0.26	−0.01	1.36	2.90	0.47	0.18

Pr = predicted, *Ac* = actual, *E* = average predicted−actual goal error, *TS* = total score in goals per game, *AAE* = average absolute error

slightly better C and AAE, but again the differences are not significant. Stefani (1987) indicates little difference for the 2435 US college and professional football games over three seasons in the top panel of Table 12.13 in comparison with James–Stein. It was also estimated that Harville's predictor would have similar performance since Harville and LSG had nearly the same differences when compared with a gambling 'line', although for a different set of games.

Stefani (1983) contains probabilities and suggested betting strategies for European soccer pari-mutuel national pools. Under European betting rules, multiple bets may be placed using one coupon, so the team associated with the most likely outcome (home win, draw or away win) can be taken in combination with the second most likely outcome to enhance the likelihood of a coupon containing at least one winning combination. Certainly, a successful pari-mutuel betting strategy requires accurately-chosen probabilities of HW, D and AW. Similarly, if accurate probabilities are available, each may be multiplied by a fixed-odds payoff to determine whether or not a prospective bet is expected to be profitable. Table 12.14 contains aggregated results showing that the LSG-based probabilities are unbiased estimators of the true probabilities.

For the 4346 Four Nations games in Table 12.14, the predicted probabilities approach the actual probabilities and the average win margin error E approaches zero. England and Italy had the lowest TS and AAE while Germany and Norway had ascending values of TS and AAE. Clearly those statistics are correlated. The statistic AAE/TS was nearly constant; apparently it is a figure of merit that is minimized by the LS process. Italy had the largest home advantage with the highest ratio of h/TS, 0.23, and the highest HW-AW difference, 0.30. The other nations had equal ratios of h/TS and HW-AW differences of about 0.20. In the case of Italy, crowd intimidation is most likely the cause of the larger home advantage.

Since the data in Table 12.13 show little difference among properly chosen predictors operating only on a margin of victory corrected for home advantage, additional accuracy most likely requires the use of more than a corrected margin of victory. For example, Dixon and Coles (1997) use Maher's method with two parameters per team (attack and defense) while making several adjustments (including home advantage) to fit better a Poisson distribution to scoring tendencies, resulting in probabilities that provide successful fixed-odds betting.

The MP approach requires summing over many Poisson-distributed scores to obtain the likelihood of HW, D and AW. Since the Central Limit Theorem in statistics, Papoulis (1965), dictates that summation over several (properly restricted) probability densities approaches a Gaussian distribution, it is reasonable to conclude that the Poisson summation approach for individual scores and the LS-Gaussian approach acting on score difference should achieve similar accuracies.

To include more data for prediction, let Y contain a row of data for each of k previous competitions, usually the differences in values for two teams for some set of evaluated performances (like shots on goal, passing accuracy and home advantage), and let Z contain the corresponding values of scoring difference in a column vector. It is of interest to find the weighting P, relating Y and Z. For example, P can be found so that YP equals Z, in some sense. The LS estimate of P is

$$P = [Y^{\mathrm{T}} Y]^{-1} Y^{\mathrm{T}} Z \tag{12.22}$$

where one set of weights is found for use by all teams.

The scoring difference for the next game by each home team is estimated by

$$\hat{z}_i^{k+1} = [\bar{y}_i^k - \bar{y}_j^k]P \tag{12.23}$$

where the overbar indicates an average over the previous k competitions. It is thus possible to determine those weighted elements of $\bar{y}$ at which team i is superior, permitting exploitation of that advantage and, similarly, to determine where team i is at risk, suggesting a defense. Croucher (1992) presented a scoring system for tennis to facilitate this type of SRS, and Goode (1976) applied this approach to US professional football and found that the statistic correlating best with positive scoring difference is yards gained per pass attempt, one of the statistics used by the NFL to rate passers.

12.7 Summary

A good sports rating system provides rankings of individuals and teams, taking into account differences in numbers of competitions, importances of the matches and other sports-specific factors. Those ordinal rankings may be of interest simply to know who is best or may be used for the seeding of tournaments and competitions. Ratings, especially those providing probability estimates, may be of value for betting purposes.

In order to provide a coherent coverage of SRSs, the natures of sports and SRSs are of interest. Sports may be divided into three classes: combat sports where each competitor tries to control the opponent; object sports where each competitor tries to control an object; and independent sports where the competitor tries to control him(her) self because interference is not allowed. Each sport is either judged, scored or measured. The taxonomy of sports rating systems includes the evaluation and weighting of each performance followed by the calculation of a rating.

For combat sports, subjective polls are available, but few organized SRSs exist. For independent sports, averages may be used to predict future performance. World rating systems exist for golf and skiing. An evaluation of Olympic winning performances shows that women improved faster than men from 1952–72, that women and men improved at nearly the same rate from 1972–96 and that both improved less during the recent period. Improvement appears to increase in events requiring progressively more upper body flexibility, arm strength and reliance on technique. Future Olympic winning performances one Olympics ahead have been predicted with average absolute percentage errors of about 2% for swimming and athletics, 3% for speed skating and 6% for weightlifting.

Many SRSs are available for object sports. World ratings exist for soccer, chess and tennis. A number of predictive schemes have been proposed in which the actual outcome (home win, draw or away win) may be predicted about 19% more often than for random chance. Predictive SRSs have been applied to US college basketball, US and Australian Rules Football and soccer, for example. The predictability of score difference appears to be strongly influenced by the ease of scoring under the rules of each sport (easier scoring provides more accurate prediction), apparently supporting the adage that it takes variance to predict variance.

Home advantage must be understood so a margin of victory may be adjusted by an SRS. The home advantage is due to such influences as tactical familiarity of the home team with playing conditions, travel fatigue and crowd intimidation acting upon the visiting team. NBA basketball and soccer have particularly high home advantages while NHL hockey and baseball show much less. Home advantage can increase with time as in Australian Rules Football and it can decrease with time as in international soccer club cup competition.

In pari-mutuel gambling, each gambler competes with the other gamblers for a profit. For fixed-odds betting, the gambler competes against odds pre-selected by a betting agency. Probabilities provided by the least squares-Gaussian SRS and by the Maher–Poisson SRS both show promise for pari-mutuel and fixed-odds gambling.

References

Ali, M. (1994) Probability models on horse racing outcomes. *Proceedings of the Section on Statistics in Sports, American Statistical Association, Joint Statistical Meeting, 1994.*

ATP (1995) *ATP Tour Computer Rankings.* Address: 200 ATP Tour Boulevard, Ponte Vedra Beach, FL 32082, USA.

Bernstein, A. (1976) *Beating the Harness Races.* New York: Arco.

Bolton, R. and Chapman, R. (1986) Searching for positive returns at the track: a multinomial logit model for handicapping. *Management Science*, **32**(4), 24.

Clarke, S. R. (1993) Computer forecasting of Australian Rules Football for a daily newspaper. *Journal of the Operational Research Society*, **44**, 753–759.

Croucher, J. S. (1992) Winning with science. *Proceedings of the First Annual Mathematics and Computers in Sports Conference*, Australian Mathematics Society and Australian Sports Commission, Bond University, Queensland, Australia, 13–15 July, pp. 1–22.

Dixon, M. J. and Coles, S. G. (1997) Modelling association football scores and inefficiencies in the UK football betting market. *Journal of Applied Statistics*, **46**, 265–280.

Efron, B. and Morris, C. (1973) Stein's estimation rule and its competitors – an empirical Bayes approach. *Journal of the American Statistical Society*, **68**, Theory and Methods Section, March 1973.

Efron, B. and Morris, C. (1975) Data analysis using Stein's estimator and its generalizations, *Journal of the American Statistical Society*, **70**, Applications Section, June 1975.

Efron, B. and Morris, C. (1977) Stein's paradox in statistics. *Scientific American*, May, pp. 119–127.

Elo, A. E. (1978) *The Rating of Chess Players Past and Present*. London: Batsford.

FIFA (1995) *FIFA/Coca Cola World Football Ranking*. Address: PO Box 85, Hitziweg 11, CH-8030, Zurich, Switzerland.

FIS (1995/96a) *Rules for the Alpine FIS World Cup 1995/96*. Address: FIS, Blochstrasse 2, CH-3652, Oberhofen/Thunersee, Switzerland.

FIS (1995/96b) *Media Guide FIS World Cup Cross Country 1995/96*. Address: Blochstrasse 2, CH-3652, Oberhofen/Thunersee, Switzerland.

Goode, M. (1976) Teaching statistical concepts with sports. Presented to American Statistical Association, Boston, August 1976.

Greer, T. (1995) *Sony Ranking (Sanctioned by the Championship Committee of the Royal and Ancient Golf Club of St. Andrews)*. Address: Pier House, Strand on the Green, London W4 3NN, UK.

Harville, D. (1977) The use of linear system methodology to rate high school or college football teams. *Journal of the American Statistical Society*, **72**, Applications Section, June 1977.

Harville, D. (1980) Predictions for national football league games via linear-model methodology. *Journal of the American Statistical Society*, **75**, Applications Section, September 1980.

Harville, D. and Smith, M. (1994) The home-court advantage: how large is it and how does it vary from team to team? *American Statistician*, **48**(1), 22.

Information Please Sports Almanac (1994) Boston: Houghton Mifflin.

James, W. and Stein, C. (1961) Estimation with quadratic loss. *Proceedings of the Fourth Berkeley Symposium on Mathematical Statistics and Probability*, vol. 1, pp. 361–379.

Larkey, P. (1990) Fair bets on winners in professional golf. *Chance*, 3(4), 24.

Lee, C., Jean, B.-S., Kwon, I.-W. and Yun, T.-S. (1997) An empirical study of boxing match prediction using a logistic regression analysis. *Proceedings of the Section on Statistics in Sports, American Statistical Association, Joint Statistical Meeting*, Anaheim.

Maher, M. J. (1982) Modelling association football scores. *Statist. Neerlandica*, **36**, 109–118.

Official World Encyclopedia of Sports and Games (1979) New York and London: Paddington.

Olympic Review (1995) number XXV-4, August–September 1995, p. 72.

Papoulis, A. (1965) *Probability, Random Variables and Stochastic Processes*. McGraw-Hill, pp. 215–219.

Passer, D. (1981) *Winning Big at the Track With Money Management*. New York: Playboy.

Pollard, R. (1986) Home advantage in soccer: a retrospective analysis. *Journal of Sports Sciences*, **4**, 237–248.

Quinn, J. (1987) *Best of Thoroughbred Handicapping*. New York: Morrow.

Stefani, R. T. (1977a) Football and basketball predictions using least squares. *IEEE Transactions on Systems, Man and Cybernetics*, **SMC-7**, 117–121.

Stefani, R. T. (1977b) Trends in Olympic winning performances. *Athletic Journal*, December, pp. 44–46.

Stefani, R. T. (1980) Improved least squares football, basketball and soccer predictions using least squares. *IEEE Transactions on Systems, Man and Cybernetics*, **SMC-10**, 116–123.

Stefani, R. T. (1982) Olympic winning performances: trends and predictions. *Olympic Review*, No. 176, pp. 357–364.

Stefani, R. T. (1983) Observed betting tendencies and suggested betting strategies for European football pools. *The Statistician*, **32**, 319–329.

Stefani, R. T. (1985) Trends and predictions of Olympic winning performances. *Olympic Review*, No. 215, pp. 554–559.

Stefani, R. T. (1987) Application of statistical methods to American Football. *Journal of Applied Statistics*, **14**(1), 61–73.

Stefani, R. T. (1989a) Trends and predictions of Olympic winning performances (1952–1992) Part 1. *Olympic Review*, No. 258, pp. 157–161.

Stefani, R. T. (1989b) Trends and predictions of Olympic winning performances (1952–1992) Part 2. *Olympic Review*, No. 259, pp. 211–215.

Stefani, R. T. (1992) Applying least squares to team sports and Olympic winning performances. *Proceedings of the First Annual Mathematics and Computers in Sports Conference*, Australian Mathematics Society and Australian Sports Commission, Bond University, Queensland, Australia, 13–15 July, pp. 43–59.

Stefani, R. T. (1994a) The best performances in athletics. *Olympic Review*, No. 321, pp. 357–360.

Stefani, R. T. (1994b) Athletics, swimming and weightlifting: improvement in performances. *Olympic Review*, No. 323, pp. 470–474.

Stefani, R. T. (1994c) Athletics, swimming and weightlifting from Barcelona to Atlanta. *Olympic Review*, No. 326, pp. 598–603.

Stefani, R. T. (1997) Survey of the major world sports rating systems. *Journal of Applied Statistics*, **24**, 627–638.

Stefani, R. T. and Clarke, S. (1992) Predictions and home advantage for Australian Rules Football, *Journal of Applied Statistics*, **19**, 251–259.

World Almanac and Book of Facts (1995) New Jersey: Funk and Wagnalls.

WTA (1997) *WTA Tour Ranking System*. Address: 133 First Street NE, Third Floor, St Petersburg, FL 33701-3383, USA.

13

Multiplicity, Bayes and Hierarchical Models

Carl N. Morris
Harvard University, USA

13.1 Introduction

The two-level hierarchical model described in this chapter is valuable for evaluating players and teams from data on many sports, because it helps account for small sample sizes and for multiple opportunities to observe exceptional performances. The examples that follow emphasize baseball statistics because that has been the focus of most published sports research which uses this method, and because abundant baseball data are widely available. Most readers who do not have special knowledge about baseball will still be able to follow the statistical ideas. Readers are invited to consider how this approach applies to other sports. Some references to its application in other sports are given.

13.2 Multiplicity

Highly unlikely events occur, given multiple opportunities. One must account for 'multiplicity' when considering whether an event is extraordinary. A .300 hitter has one chance in 301 of making 16 consecutive outs. That such a streak will start with the player's next at-bat is highly unlikely. That it will happen some time during a season to a .300 hitter who plays every day is predictable by randomness and need not be considered as a 'slump' when it happens. Randomness alone suggests there will be about two no-hit games in the major leagues each season, even if the skills of pitchers and batters do not vary. (See Frohlich, 1994, for a discussion of the rate of no-hitter occurrences.) Thus, there must be more than two no-hitters per season to conclude that pitchers are unusually dominating in such games. Otherwise, these unusual events are interesting to observe and record, but they would indicate good or bad luck rather than exceptional athletic skill.

Seymour Siwoff (Siwoff *et al.*, 1985) and Bill James (1986, and other years) seem to take opposing positions on the multiplicity issue, especially with respect to 'small-sample', 'situational' or 'breakdown' statistics. Siwoff wants to report what actually happened, without making inferences. His books record how each hitter fared against each pitcher, and in which situations. This creates many small-sample situations, and thus many extreme rates occur in small samples. How seriously should we take them? Are small sample results predictive? In 1997, Brady Anderson got his ninth hit in ten career at-bats against pitcher John Burkett, something a .275 hitter (Anderson's six-year average since becoming a regular in 1992) could do once in 15 000 chances. We do not think .900 is the best estimate of Anderson's future success rate against Burkett, but attribute this result to multiplicity. There are many combinations of regular batters and pitchers, of the order of 10 000 of them, and Anderson's performance is the most impressive among such match-ups by that standard. It is unsurprising that some batter performed at this level against some pitcher.

In contrast, Bill James wishes to make inferences and predictions from baseball data, so rare events attributable to multiplicity are not useful to him. James almost certainly would estimate Anderson's future performance against Burkett to be closer to Anderson's recent average against all pitchers of Burkett's caliber. There need be no Siwoff/James controversy; the two simply have different objectives.

The media often manipulate statistics to exaggerate extreme performances, as in the Anderson example. Casella and Berger (1994) showed how to correct such reporting biases, assuming a Binomial distribution model (making inferences always requires some distributional assumptions) for the player's batting performance. They showed that a media report made late one season, saying that a player (David Winfield) had eights hits in his last 17 at-bats (.471), is most consistent with a .359 (the maximum likelihood estimate in their model) batting average, not .471. Winfield must have been 0 for 2 just before the streak, for otherwise the media would have reported nine or ten hits for 19 at-bats. Inferences inevitably involve uncertainty (unless they are mere deductions, such as 8 for 17 implies no streak of ten outs). So inferences not only require probability models, but they require their own probability distributions to describe their uncertainty.

Managers and owners must make inferences about future performances based on past data, so they can decide for whom to trade, what free-agents to sign, what salaries to pay, which players to use, which players will be best in pinch-hitting and relief pitching situations, and so on. We might even wish to make inferences about past events. We will ask later whether Ty Cobb, who batted over .400 in three of his 24 seasons, ever had a full season in which his underlying 'true' batting average exceeded .400 (as opposed to having a true average of .399 or less, with his observed average exceeding .400 only by good luck).

The multiplicity issue arises in many ways, especially when identifying the best or worst performers and teams. The 'sophomore slump' is said to hit batters in their second year, following a glowing rookie season. In the ten years of 1987–96, 72% of the teams that won their division one year had a lower winning percentage the next year. Should we conclude that players and teams allow their success in one year to go to their heads, losing their competitive edge? The statistical models and methods of this chapter provide a different explanation. The rookie with the highest batting average will be one who was both lucky in the first year, and who was good. First-year luck cannot be expected to persist, but the ability will. Such a player, more often than not, will have a lower average in his second year. The same argument applies to successful teams in any sport.

One sometimes can account for multiplicity by making calculations that only assume the (level-1) randomness of the observed performance data, given a completely specified distribution for the player's or team's skill. We did that above by assuming a 'true' .275 average for Brady Anderson, and most analyses of sports statistics make such an assumption. When one cannot specify a player's skill, a two-level model is needed, the second level providing a probability distribution for the player's skill. For example, if Anderson's .275 was not assumed constant for all games, we might also specify a level-2 'prior' probability distribution for his true batting skill, perhaps having mean .275 and some standard deviation. When this level-2 prior distribution is known, one may use Bayes' theorem to make statistical inferences. If the level-2 distribution is also unknown, then sometimes, given appropriate data on other players or teams, one may carry out an empirical Bayes or a fully Bayesian analysis of the two-level 'hierarchical model.' A few published analyses in the statistical literature which do this are central to this chapter. Because such analyses are complicated, we start with simpler Bayesian analyses of sports data for which the level-2 distribution is fully specified.

13.3 Modeling regression-to-the-mean

The tendency to do less well after a successful period, or to improve after a poor one, is predicted by a statistical phenomenon called 'regression-to-the-mean'. Galton (Stigler, 1986) first studied this, noticing that the sons of tall fathers, while tending to be taller than average, are usually shorter than their fathers. He analyzed this by plotting the heights of sons (vertical axis) against the heights of their fathers (horizontal axis) for many father–son pairs, and obtained the least-squares regression line. The now widely used least-squares method that Galton developed for fitting his straight line is still called 'regression' because of this pioneering application. If the correlation between a father's height and his son's height is about $r = 0.5$, then the best predictor of a son's height is about half way between his father's height and the average height. This same explanation applies to players and to teams, although we now replace Galton's scatter-diagram with one that plots second period performance rates y_2 for teams or players (vertical axis) against first period performance rates y_1 (horizontal axis). The linear regression line, given y_1, is

$$E(y_2) = \mu + b(y_1 - \mu) \tag{13.1}$$

with μ the mean of y_2 and of y_1 (assumed the same for both time periods) and b, the regression coefficient, as determined from least squares theory. If y_1 and y_2 have the same variances, then the slope b is simply r, the correlation between y_2 and y_1. Correlations must be less than 1, and given y_1, this forces $E(y_2) < y_1$ whenever $y_1 > \mu$. Thus, Galton's regression-to-the-mean observation follows from linear regression theory.

There are 'two levels' of variation in the baseball situations described above, and in Galton's. Suppose that y_2 is the observed winning percentage (games won/games played) for a team in year 2, and y_1 is the observed percentage for the first year. Assume p_2 and p_1 are the 'true' percentages for each year (idealized as the percentage that would be known if the teams had been able to play millions of games each year). Let N_1 and N_2 be the number of wins (or hits) in n_1 and n_2 games (or opportunities), so that the observed rates are $y_1 = N_1/n_1$ and $y_2 = N_2/n_2$. Given p_1 and p_2, the probabilities of success, we assume that N_1 and N_2 are independent and follow the Binomial distribution:

Level 1: $N_1 = n_1 \times y_1 \sim \mathrm{Bin}(n_1, p_1)$ and $N_2 = n_2 \times y_2 \sim \mathrm{Bin}(n_2, p_2)$ (13.2)

Assume, for simplicity only, that $p_1 = p_2$ (the team's skill stays the same for the two years), and denote the common (unknown) value as $p = p_1 = p_2$. Because of the central limit theorem, both Binomial distributions are reasonably well approximated for large n_1 and n_2 by Normal distributions. Thus, approximately,

Level 1: $y_i \sim N(p, V_i)$ independently for $i = 1, 2$ (13.3)

with $V_i = p(1 - p)/n_i$, the variance of y_i.

These Binomial distributions describe one level of variation, the variation of a team's or player's averages around its or his own 'true' average p.

The second level of variation involves p. Because we do not know p, we may describe its possible values with another distribution. We might do so because the team or player was taken at random from a group of teams or players and the true abilities of teams or players vary in the group. Thus,

Level 2: $p \sim N(\mu, \tau^2)$ (13.4)

That is, p is thought of as having a Normal distribution with mean and standard deviation being μ and τ. (The Normality assumption is unnecessary, and is assumed here only for convenience of calculation.)

In our first example, we assume we know μ and τ after careful thought. Later, we will discuss how these values can sometimes be estimated from the data for all players or teams, so that μ and τ do not have to be known. The latter approach is complicated, and the main concepts are captured in this simpler case.

Between 1987 and 1996, there were 46 division winners in Major League Baseball. The regression-to-the-mean phenomenon predicts that most of these 46 teams would have had a lower winning percentage in the following year, and in fact 33 (72%) of these teams did less well. While individual teams varied, their winning percentage averaged about 0.585 in their division-winning year, and then dropped to an average of about 0.545 in the following year. We will see that this 0.040 drop is about what is predicted from the two-level model.

For the level 2 model (13.4), with p referring to the (hypothetical) long-run winning percentage for a given team, we choose $\mu = 0.500$ as the mean, because this is the average for all teams. We will take the standard deviation of p to be $\tau = 0.050$ on the grounds that we believe true winning percentages exceed 0.600 only about 2% of the time ($p > 0.500 + 2 \times 0.050 = 0.600$). If 2% seems too small, remember that we are evaluating the variability of the true (unobserved) p in formula (13.4), and not of the observed winning percentage y for a team. In our given ten-year span, only about 5% of all teams won at least 60% of their games, and many of those probably had $p < 0.600$, but were lucky enough to have $y_1 > 0.600$.

The level 1 distribution, formula (13.3), is easier to consider. The Normal distribution applies because 162 games in a baseball season represents a large number of trials. We assume that y_1 and y_2, the fractions of games a team won in consecutive years, are unbiased estimates of the unknown p, and that the variances are $V_1 = V_2 = 0.5 \times (1 - 0.5)/162 = (0.0393)^2$. (Teams ordinarily play $n_1 = n_2 = 162$ games, except in strike years, and we are using the traditional variance approximation here for the Binomial.)

Bayes' Theorem now provides the distribution of the unknown p, given the observed value of y_1. The well-known result is that

$$p \mid y_1 \sim N((1 - B)y_1 + B\mu, V_1(1 - B)) \qquad (13.5)$$

where $B = V_1/(V_1 + \tau^2)$. B is the portion of the total variance $(V_1 + \tau^2)$ of y_1 that is attributable only to sampling variation (V_1). The mean of p, given y_1, is $(1 - B)y_1 + B\mu$. This is a weighted average of the observed winning percentage (y_1) and of the mean for all teams $E(p) = \mu = 0.500$. B is often called a 'shrinkage factor' because it determines whether the (Bayes) estimate of p, the minimum mean-squared-error estimate $E(p \mid y_1)$, is closer to y_1 or to μ. If τ is large, B is near 0, and y_1 is nearly the Bayes estimate. In our case,

$$B = (0.0393)^2/((0.0393)^2 + (0.050)^2) = 0.382$$

so a 38.2% regression-to-the-mean is expected.

Taking $y_1 = 0.585$, the average winning percentage for the 46 division winners, we calculate the variance in (13.5) to be $(0.0393)^2(1 - 0.382) = (0.031)^2$, and so

$$p \mid y_1 \sim N(0.553, (0.031)^2)$$

The expectation is $E(p \mid y_1 = 0.585) = 0.553$. This means that the typical division winner in this period is estimated to have a true winning percentage of 0.553, not 0.585.

We now ask about the distribution of y_2, next year's winning percentage, given y_1, still assuming p to be the same for both years. Given y_1, p has the distribution in (13.5), and y_2, given p, has the $N(p, V_2)$ distribution in (13.3). Thus, given y_1,

$$y_2 \mid y_1 \sim N((1 - B)y_1 + B\mu, V_2 + V_1(1 - B)) \qquad (13.6)$$

This is the 'predictive posterior' distribution, so called because it predicts y_2 (which we eventually can observe), assuming y_1 is known. In (13.6) the variance of y_2 exceeds that of p by the amount of its variance V_2 around p, and so the square root of the variance in (13.6) is calculated to be 0.050, noticeably larger than the 0.031 for the variance of p. The V_2 term is required because even if we knew p (i.e. even if $V_1 = 0$), a team's winning percentage y_2 for the next year would still vary around p by this amount. The actual average winning percentages for the 46 division winners in the year following their division championship was 0.545, 0.008 below (about 1.1 standard errors below) the prediction of $0.553 = E(y_2 \mid y_1 = 0.585)$ computed from (13.6).

Assume that a team has $y_1 > 0.500$, so that downward regression-to-the-mean is expected in year 2, i.e. $E(y_2 \mid y_1) < y_1$. We use $V_1 = V_2 = (0.0393)^2$

and $B = 0.382$ in (13.6), with y_1 subtracted from the mean, to calculate the distribution of the difference of the winning percentages in successive years:

$$(y_2 - y_1) \,|\, y_1 \sim N(-0.382(y_1 - 0.500), (0.050)^2) \qquad (13.7)$$

The probability that $y_2 - y_1 < 0$ is easily obtained from (13.7) and the standard Normal cumulative distribution function $\Phi(z)$ as

$$\Phi(0.382(y_1 - 0.500)/0.050)$$

Averaging these probabilities, calculated for each of the 46 division winners, predicts that 74% of them will regress to the mean in the next year, which is close to the 72% (33/46) that actually did.

The predictive probability of regression-to-the-mean just calculated as 74% is less than the probability that y_1 exceeded p, the team's long-run winning ability, which is $P(p < y_1 \,|\, y_1)$, computed from (13.5). The average of these latter values for the 46 division winners (the average of 46 values given by $\Phi(0.382(y_1 - 0.500)/0.031)$) is 84%. Predicting a future average over a limited number of trials, y_2, is always less certain than predicting for an infinite number of trials, p.

Among all teams that finished above 0.500 in year 1 (not just the division winners), the fraction expected to regress to the mean in year 2 (i.e. to have $y_2 < y_1$) can be shown to be

$$1/2 + \arcsin((B/2)^{1/2})/\pi \qquad (13.8)$$

This is $0.500 + \arcsin((0.382/2)^{1/2})/3.1416 = 0.644$ for baseball teams, assuming $\tau = 0.050$. It follows that among all better-than-average teams (not just division winners) in a given year, 64% are expected to win fewer games in the following year. Conversely, 64% of those below 0.500 should improve. Note that this probability (64%) of all better-than-average teams regressing to the mean is less than the probability for the division champions (74%), because mediocre teams are less likely to change in performance in the following year. (Formula (13.8), and a generalization of it for different sample sizes not shown here, are newly derived for this example.)

James *et al.* (1993) used simulations and a hierarchical model, much like (13.1) and (13.2), to estimate the probability that a 'good' team (one in the top 10% based on p, so $p > 0.567$) wins its division. They did this for the years before 1994, when the American and the National Leagues each still had two seven-team divisions. Their simulation accounts for the paired-comparisons structure of sports competition. They showed that 'good' teams win their division only about 36% of the time, even though most 'good' teams are the best in their division. (By definition, the best team in the division would always win in an infinitely long season.)

The preceding ideas about two-level models apply to many other situations, and can involve distributions other than the Normal, such as the Poisson, as discussed in Section 13.4. The reader may wish to use the distribution (13.6) and his or her ingenuity to investigate the magnitude of the 'sophomore slump' for second-year players who had high batting averages in their rookie years. It may be hard to determine the parameters needed for the level-2 distributions (e.g. should μ be chosen as the mean for all rookies, or what?). It is generally more difficult to choose these models once a specific player (or team) is named, because we must then think about all of his or her available information. For example, Nomar Garciaparra, the 1997 American League Rookie-of-the-Year, batted .306 (209 hits in 684 at-bats). What should we choose for μ and τ to calculate Garciaparra's probability of a sophomore slump (batting less than .306 in 1998)? Garciaparra excelled in so many offensive categories that it seems inappropriate to base the needed μ and τ for him on data for other 1997 rookies. It might be appropriate to use all past rookies-of-the-year, together with their second-year data, but that would require a much more complicated analysis and a fully Bayesian hierarchical model.

13.4 Sports statistics and hierarchical models

Given a sample from a population of players or teams, one may wish to estimate the population mean and standard deviation of true values, as with μ and τ for batting averages, or perhaps some other unknown parameters that govern the level-2 distribution. There is a large, rapidly growing statistical literature on empirical Bayes methods, hierarchical modeling, multi-level (or two-level) models, and on other closely-related topics that shows how to do this. Only a few of these published articles directly address their application to sports examples. We briefly review some of them. These models are more complicated than those of Section 13.3, but the general ideas are the same.

13.4.1 Estimating final batting averages from early season averages

Efron and Morris (1975, 1977) asked if a season's final batting average can be better estimated from the early season's batting average by shrinking each player's average, using data from other players to help estimate the shrinkage constant B. These two papers both concern the theory needed to estimate μ and τ in the level-2 model (13.4), and the shrinkages. Charles Stein (James and Stein, 1961) had shown that shrinkage estimates give a lower expected total squared error for all estimates. Efron and Morris used the following baseball data to show that the theory also works in practice.

Both Efron–Morris papers analyzed the 18 players who had 45 official at-bats on a certain day early in the 1970 baseball season (having the same number of at-bats was necessary to apply Stein's method, although the papers provided generalizations for other situations). They assumed all $k = 18$ players followed the two-level model (13.3) and (13.4). The best batter after the first 45 at-bats hit .400 (=18/45), and we anticipate regression-to-the-mean. However, that player was the great Roberto Clemente, so the concern was that combining him with a random sample of (regular) players to estimate shrinkages might be worse than keeping all the players separate and ignoring shrinkage estimators.

The goal is to estimate μ and B using data from all 18 players, and use that information with (13.5) to estimate a probability distribution for each player's true batting ability. These 18 players had an overall average of .265 after 45×18 at-bats, .265 being the estimate of μ. The variance $V_1 = V_2 = \cdots = V_{18}$, the binomial variance after 45 at-bats (the same for all players, because they batted equally often), is computed as $0.265(1 - 0.265)/45 = (0.066)^2$. A

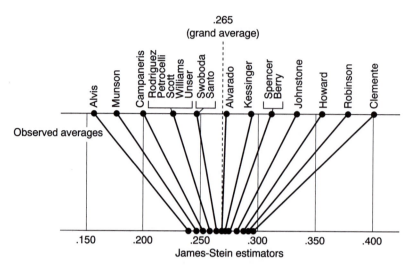

Figure 13.1   James–Stein Estimators for the 18 baseball players were calculated by 'shrinking' the individual batting averages toward the overall 'average of the averages'. In this case the grand average is .265 and each of the averages is shrunk about 80% of the distance to this value. Thus, the theorem on which Stein's method is based asserts that the true batting abilities are more tightly clustered than the preliminary batting averages would seem to suggest.

knowledgable baseball fan might guess $\tau = 0.025$ (25 batting average points) for this group of randomly sampled regular players. Then, we would have $B = V_1/(V_1 + \tau^2) = 0.874$, and so would predict the final average to be $0.126y_1 + 0.874 \times 0.267$ for each player. Instead, Efron and Morris used Stein's method, which estimates $B = (k - 3)V_1/SS$, with $k = 18$, $SS =$ the sum of squares about the mean 0.265, and $\mu = 0.265$. This is called an 'empirical Bayes' approach, since the unknown level-2 parameters μ and τ^2 are estimated by using the data. The estimates and raw data are shown in Figure 13.1 from Efron and Morris (1977).

The results were spectacular considering that little knowledge of baseball is required to make this estimate. These shrunken averages predicted 16 of the 18 player's batting averages for the remainder of the season more accurately than their individual batting averages after 45 at-bats. The overall accuracy was improved by a factor of more than three, the sum of squared errors being three times smaller than for the 18 individual estimates.

Morris (1983a, b) re-analyzed the data with a more complete fully hierarchical Bayesian analysis for these data, as did Steffey (1992). Such analyses provided posterior standard deviations and distributions for the 18 hierarchical estimates. The main conclusions are similar to those obtained by Stein's method. In Figure 13.2 (Morris, 1983a), the numbered circles plot the final season averages (vertical axis) against the early season batting averages (horizontal axis) for these batters. The linear regression line and the 68% and 95% intervals for the true averages are shown for these players using the solid lines. Note that Clemente's 95% interval barely reaches .400 at the upper end, even though he batted .400 for the first 45 at-bats. This is unlike

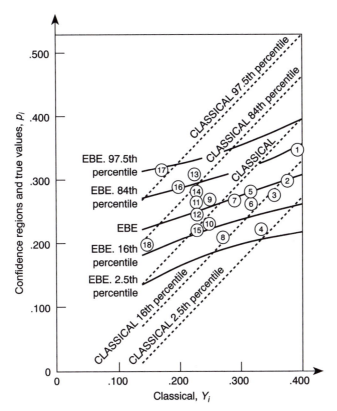

Figure 13.2 Eighteen points plotted at (Y_i, p_i). Classical $= Y_i$, $Y_i \pm V^{1/2}$ and $Y_i \pm 1.96V^{1/2}$ (dashed lines). EBE $= \hat{p}_i =$ empirical Bayes estimate, EBE $\pm s_i$ and EBE $+ 1.96s_i$ (solid curves), s_i is the (posterior) standard error of p_i.

more traditional statistical methods based on data for 45 at-bats (shown by the dashed lines) because this analysis predicts substantial regression-to-the-mean. Clemente's .346 average for his remaining 367 at-bats is well within his prediction interval.

The hierarchical modeling formula of Stein's, described above, applies only when all the players (or teams, etc.) have the same level-1 sampling variances V_1 (and sample sizes). That does not hold for most situations. The applications that follow all require more complicated methods for handling unequal sample sizes and/or other distributions than the Normal. Many such methods are now available in the statistical literature, and some will be described here.

13.4.2 The .400 skill level

A Bayesian statistical analysis allows probabilities to be calculated that cannot be evaluated by non-Bayesian approaches. In the next example, we ask about the probability that Ty Cobb was ever a 'true' .400 hitter for a full season. Ty Cobb played from 1905 to 1928 and attained the highest career batting average

(.366) in professional baseball history. Cobb batted .400 in three different seasons but his batting averages in those three years were only 1.05, 0.50 and 0.05 standard deviations above a true .399 skill level. Since Cobb must have batted at least this many standard deviations above his true mean several times in his 24 year career, his .400 years could have been achieved by combining good luck with less than .400 skill. However, we must account for year-to-year variability in Cobb's true skill resulting from changes in his age and various unpredictable components. Accounting for these level-2 variations requires a Bayesian hierarchical model.

Morris (1983a) used this model to analyze Cobb's 24 years of batting data. The model assumes that Cobb's 24 true means, $p_1, \ldots, p_{24}$, varied randomly about a quadratic curve that (as fitted) rose in his early years and diminished later. The shrinkage factors, B_i for year i, must now vary to provide more shrinkage in years with fewer at-bats. They were estimated to determine how much each season's true batting skill should be regressed toward the fitted quadratic curve.

Figure 13.3 (Morris, 1983a) shows the results, including Cobb's career batting data (dashed line), the fitted quadratic curve (smooth solid line), and the estimates (dotted line). In no year do we estimate that Cobb's batting skill

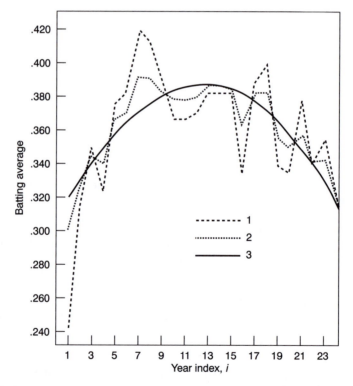

Figure 13.3 Ty Cobb's batting averages, 1905–28. 1 = observed average; 2 = empirical Bayes estimate, shrinkages B_i range from 56 to 79%; 3 (dark line) = regression estimator ($r = 3$ quadratic).

level that year exceeded .400. The highest probability of his exceeding .400 is 34% in 1911 (his seventh year) when he hit .420. On the other hand, we estimate that Cobb probably did achieve a skill level above .400 in some (not identifiable) year, because the probability that $p_i < .400$ in all 24 of his playing years is 12%, a figure obtained approximately (ignoring correlations in these estimates) by multiplying the 24 easily calculated values $P(p_i < .400)$. Thus, there is an 88% chance that Cobb's true batting skill exceeded .400 at least once. If so, it might have been in one of his three .400 years, but it also might have happened in a year when he batted under .390. This 88% probability can be calculated more accurately (without ignoring the correlation between years) by using simulation methods. One approach would be to use a Gibbs sampling method like that of Albert's (1992) home run analysis. Woodburn (1987) did this calculation several ways by using bootstrap simulation and got similar results.

If there ever were a 'true' .400 hitter for a full season, it almost only could have been Cobb or Rogers Hornsby. Hornsby had a combined average of .402 for five consecutive years, 1921–25. No one else would come close, by this analysis, even though about ten other players have batted .400 for a full season. Tony Gwynn, the best of today's hitters (for batting average), is good enough to have a chance to bat .400. If so, it will take lots of luck, as he'll have to bat a standard deviation or two (20 or 40 points) above his true skill level.

13.4.3 Situational statistics

Jim Albert (1994), in a paper with far-reaching implications for baseball, used a hierarchical model to study how much batters differ from one another with respect to 'situational statistics'. Using situational data for 154 batters from the 1992 season (Cramer and Dewan, 1992), Albert fitted a hierarchical model similar to that just described for the Ty Cobb analysis, but generalized for his more complicated setting. In eight separate analyses, Albert studied each of eight distinct batting situational effects. In each case, Albert estimated the main effects (the mean batting average differences between two alternative situations) for these 154 players. The eight pairs of situations, with estimated main effects in parentheses, are: grass versus turf (-0.003); runners in scoring position versus no base runners and no outs (0.000); day game versus night game (0.002); before the All-Star break versus after (0.003); home games versus away (0.008); ground-ball pitcher versus fly-ball pitcher (0.010); opposite-armed pitcher versus same (0.020); and ahead in the count versus two strikes on the batter (0.123). Level 1 of Albert's model specifies that the (logistically-transformed) batting averages for each player follow Normal distributions independently, and estimates the variances V_{ij}, for each player i in both situations ($j = 1$ and 2). Level 2 of each model specified that each player has his or her own true means (true batting skills) in both batting situations. The level-2 means and variances then were estimated for each player by using the data for all 154 players.

As noted above, the average advantage for a right-handed batter facing a left-handed pitcher, relative to his facing a right-handed pitcher (and conversely for left-handed batters), is 20 points (0.020). Albert asked whether some of the 154 batters' skills deviate from this main effect because their relative advantage is not 20 points. Siwoff *et al.* (1985) believe that substantial differences exist for many batters based on the data in the two situations. Albert gets the opposite result, saying that the variation of these differences for the 154 players is completely consistent with what randomness suggests. On average, a right-handed batter who had a season's average of .300 overall, who has faced 'righties' for 70% of his at-bats, and 'lefties' for 30%, would be estimated to have had a .294 skill against the righties and .314 against the lefties. Interpreting Albert's results in the strongest way, no other adjustment is necessary for any of these players, regardless of how much the player's actual performance differed against lefties and righties.

In recognition of the enormous strategic emphasis baseball managers place on right-hand/left-hand match-ups, Albert's findings are shocking. One will make much better inferences from batting data by ignoring situational data compiled for individual batters than by putting much stock in them. However, Albert's paper does demonstrate a relatively large general effect of 20 points with respect to handedness. So, the situational data have value when applied over all baseball players as a group. Thus, platooning players with relatively equal batting skill but opposite handedness does make sense. Similar results hold for the other seven types of situations in Albert (1994). The major exception involves Tony Gwynn, whose average drops very little with a two-strike count versus when he is ahead in the count, whereas the average difference for the other 153 players is a whopping 122 points (0.122).

We note that Albert's main effect conclusion fails for home–away comparisons involving Colorado players after 1993. The Colorado Rockies joined the National League in 1993, and so escaped Albert's 1992 analysis. All hitters – the Rockies' and the visiting teams' – have an enormous advantage at Coors Field, which makes Rockies hitters seem to be better than they are (and their pitchers seem worse). This effect is generally attributed to the stadium's location in Denver, Colorado, which places Coors Field at a substantially higher altitude level than all other Major League Baseball stadiums. Rockies players batted .330 as a team in home games in 1995–96 (combined) and only .237 on the road – a whopping 93 point advantage (rivaling the 0.122 difference that Albert estimated for pitch count situations). This 0.093 dwarfs the overall 0.008 home–road advantage that Albert estimated. For comparison, the Boston Red Sox had the second best home field advantage (0.031).

13.4.4   Other applications

Hierarchical statistical models have been applied to sports data, usually to baseball, in other papers. Albert (1992) analyzed the seasonal home run rates of 12 of the greatest career home run hitters, using a hierarchical model for Poisson data (home runs are relatively rare events, so they have Poisson distributions, approximately). His model is similar to the hierarchical models for Normally distributed data described earlier. Key results included estimates of the peak seasonal rates for each of the players. For example, the estimated 0.101 home runs per at-bat for Ruth's peak season exceeds by far the highest for anyone else ever to play before 1992. However, Mark McGwire's amazing recent home run performances (0.116 home runs per at-bat from 1995 to 1997) achieved since Albert's analysis would change that.

Morris and Christiansen (1996) showed how to use a hierarchical model to rank baseball teams by their runs per game using 1993 data for the 14 American League teams. They calculated the probability that each team was best (and worst) at scoring runs, and also the predictive probability that each team would have the highest rate in the next year (assuming the rate stays constant for the two years). The following technical result developed for their analysis should be useful in other baseball applications. If a team's mean scoring rate is μ runs per game, then a very accurate theoretical approximation to the variance of runs the team scores in one nine-inning game is $(11 \times \mu/8)^{1.2}$.

Less has been done with applications of hierarchical models in sports other than baseball, but that will change, as understanding and software for these models continues to improve. Computer simulation methods for fitting hierarchical models allow quite complicated, quite realistic situations to be analyzed. An early application of hierarchical models to football was by Harville (1980), who used them to rank professional football teams for several years, based on the schedules and on the scores of games. Components of variation are used at level 2 to reflect changes in team strength over seasons. Glickman and Stern (1998) have a related application for the 1988–93 seasons. They allow even more detail, as permitted by advances in computation over two decades, and their methods outperformed the Las Vegas betting line over the last half of the 1993 season. Other sports applications that used hierarchical models are by McArdle and Hamagami (1994), who predict graduation rates of college athletes, and by Ridder *et al.* (1994), who use a Poisson model for soccer scores to determine how much the expulsion of a player for a rules infraction affects the score.

13.5 Conclusions

Applications to many sports clearly are possible, and these models will be most valuable for assessing the strengths of teams, for predicting future outcomes, and for assessing players, e.g. to establish trade value. These methods also can be effective in improving evaluations from the past. For example, they were used by the author in unpublished work to show that Mark McGwire's 39 home runs in 317 at-bats in 1995 was more extraordinary than Albert Belle's league-leading 50 home runs that year in 546 at-bats. Although there will be objections to using complicated statistical methods to decide championships and other awards, the result just given did predict McGwire's amazing home run performance in the following two years. Hierarchical models are most necessary for making good predictions, and that is valuable in many sports for personnel and economic decision making.

Acknowledgments

Research for this chapter was supported by grant DMS-9705156 from the National Science Foundation.

References

Albert, J. (1992) A Bayesian analysis of a Poisson random effects model for home run hitters. *The American Statistician*, **46**, 246–253.

Albert, J. (1994) Exploring baseball hitting data: what about those breakdown statistics? *Journal of the American Statistical Association*, **89**, 1066–1074.

Casella, G. and Berger, R. L. (1994) Estimation with selected binomial information or do you really believe that Dave Winfield is batting .471? *Journal of the American Statistical Association*, **89**, 1080–1090.

Cramer, R. and Dewan, J. (1992) *STATS 1993 Player Profiles*. Skokie, IL: STATS, Inc.

Efron, B. and Morris, C. (1975) Data analysis using Stein's estimator and its generalizations. *Journal of the American Statistical Association*, **70**, 311–319.

Efron, B. and Morris, C. (1977) Stein's paradox in statistics. *Scientific American*, **236**(5), 119–127.

Frohlich, C. (1994) Baseball: pitching no-hitters. *Chance*, **7**(3), 24–30.

Glickman, M. E. and Stern, H. S. (1998) A state-space model for National Football League scores. To appear in *Journal of the American Statistical Association*.

Harville, D. (1980) Predictions for National Football League games via linear-model methodology. *Journal of the American Statistical Association*, **75**, 516–524.

James, B. (1986) *The Bill James Baseball Abstract*. New York: Ballentine Press.

James, B., Albert, J. and Stern, H. (1993) Answering questions about baseball using statistics. *Chance*, **6**, 17–22.

James, W. and Stein, C. (1961) Estimating with quadratic loss. *Proceedings of the Fourth Berkeley Symposium*, Vol. I. Berkeley: University of California Press, pp. 361–379.

McArdle, J. J. and Hamagami, F. (1994) Logit and multilevel logit modeling of college graduation for 1984–1985 freshman student-athletes. *Journal of the American Statistical Association*, **89**, 1107–1123.

Morris, C. (1983a) Parametric empirical Bayes inference: theory and applications (with discussion). *Journal of the American Statistical Association*, **78**, 47–65.

Morris, C. (1983b) Parametric empirical Bayes confidence intervals. *Scientific Inference, Data Analysis, and Robustness*. New York: Academic Press, pp. 25–50.

Morris, C. and Christiansen, C. L. (1996) Hierarchical models for ranking and for identifying extremes, with applications. In *Bayesian Statistics 5*, J. M. Bernardo, J. O. Berger, A. P. Dawid and A. F. M. Smith (eds). Oxford: Oxford University Press, pp. 277–296.

Ridder, G., Cramer, J. S. and Hopstaker, P. (1994) Down to ten: estimating the effect of a red card in soccer. *Journal of the American Statistical Association*, **89**, 1124–1127.

Siwoff, S., Hirdt, S. and Hirdt, P. (1985) *The 1985 Baseball Analyst*. New York: Collier Books, Macmillan.

Stats Major League Handbook 1997 (1996) Skokie, IL: STATS Publishing.

Steffey, D. (1992) Hierarchical Bayesian modeling with elicited prior information. *Communications in Statistics*, **21**, 799–821.

Stigler, S. (1986) *The History of Statistics.* Cambridge, MA: The Belnap Press of Harvard University.

Woodburn, R. L. (1987) Applications of the bootstrap technique to linear regression models in the empirical Bayes framework. Master Degree Thesis, The University of Texas at Austin, December 1987.

Glossary

This glossary provides general definitions of sports terminology used in the book. It is provided as an aid in understanding the material in the chapters and not as a standard reference of sports terms. The degree to which sports are covered in the glossary depends on three factors: (1) the level of detail in the chapter, (2) general familiarity with the sport internationally, and (3) whether the term is defined in the course of the chapter. Using these criteria, certain sports (e.g. baseball, cricket) are heavily represented, while others (e.g. soccer) have few if any definitions.

The glossary is organized alphabetically by the term. Each entry includes the sport to which the term pertains as well as the definition. This is necessary because of the overlap in usage of certain terms (e.g. assist).

Ace	Tennis	A service the opponent cannot even touch
All rounder	Cricket	A player who excels at both batting and bowling
Alternate shot	Golf	A form of two-person team competition where team members alternate shots on a single ball. It is usually played in a match format
American League (AL)	Baseball	The younger existing major professional baseball league in North America, established in 1901
Assist (A)	Baseball	A fielding play in which the fielder is instrumental in getting an out by throwing the ball to another fielder
Assist	Ice hockey	Up to two players on the goal-scoring team who handle the puck immediately prior to a scored goal may be credited with assists
Association of Tennis Professionals (ATP)	Tennis	A players' union (men's tennis) that runs tour events outside of the Grand Slams, Grand Slam Cup and Davis Cup

At-bat (AB)	Baseball	Plate appearance that results in a hit or out
Australian Rules Football	Australian Rules Football	A continuous action sport where the players may align themselves anywhere along an oval field more than 100 m long. The ball may be advanced forward by running, punching or kicking. There are four vertical goal posts. If a ball is kicked between the two centre goal posts a six-point goal is scored. If the ball is kicked between a centre and an outer goal post a one-point 'behind' is scored. More than 200 points are scored in a typical game. In 1997, the Australian Football League consisted of ten teams in Victoria and six teams in other states
Average	Cricket	Batting average = total runs scored divided by number of times dismissed. Bowling average = number of runs conceded divided by number of batsmen dismissed
Backhand court	Tennis	The left-hand court
Ball	Baseball	A pitch outside the strike zone at which the batter did not swing
Base on balls (BB)	Baseball	Batter goes safely to first base on four pitches called balls by the umpire. Also, called a walk
Batting average (BA)	Baseball	Hits divided by at-bats
Best ball	Golf	The four-person version of Better ball in which one, two or three of the four scores may count for the team
Better ball	Golf	A form of two-person team competition where each team member plays their own ball and, on each hole, only the lowest of the two scores count for the team. It can be either stroke or match play between the teams
Birdie	Golf	One stroke less than par
Bogey	Golf	One stroke over par
Break point	Tennis	A rally in which the player receiving service is within a point of winning the game
Bunt	Baseball	Batter taps the ball lightly in the infield, generally to advance runners
Bye	Cricket	A run taken when the batsman has not made contact with the ball with his body or the bat

Caught stealing (CS)	Baseball	Runner tagged out while attempting to steal a base
Century	Cricket	A score of 100 or more runs
Class	Horse racing	Each horse may be assigned to a class depending on that horse's recent win–loss record and money won. In betting, a horse moving to a higher class is sometimes discounted and a horse moving down to a lower class may be assigned an enhanced chance of winning
Complete game	Baseball	A game in which the starting pitcher pitches every inning
Count	Baseball	The current number of balls and strikes on a batter. A batter is ahead in the count when he has more strikes than balls
Declaration	Cricket	The captain of the batting side can declare his innings closed at any stage. Usually done if the team is in a strong position, to give time to dismiss the opposition
Double (2B)	Baseball	Batter hits the ball and stops safely at second base. Counts as two bases towards total bases
Double bogey	Golf	Two strokes over par
Double eagle	Golf	Three shots under par; a score of 1 on a par four or, more common, 2 on a par five
Double fault	Tennis	Both serves are faults, leading to a loss of point for the server
Double play (DP)	Baseball	A single play in which two outs are recorded
Doubles	Tennis	Two players against two other players
Draft	Baseball	The procedure in which the rights to players leaving high school and college sports to enter professional sports are picked by teams
Duck	Cricket	A score of zero runs
Eagle	Golf	Two strokes less than par
Earned run (ER)	Baseball	A run which is scored without the aid of a fielder's error
Earned run average (ERA)	Baseball	Earned runs allowed by a pitcher divided by the number of innings pitched, then multiplied by nine
Economy rate	Cricket	For a bowler, number of runs conceded per 100 balls bowled, or sometimes the number of runs conceded per over

Enforcer	Ice hockey	A player whose explicit function is physically to discourage opposing teams' players from abusing his own team's better players
Error (E)	Baseball	Misplay by a fielder which should have resulted in an out or which allowed runners to advance
Even strength	Ice hockey	Teams are said to be playing at 'even strength' when both teams have the same number of players on the ice. Even strength may be at four, five or six players each
Fair ball	Baseball	A ball hit by the batter into the 90° arc in front of home plate
Fairway	Golf	The closely mowed landing area between tees and greens
Field goal	American football	An attempt by the offense to kick the football from its position on the field toward the goal posts, worth 3 points if successful (between the goal posts and above the crossbar)
Field goal	Basketball	A shot attempt which is worth 3 points if made outside the 3-point line and 2 points otherwise
First innings win	Cricket	Obtained by the team with the most runs after both have completed their first innings. In some tournaments, a team is awarded points for this win
Fly ball	Baseball	A ball hit in the air generally into the outfield
Fly ball pitcher	Baseball	A pitcher who gets outs by inducing the batter to hit easily fielded balls into the air
Forehand court	Tennis	The right-hand court
Foul	Basketball	A penalty which results in free throws or loss of possession
Foul ball	Baseball	A ball hit by the batter that does not stay in fair territory
Free agency	Baseball	The player's right to change teams. Until the mid-1970s, a player could only change teams if his team traded or sold his contract
Free throw	Basketball	An uncontested field goal worth 1 point if made. Awarded as a result of a foul
Game	Tennis	To win a game, a player must win four points unless the score reaches three each (deuce) in which case a two-point lead must be established

Goal	Ice hockey	A goal is counted for a team when a player on that team shoots the puck into the opposing team's net
Grand Slam Event	Tennis	These events consist of the championships of Australia, Wimbledon, France and the USA
Green	Golf	The closely mowed area where the pin is located
Ground ball	Baseball	A ball hit so that it bounces on the ground
Ground ball pitcher	Baseball	A pitcher who gets outs by inducing the batter to hit easily fielded balls on the ground
Half-inning	Baseball	The portion of an inning in which one team bats
Hit (H)	Baseball	Single, double, triple or home run
Hit by pitcher (HBP)	Baseball	Batter goes safely to first base because he was hit by a pitched ball
Home run (HR)	Baseball	Batter hits the ball, touches all the bases, and arrives safe at home plate: typically, when the batter hits a ball on a fly in fair territory beyond the field of play. Counts as four bases towards total bases
Infield	Baseball	The area of the playing field encompassed by home plate and the three bases
Inning (ING)	Baseball	Each game has nine innings in which each team gets to bat (has the opportunity to score runs). If a game is tied at the end of nine innings, extra innings are played until one team is ahead at the end of a complete inning
Intentional base on balls (IBB)	Baseball	A base on balls in which the pitcher prevents the batter from hitting by purposely pitching four balls
Leg bye	Cricket	A run taken after the ball has hit the batsman without hitting his bat
Lie	Golf	Position of the ball such as tight lies (sitting down), fluffy lies (sitting up in grass), buried lies (embedded usually in a bunker), and unplayable lies (impossible position with a stroke penalty assessed to move it to a playable lie)
Maiden	Cricket	An over off which no run is scored by the batsmen, and no wide or no ball is delivered

Major League Baseball (MLB)	Baseball	The major North American professional baseball league, composed of the American and National Leagues
Major Tournaments	Golf	The Masters, the US Open, the British Open, and the PGA Championship
Match play	Golf	A form of individual or team competition in which the winner is decided on the basis of holes won, with fewest strokes determining the winner
Match point	Tennis	A rally in which one player is within one point of winning the match
Medal round	Golf	A round (or more) of stroke play to determine which subset of players in a large field go forward to match play
National League (NL)	Baseball	The oldest professional league in North America, established in 1876
Net-rusher	Tennis	A player who attempts to come into the net as soon as possible during a point
No ball	Cricket	A ball declared by the umpire to be delivered illegally (e.g. by overstepping or by throwing)
Out	Baseball	Batter does not reach base safely or a base runner is tagged by a fielder with the ball while trying to advance
Outfield	Baseball	The area of the playing field beyond the three bases (i.e. area past the infield)
Outright win	Cricket	Obtained by the team with the most runs after both teams have completed two innings. (As in baseball, the team batting second may not have to complete its final innings if it is ahead.) In Test cricket, this is the only victory that counts
Over	Cricket	A set of (usually six) balls bowled consecutively by one bowler
Par	Golf	The score that an expert golfer would be expected to make for a given hole. Errorless play without flukes and under ordinary playing conditions, allowing two strokes on the putting green
Partnership	Cricket	The period, or score made, by two batsmen at the crease together

Penalty	Ice hockey	A player 'assessed a penalty' cannot play for two to five minutes depending on the severity of the infraction. During this time, his team must play one player short (see Short handed, Penalty minutes and Power play)
Penalty minutes	Ice hockey	The number of minutes a player is off ice as a result of penalties
Platooning	Baseball	A manager's line-up strategy in which he alternates starting players according to their handedness. The left-handed (right-handed) batter starts against right-handed (left-handed) pitchers in order to optimize the use of batting skills
Plus/minus	Ice hockey	When a team scores an even strength goal, every player on the ice for that team gets a 'plus' and every player on the ice for the opposing team gets a 'minus'. The net difference between a player's pluses and minuses is referred to as his 'plus/minus'
Point	Ice hockey	The sum of goals and assists
Point-after-touchdown	American football	A field goal attempt from the two-yard line after a touchdown, worth 1 point if successful
Power play	Ice hockey	When one team is playing short-handed, the other team is said to be on a 'power play'
Putout (PO)	Baseball	A fielding play in which the fielder puts the batter or runner out
Rally	Tennis	The exchange of shots between the beginning and end of a point
Rebound	Basketball	Obtaining possession of the ball after a missed field goal. Defensive if the field goal was missed by the opposing team and offensive otherwise
Run (R)	Baseball	A run is scored when a base runner is safe at home plate
Run batted in (RBI)	Baseball	A batter is credited with a run batted in for each run scored as a direct consequence of his at-bat
Run rate	Cricket	For a batsman or team, the runs scored per ball, or sometimes runs scored per over
Sacrifice fly (SF)	Baseball	A fly ball out long enough to allow a runner on third base to score
Safety	American football	Two points for stopping the opposing team behind its goal line

Scoring position	Baseball	A situation in which runners are on second and/or third base
Scramble	Golf	A form of team competition where each team member takes each shot, but successive shots are taken from the most advantageous position resulting from the previous shot. Teams are usually two person match play or four person stroke play
Short handed	Ice hockey	When a player is assessed a penalty, the player is temporarily forced off the ice. His team may not replace him at this time and must continue play with one less player on the ice. The team is said to be playing 'short handed'
Shot-on-goal	Ice hockey	A shot at a net requiring that the goalkeeper stop the puck to prevent it entering the goal
Shutout	Baseball	A game in which a baseball team scores no runs in a game
Single (1B)	Baseball	Batter hits the ball and stops safely at first base. Counts as one base towards total bases
Single	Cricket	One run
Singles	Tennis	One player against one other player
Slugging percentage (SLG)	Baseball	Total bases divided by at-bats
Stolen base (SB)	Baseball	Runner advances safely to next base without the ball being hit
Strike	Baseball	A pitch at which the batter swung and missed or fouled, or one in the strike zone at which the batter did not swing
Strike	Cricket	The batsman facing the bowler is said to be on strike
Strikeout (K)	Baseball	Batter called out on three strikes
Strike rate	Cricket	For a bowler, the balls bowled per wicket taken
Strike zone	Baseball	The volume over home plate, above the batter's knees, and below his shoulders (as judged by the umpire)
Stroke play	Golf	A form of usually individual competition in which the winner is decided in terms of fewest total strokes
Sundry/extra	Cricket	A bye, a leg bye, a wide or a no ball. As a run not scored off the bat, it is credited to the team, but not credited to any batsman

Test Match	Cricket	An official match between two countries of Test Match status. Usually scheduled over five days, each team would have two innings
Tie-breaker	Tennis	A method of deciding the winner of a set when the game score reaches 6-6. The player due to serve next serves the first point and the players then serve two consecutive points in turn. The first player to score seven points wins the game and the set
Total bases (TB)	Baseball	Bases accumulated from hits
Triple (3B)	Baseball	Batter hits the ball and stops safely at third base. Counts as three bases towards total bases
Two point conversion	American football	A bonus play run from the two-yard line after a touchdown, worth 2 points if the goal line is crossed
Umpire	Baseball	An official designated to make rulings of play
Walk	Baseball	See Base on balls
Wickets in hand	Cricket	The number of wickets to fall to complete the innings (i.e. ten less the number of batsmen dismissed)
Wide	Cricket	A ball declared by the umpire to be too wide or high for the batsman to hit
Wimbledon	Tennis	An annual championship played on grass in London. A much coveted title
Winning percentage (WPct)	Baseball	Wins divided by the sum of Wins and Losses

Index